Implementing and Upgrading for Profitability

Telecommunications Billing Systems

Implementing and Upgrading for Profitability

Jane M. Hunter
Maud E. Thiebaud

McGraw-Hill
New York Chicago San Francisco Lisbon
London Madrid Mexico City Milan
New Delhi San Juan Seoul
Singapore Sydney Toronto

The McGraw-Hill Companies

Cataloging-in-Publication Data is on file with the Library of Congress

1 2 3 4 5 6 7 8 9 0 DOC/DOC 0 9 8 7 6 5 4 3 2

P/N 141282-4
Part of
ISBN 0-07-140857-6

The sponsoring editor for this book was Judy Bass and the production supervisor was Pamela A. Pelton. It was set in Century Schoolbook by Patricia Wallenburg.

Printed and bound by R.R. Donnelley & Sons Company.

This book was printed on recycled, acid-free paper containing a minimum of 50% recycled, de-inked fiber.

Contents

Section

I

What Does a Billing System Do?

Telecommunications billing systems—or billing "solutions"—do a lot of pretty complicated things. Sometimes a single "system" performs these functions. More often, the functions are modularized and may come from different vendors. Sometimes a system is designed to handle only specific types of billing and markets (e.g., consumer or "residence," business, wholesale). The "Billing Functions" figure shows the functions of a typical billing system architecture, as well as key interfacing functions. This figure places system functions within the business organizations or processes where they are typically performed. As we walk through what billing systems do, we will trace process and information flows between these functions.

Billing systems get a whole lot of information about a customer and the services and products the customer orders. They get some information about how and when the telecommunications provider delivers those services and products. This information originates in the sales and provisioning process. They get information about how and when the customer uses those services ("usage data") or if the services fail or need repair. This information originates in the network itself.

Within the billing environment, billing systems process large quantities of time- and location-specific usage data into customer-specific usage information. They figure out how much to charge for the usage, as well as for other non–usage-based products and services. They apply relevant federal, state, and local taxes to the billed services. They generate a readable bill format and print or electronically distribute the bill.

Billing systems also get information on payments and apply those payments to customer accounts. They track delinquencies and flag accounts for "treatment"—collection activities including service suspension. They may generate letters to effect collection or give notice of discontinuance of service.

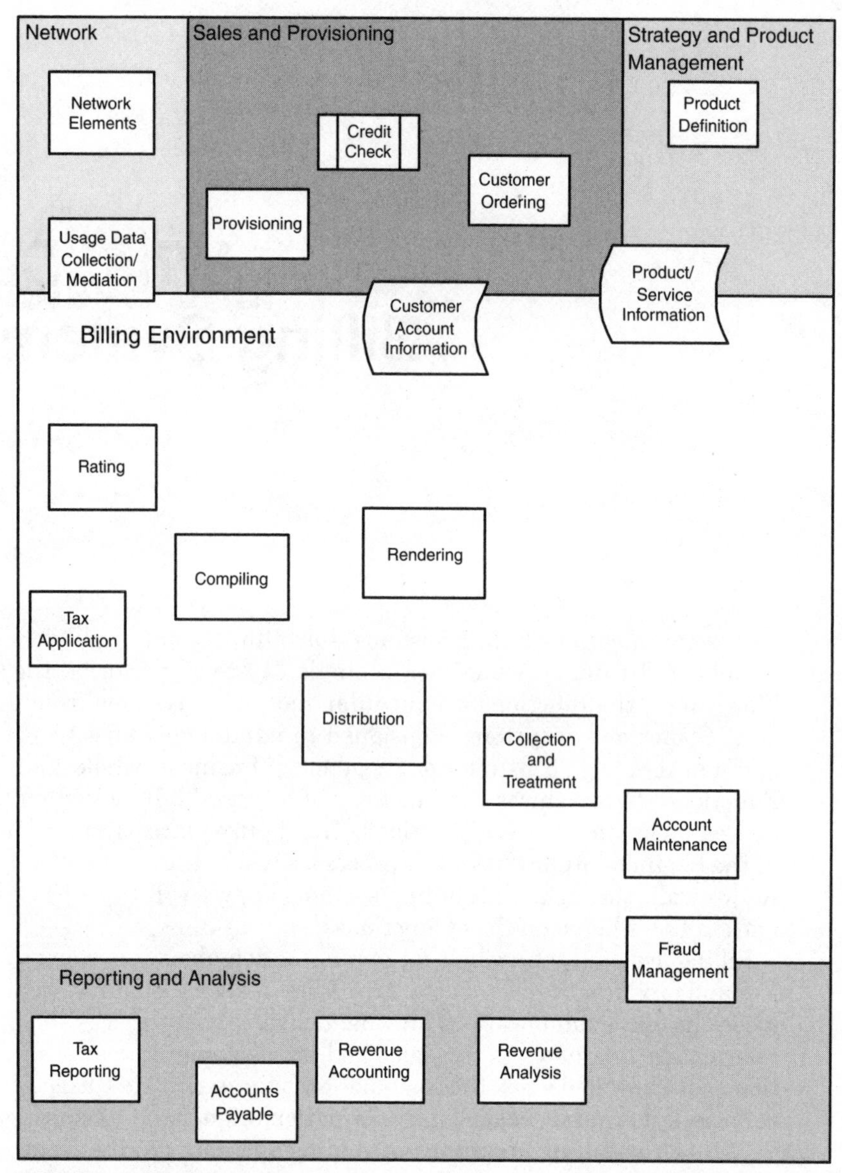

Figure I-1 Billing functions.

Billing systems also do a lot of record keeping, which supports business and financial reporting and analysis. They keep track of all taxes collected from customers and often create reports to appropriate taxing authorities, as well as collecting and reporting all types of general and specific product and service information to support sales and marketing efforts. They also direct receivables, payment and tax information to corporate accounting systems.

Chapter

1

Introduction to What Billing Systems Do

This chapter provides an overview of the major information flows in a telecommunications billing environment—billing operations, information management, and financial management. For simplicity, this chapter focuses primarily on the conventional telecommunications environment. Subsequent chapters will address unique features of wireless services, content-delivery services, IP-based services, Internet-specific services, cable services and the intersection of all of the above, which is often labeled "convergence."

Billing Operations

Billing operations identify and calculate all the charges for a specific customer account, generate the actual bill and apply payments and credits against that bill. The billing operations information flow is illustrated in Figure 1-1. In this overview of billing operations, we refer to various types of data that are used in the development of a customer bill. These data are surveyed in "Information Management" later in this chapter.

Obtaining usage

In traditional usage-based telecommunications billing (toll calling), the originating toll switch or handling operator service system (Network Elements) will create a data record for each call. The records created at each toll switch are generally referred to as "Call Detail Records" or CDRs. Call detail records generally include: the originating number, the terminating number, the billed number if different from the originating number, the time (at the originating location) the call was connected, the time (at the originating location) of hang-up and a flag for direct dial or operator handled. Call detail records can also

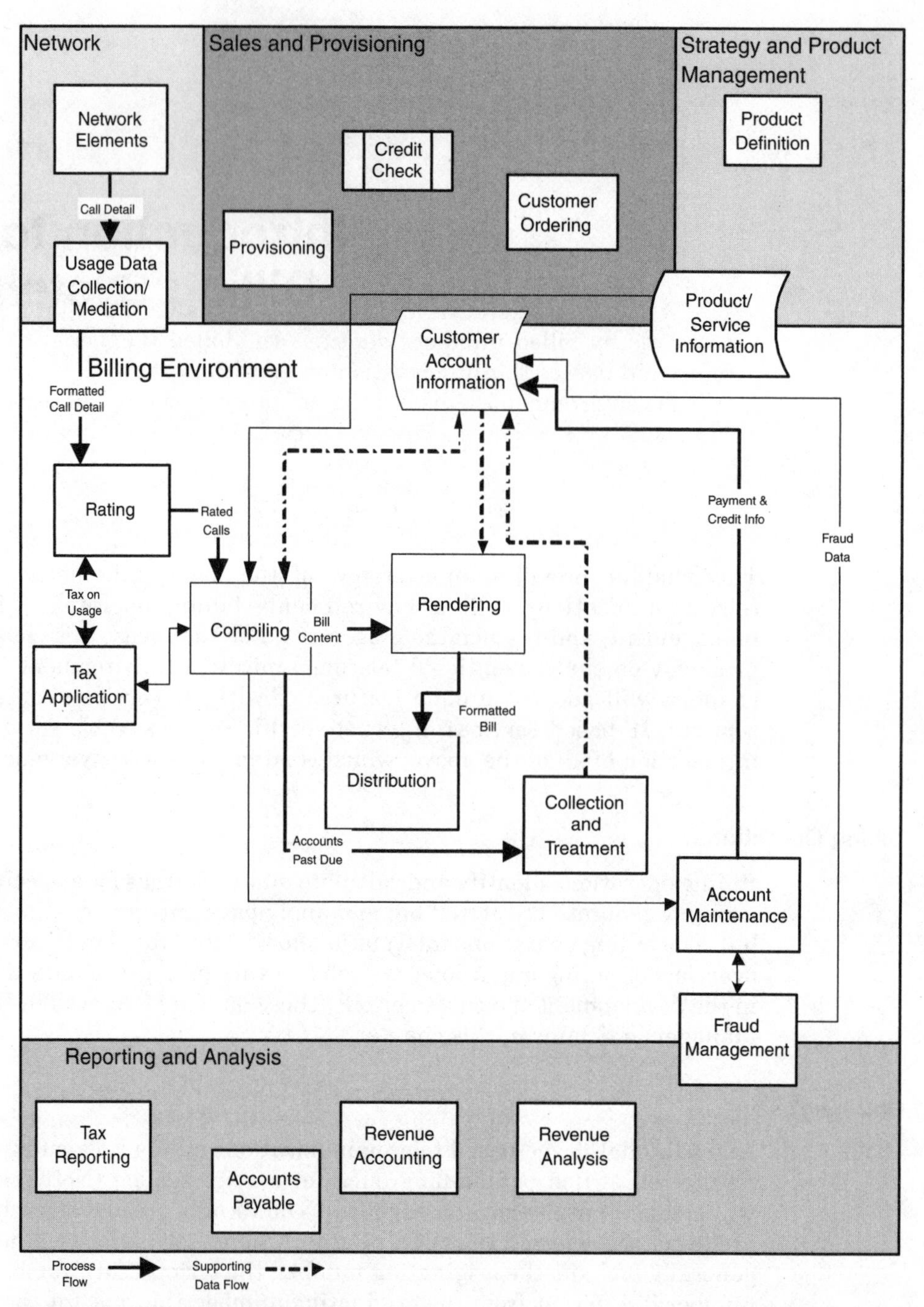

Figure 1-1 Billing operations flow.

contain information on transaction-based services accessed (such as "*69" or directory assistance) and lots of fixed and configurable data flags.

These records are then collected by Usage Data Collection and Mediation systems and transmitted to the billing system. The format produced by the Network Element may or may not be the format expected by the billing environment. The Mediation function formats call detail information as required by the specific billing environment to receive the record. The forwarding of formatted call detail information to the billing environment may be on a periodic basis (hourly or daily, etc., depending on call volume and company strategy) or it may be triggered as a call record is available. The billing system sorts the records by billed number, identifies the billed number with a customer account and passes that information on to rating.

In some environments, this is a "batch" process, where call records are gathered at a data collection site for some period of time (usually one day) and then transmitted as a "batch" to rating and billing. Newer environments provide a more "real time" approach—call records are transmitted to the biller, processed for rating and also reviewed for any suspected fraudulent conditions as the call is being processed.

Most commonly, the billed number is the originating telephone number. If you call California from your home telephone in New Jersey by dialing 1 plus a California area code and the local number in California, that call will be billed to your home telephone number—the originating number.

Of course, nothing is ever easy. Telephone companies have spent a century figuring out ways to make billing more complicated. Many different telephone services are simple voice telephone calls with very complicated billing arrangements! For example:

> For a "collect" call, the billed number is the terminating (called) number.
>
> In toll-free billing—still referred to as "800" service, although now using a wider range of area codes—the billed number is actually the terminating number, which isn't the 800 number, but a regular telephone number associated with the 800 number! And, if the call is to be rated differently in one period of the day versus another, the time of connection used is that at the terminating location instead of the originating location.
>
> In billing to a telco calling card, the billed number is the main or home phone number of the calling card holder or a unique telephone number format number assigned to that account.
>
> In "Call Home" card billing, the billed number is the called number—usually the home or office phone number of the purchaser of the calling card, not the holder of the card.

Rating calls

The rating process begins with the Call Detail Record (CDR). It looks at the type of call and calculates the length of the call from the connect time and hang-up time. The rating software then looks up customer-specific rate information, determines the appropriate rate table by customer information (and, possibly, time of day, and/or date) and assigns a price to each call.

For residential customers, this is usually pretty straightforward, although various marketing promotions such as "free minutes" or "holidays free" or 10 percent off for the first 6 months of service make even residential rating moderately complex. For business and wholesale customers, rates may be determined by individual negotiations and therefore may be unique on a customer-by-customer basis.

Aggregating usage

Each rated call, now linked to the customer's account, is held by the billing system until time to compile all charges for billing. During the period between the billing system's receipt of the CDR and bill compilation, the rated call information is available for inquiry by the company or the customer.

Compiling charges

In compiling a bill for a specific customer, the billing environment will go to the customer account information to determine what services should be included on the bill. It will then collect the usage charges for all the usage-based services, any recurring and one-time charges and all relevant taxation information. Information about recurring and one-time charges is generally captured in the data about specific services that are linked to the customer account record by service identifier or service order identifier. We look at the customer account record briefly as we discuss "Information Management" and in much more detail in Chapter 2.

Applying taxes

Telecommunications services and products are subject to federal and state and, sometimes, local taxes. For usage-based charges, taxes are applied based on the cost of an individual call, the originating location and the relationship of the originating location to the terminating location. Interstate calls are subject to both federal and state taxation while intrastate calls generally are subject to state taxation authorities. Products and services delivered at a specific location are subject to state taxes in that location, except when the products or services are provided in conjunction with an interstate service. This information is compiled to create the tax totals on individual bills.

Rendering bills

Once all the charges are determined, those charges are gathered together with any applicable late charges, payments, and credits for a specific bill. Billing will identify promotional or informational materials designated for a type of customer or for a subscriber to a particular service or for customers in a specific geographic area. This may become part of the bill format or be included in the bill in some other way, such as billing inserts for paper bills or banner ads on Web-based billing. The bill will then be formatted for the appropriate distribution media (based on system parameters and/or customer account information) and will be distributed to the customer or billing agency.

Maintenance of the customer account

The customer account information and/or the data tied to it (such as CDRs and payments to be reflected on the next bill) may change daily, and each of those changes must be available to a number of clients of the billing system. For example, any payment received will be deducted from the previous amount due and notification will be provided to financial organizations. Also, information will be made available to sales and collections groups. In the event of network failures (e.g., due to weather outages or central office disasters), the billing system may need to make wholesale adjustments to accounts and provide subsequent reports to customer contact personnel.

Information Management

As we take a look at information gathered, stored, managed and used by billing systems, notice that some information can be customer specific, service specific or business and system specific. For instance, billing date and billing cycle can be established by customer or can be standard across the entire system. Pricing can be customer specific, fixed for a specific product or service or some hybrid of the two. The levels at which data are received and stored by billing systems are determined by the telecommunications company policies and system standards, which are addressed in Section II.

Establishing customer account information

In order to bill a customer, a billing system must know about that customer. All billing systems have some mechanism to establish the required customer account information.

What information? At a minimum, a billing system must have a mechanism or mechanisms for collecting, validating, and storing the following customer specific data:

- To whom the account belongs—Account identification (customer name and account number)
- Where the bill goes—Billing address (mail, email, Web identity or credit card account)
- What is being billed (service description or product catalog identifier, quantity)

- Service identifier (e.g., telephone number or circuit number)
- Service address/location (where the service is physically)
- Service start date or product delivery date
- Customer tax exempt status

Some other customer-specific information that is frequently required includes:

- Credit status, including information on deposits to secure service
- Payment terms and charges
- Billing cycle (on what date the bill is produced, payment due, charges assessed, etc.)
- Pricing plan or package
- Promotional information
- Related account(s) and bill consolidation rules
- Special pricing overrides
- Customer category or market segment (e.g., business or residence)
- Billing delivery media/agent (bill print centers, credit card companies, EDI interfaces, Web-based billing systems, email billing systems, other electronic media)

Where does it come from? Figure 1-2 provides a high-level view of customer account information flow. Generally, customer account information is created during the service ordering and provisioning process.

Customer ordering systems. The sales force may use a customer order system to collect customer and order information and to support the sales and post-sales process. Generally, the ordering system interfaces with external and internal credit management and reporting resources.

In a traditional sales scenario, a salesperson (this may be someone in sales, a customer service representative or an account manager) may discuss with the customer that customer's service needs and requirements. If the customer is a new customer, the salesperson will collect basic customer information, establish a customer account and create a service order. If the customer is an existing account—adding, changing or discontinuing services—the salesperson will generally verify customer, account and billing information already established and then create a service order. In e-business scenarios, the customer may enter the basic information to establish or access account information online—either from a public Web site or from customized customer access to ordering systems.

Once an account is created or accessed and order information is entered, a check is made for customer credit-worthiness (e.g., "credit limit"). This may be

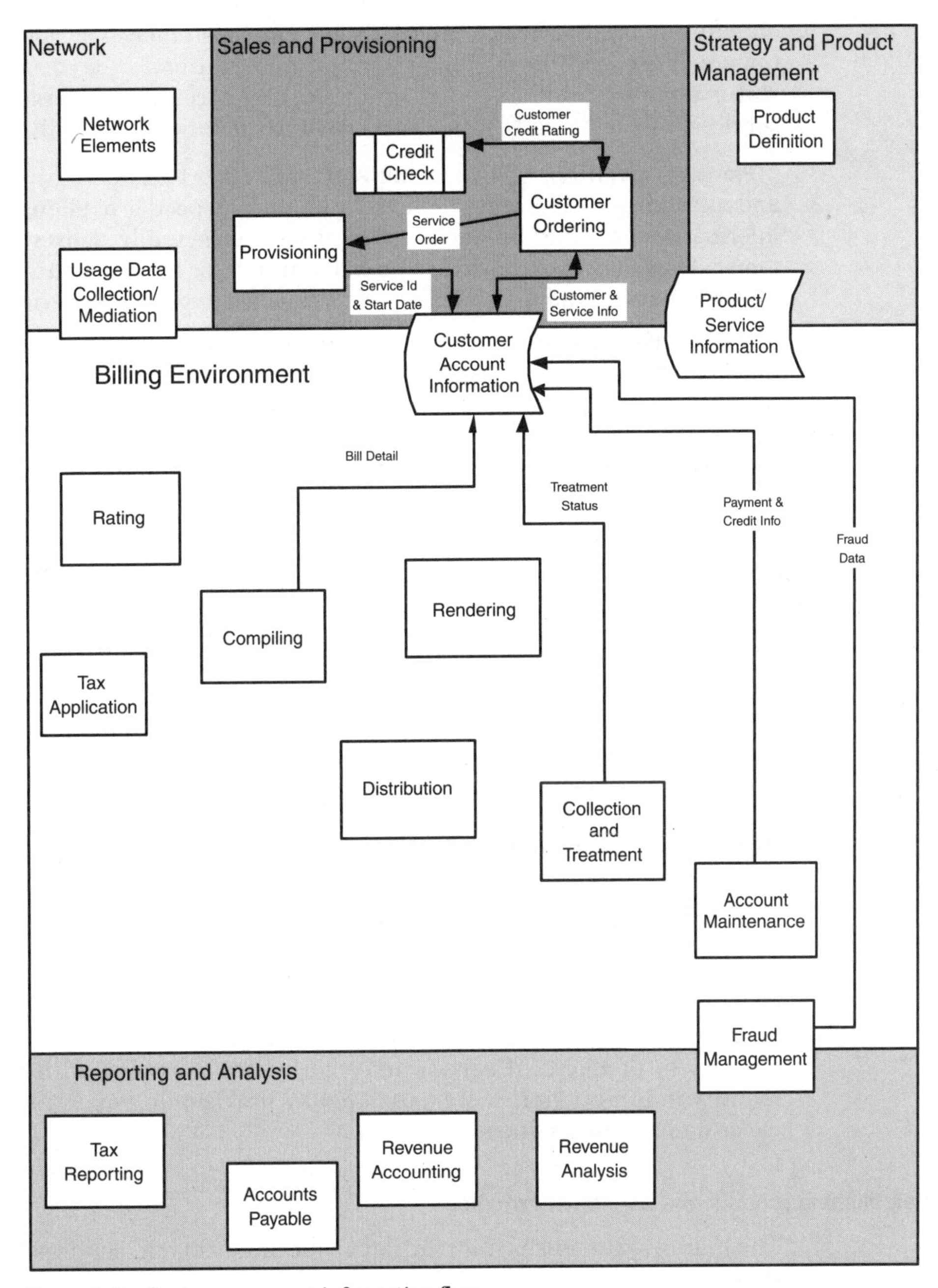

Figure 1-2 Customer account information flow.

an external credit check—especially for new customers—or, for existing customers, a check of the customer account information for payment history and status and any fraud information for existing accounts. For credit card billed services, this is literally a check on credit availability on the billed credit card.

Provisioning systems. Provisioning systems support the process of configuring and installing services and products. From the perspective of billing, they provide information on when products and services are actually delivered to the customer. They may also provide some configuration or service identity information that will be required to determine services billed (e.g., telephone number).

Billing. The billing environment creates and updates information on a customer's billing and payment history. It also supplies information on any fraud investigations associated with the customer's account.

How does it get to billing? Since most customer account information creation transactions normally take place in ordering and provisioning systems—not the billing system—that information must be made available to the billing environment. Some architectures share this information directly with the billing system by using common databases. However, ordering systems also collect other non–billing-related information for installation and maintenance purposes and an efficient ordering database may not be an efficient billing database. Most provisioning-to-billing architectures involve formatting the billing-related provisioning information specifically for billing input and passing that formatted data to the billing system. Interface architecture will be covered in more detail in Section III.

When does it get to billing? Customer account information may be entered in the billing system at any time prior to service turn up—or upon service turn up. Typically, customer identity information will be forwarded to the billing system when the information is gathered or following customer credit verification. If the identity information is forwarded immediately, it will be updated upon credit verification or credit level assignment. Service information may be forwarded with customer identity information or upon turn-up of service or delivery of products. If service information is forwarded with the customer identity information, it will be updated by provisioning upon turn-up of service or delivery of products.

Establishing product and service information

The billing system uses information about products and services in generating customer bills.

What information? At a minimum, a billing system must have a mechanism or mechanisms for collecting, validating and storing the following product and service-related data:

- Product or service name as the customer knows it
- Product or service identifier for use in processing (may be the same as "name" above)
- Type/basis of charge (e.g., one-time fixed, one-time variable, monthly recurring, annual recurring, billed in advance, or billed in arrears, usage-based)
- Source of usage data (if applicable to the type of service)
- Pricing plans
- Jurisdiction availability
- Market

Some other product and service information that is frequently required includes:

- Product or service information for use in quality-checking bills (e.g., identification of which services are not compatible with other services or products)
- Promotional information
- Billing delivery media/agent (bill print centers, credit card companies, EDI interfaces, Web-based billing systems, email billing systems, other electronic media)
- Accounting classification information, used to aggregate money to a product, service or customer level

Where does it come from? Generally, product management teams in the marketing organization supply product and service information manually. This information may be consolidated into periodically released "product and service catalogues" that are then input to ordering, provisioning, billing, and network environments. Figure 1-3 illustrates this information flow. The implementation of this flow can range from a paper or email memo from product management being manually entered to each environment to a sophisticated electronic feed with automated implementation in one or all environments. It is critical that updates to these environments be synchronized.

In some integrated solutions—whether single-vendor or middleware integrated—product and service descriptions are entered in the sales, customer service or provisioning systems and the descriptions are passed through a predefined interface to the billing system. This has the advantage of ensuring synchronization between provisioning and billing, which minimizes fallout in the provisioning and/or billing process. Often, even integrated solutions require the manual entry of billing-specific information to the billing system.

When must it get to billing? Product or service information must be properly configured in the billing system(s) prior to orders for those products or services being submitted to the billing system. If the information is not properly con-

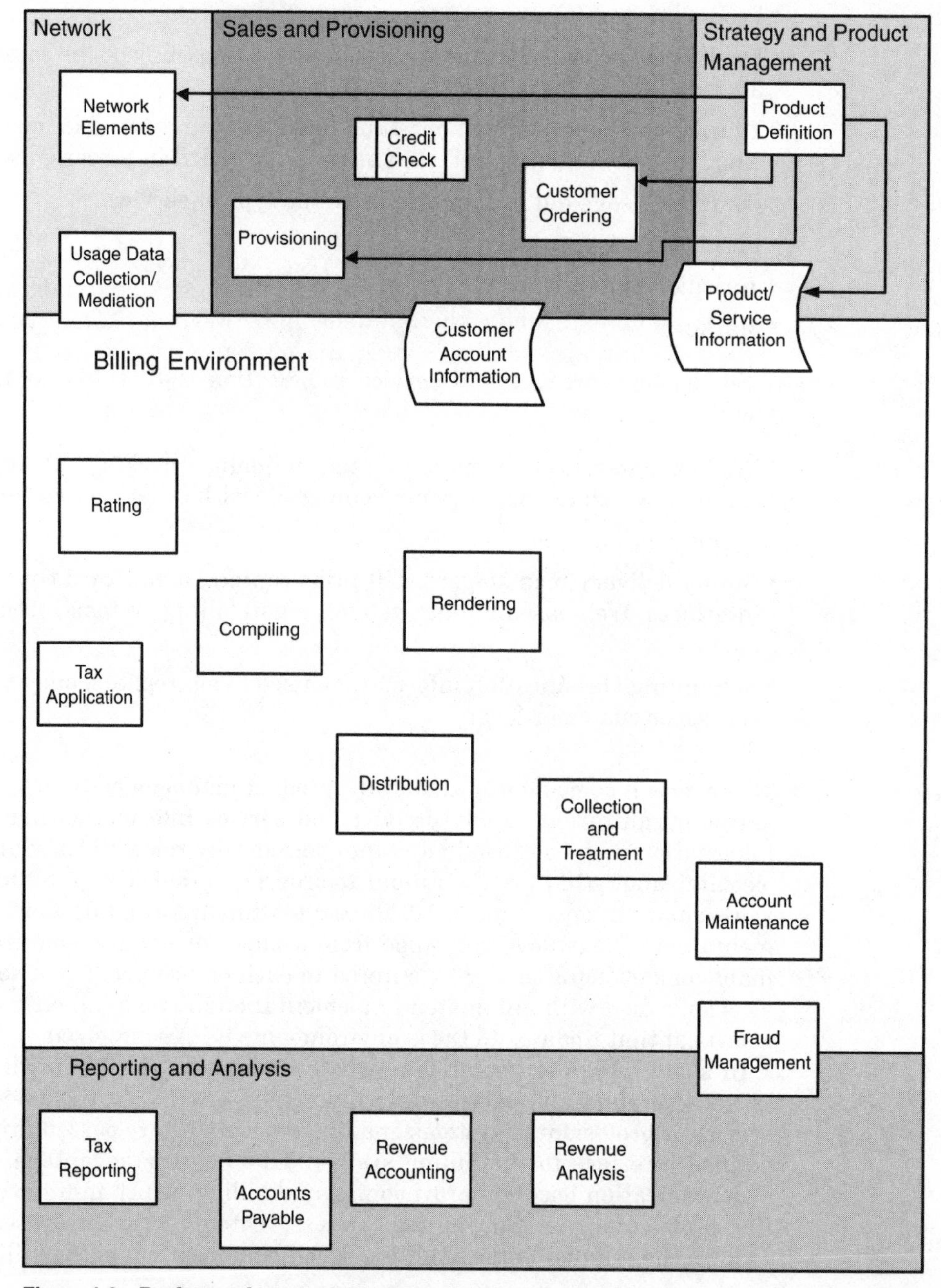

Figure 1-3 Product and service information flow.

figured, order and related charge information will "fall out"—create a processing error—or, even worse, allow a service to be provided without an appropriate bill.

For proper processing of usage data, the billing system must "know" when and if a particular service configuration is available in the network.

Establishing business and operational information

Billing systems store and use information about the company's business rules and billing operations. This information may include standards for credit quality and credit limits, the relationship between date of usage and date of billing, definitions of billing cycles and official billing dates. It may include report generation dates, as well as payables "aging" rules, any discounts for early payment, penalties for late payment and schedules for collection activities. It may also include information on the timing and synchronization of various portions of the billing environment and its external interfaces.

In complex billing implementations, these data may populate very extensive databases. In simple environments (limited product catalog, homogenous customer base), these data may be simply represented in control code or automatic scheduling routines (e.g., UNIX cron files). For instance, if the company has a small customer base, all bills may be dated and prepared on the 15th of the month. In that case, a function at the system level must know to run billing information on the 15th!

This type of information is generally created manually within the billing environment on the basis of direction by the Chief Financial Officer's organization and product management.

Financial Management

The financial management capabilities of a billing system are critical to its success. The billing system must be able to maintain and display information on all aspects of customer accounts.

Managing payment history

Billing systems track activity from the month charges are incurred through the month in which the charges are cleared. Among the items tracked are the types of charges, how long charges have been outstanding, when charges are cleared and how they are cleared (i.e., paid, adjusted, or written off). This allows the billing system to input information to corporate accounting and to provide an audit trail to answer any customer questions regarding outstanding balances.

Identifying collectables and treatment

Using the parameters given by the product management and/or customer service organizations, customer accounts that are overdue for payment are identified by the billing system. In the most simple situations, the accounts may be identified to be referred to a customer service representative for personal contact. Larger systems provide an array of actions that are keyed to the length of time that the account has been unpaid. These actions include production of standard letters requesting payment of overdue amounts, letters advising of service discontinuance, reference to a customer service representative for personal contact and, finally, communication with the provisioning system for cancellation of the service.

Reporting and paying taxes

As discussed earlier, the billing environment determines which taxes are applicable on a per-charge basis, and which are the appropriate taxing authorities. The billing environment then tallies amounts by taxing authority, creates reports of tax liability by taxing authority and forwards the appropriate information to the corporate accounts payable process for payments of collected taxes to taxing authorities.

Summary

In this chapter, we provided a high-level view of the world of billing. We took a quick tour of the entire billing operations process. We reviewed critical areas of information management in the billing environment and we briefly turned the spotlight on the relationship of billing to overall financial management.

In subsequent chapters, we will examine the content, acquisition and management of customer account information, product descriptions and business and operational information. Then, we will look at each of the functions of the billing environment to create a complete picture of the billing process end to end.

Chapter

2

Customer Accounts

In Chapter 1 we took a quick look at the process of establishing customer accounts. Chapter 2 will cover some of the ways customer account information is used, a more detailed description of types of customer account information and issues in acquiring, creating, managing and retaining account information.

Using Customer Account Information

Customer account information is used in many ways throughout the billing environment. It is used to identify charges and link them to the account. It provides the basis for determining taxing authorities. It provides instructions for compiling and addressing the bill and it is used in the determination of credit worthiness.

Identifying charges

The customer account information links specific products and services to a customer and to a billing address and specific person to contact. Generally, the customer account record contains a customer or account identifier that is then linked to order, service and product identifiers. This allows the billing process to start with records of charges incurred and then link them to a specific customer billing account.

The actual prices of products and services may be stored in the customer account information. This would most commonly occur for "negotiated price" contracts for large business customers. More commonly, the customer account information will include indicators for "price plans"—a set of predetermined prices for specific types of customers. An indicator in the customer record would tell the billing system which price table to use for that customer. Some companies decide to charge all customers the same price for products and services. In that case, no unique pricing information is required in the customer record.

Products—whether leased or sold outright—may be identified uniquely by serial number, but often are simply designated by a product code that applies to all products of that type. Services such as installation, or other forms of support such as contracted testing with other vendors, are generally designated only by a non-unique product code, although they may be linked to a specific unique service identifier.

Telecommunications services always are identified by a unique service identifier—either in telephone number format (TN) or in some circuit number format. Many companies use the Common Language Circuit Identifier (CLCI) formats—one of several naming standards maintained and administered by Telcordia (see Standards in Section III). These unique identifiers will be tied to the customer account identifier within the billing system. The service provisioning system will provide the appropriate unique service identifier and will link that identifier to information on the service order. The billing system will receive the service identifier, linked either to the service order number or the customer account identifier or both, from either the sales order system or the service provisioning system.

One-time charges. One-time charges, such as installation and set-up fees, may be related to specific completion or delivery dates and notifications. Information on the work done is generally provided to the billing system by a service provisioning system where the installer or a dispatcher logs completion of the work, an identifying code for the work ("service code") and any time and materials charges. The service provisioning system links such charges to a service order number, which is linked to the customer account identifier and thence to the rest of the customer account information.

Direct purchase of equipment can also be a one-time charge. Direct purchase charges can be invoiced upon shipping or upon installation. If they are invoiced upon shipping, the charge information may be supplied directly by the warehouse and shipping system or it may be fed to the service provisioning system and then to the billing system.

Fixed recurring charges. Fixed recurring charges, such as equipment rentals or permanent fixed bandwidth ("private lines"), are determined at the time of the customer order or contract. However, the charges usually do not commence until the customer confirms that the equipment is received or the dedicated service is up and running. From the perspective of the billing system, this means that the fixed charge amount (or code used to identify the charge) is provided by the sales order system, but the activation of billing is provided by the service provisioning system. Leased equipment without installation may be billed based on advisement from the warehouse and shipping fulfillment system.

Usage-based services. As noted above, the customer account identifier will be linked to a unique service identifier. The most familiar unique service identifier is a standard 10-digit telephone number. Usage data are collected according to unique service identifiers and then linked to a customer account through the customer account identifier.

While this process is somewhat analogous to commercial credit card billing of diverse charges, the volumes of transactions and the potential number of entities that handle a usage record between its origin and final appearance on a customer's bill contribute to the uniqueness of telecommunications billing. In addition, usage is calculated on a variety of parameters, including time, distance, configuration, quality and others.

Determining taxing authorities

Any individual charge on a telephone bill may be subject to multiple taxing authorities—local, state, federal and/or international. Which authorities apply and which tax rates apply depend on a number of conditions, some of which are identified in the customer account record and some of which come from the item to be billed.

Type of product or service. Many telecommunications services and product leases are taxed according to special tax laws that apply only to telecommunications. Other equipment and services—particularly in the Internet and "convergence" arena—may be taxed only according to local sales and use tax laws, if any. Sometimes quite similar types of equipment or services have different taxation authorities depending upon the context of sale and use. The customer account record will contain indicators (such as "type of customer" and service location) that will determine the appropriate taxation standard.

Location of customer or origination. Most often the taxing authority is determined, at least in part, by the geographic location of the customer or of the originating telephone number. That location and the relevant jurisdictions will be identified in the customer account record or the call detail record. In some locations, municipalities or counties impose a tax on some telecommunications services. In most states, a state tax applies to some or all telecommunications services.

Relationship of origination and termination location. The relationship of the originating and terminating locations of a telecommunications service or call usually determines which regulatory and taxing agencies have authority over that service or call. Interstate services are subject to federal telecommunications taxes whereas a call that originates and terminates within the same state usually is subject only to state taxation requirements.

Legal boundaries—LATAs, states, countries, et al. In some cases, the determination of legal boundaries and jurisdictions is pretty straightforward. For instance, a call or a dedicated service that has one end in New York City and the other end in San Francisco, California, would intuitively appear to be an "interstate" (between two different states) service. A call or dedicated service that has one end in lower Manhattan and the other end in midtown Manhattan would intuitively appear to be an "intrastate" (within one state) service.

There are exceptions to these intuitive classifications, however. A dedicated service with both ends in Manhattan might be considered "interstate" if the majority of the traffic it carries is destined for transport on interstate services. A call between New York City in New York State and Newark, New Jersey, may be considered "intrastate" if New York City and Newark have been defined as part of "metropolitan area" traffic. Similarly, some U.S. border "metropolitan areas" include non-U.S. locations and the taxing authority is generally that of the originating telephone location. Exceptions such as these are usually identified and addressed within tax processing modules in the billing environment.

Compiling the bill

A bill will be compiled for each account in the system, a total probably fewer in number than the services being billed. Customer account information links specific services to a customer account. For instance, if you have three telephone lines at home you may elect to receive three separate bills, two bills (one bill will show combined charges and one will have the charges for only the third line) or one bill containing all charges. This is dependent on how you order the services and how the salesperson establishes the accounts: as one account, two accounts or three accounts.

Generally, billing and account folks refer to a "telephone number," "service number" or "circuit number" when referring to the designation of a specific single service. Additionally, each service will be assigned a "billing telephone number" or BTN that is linked to the account number. For single-line residential service, the service telephone number, the BTN and the account number are usually the same.

To emphasize, a BTN is not the same as an account identifier (even though confusingly enough, they may be exactly the same number). For instance, a company could have one account, but have separate BTNs for each branch office. Large business customers often have custom billing requirements to support their internal bill processing. These requirements will be part of the customer account information in the billing system.

"Addressing" the bill

Customer account information includes how the customer is to be billed and where the compiled bill information is to be sent. In traditional bill printing and mailing, the customer's postal address is part of the customer record. It is printed on the paper bill as a mailing address—including bar coding for postal handling. It is also used to direct bill print processing for automated envelope enclosure, mail sorting and bundling.

Not all customers receive traditional paper bills. Some customers receive their telephone and/or Internet bills on compact disc media, by email or on the Web. Others handle their bills through credit card billings. The modern billing system must be able to identify how a specific customer chooses to be billed and then be able to direct appropriate billing information to that billing medium. This also increases the need for multiple differentiated billing streams, with total amounts being sent to credit cards or other online payment vendors, but billing details being made available for customer query on Web sites.

Extending credit

When a customer initially establishes an account for telecommunications services, an initial credit rating is established—usually by performing a credit check on the individual or business. This credit rating can be used to support a request for a preservice deposit, control payment terms where not controlled by regulatory guidelines and to flag usage beyond certain limits.

As the provider gains direct experience of the customer's service activity and payment history—based on information within the billing system—the customer credit rating may be modified.

Acquiring Customer Account Information

Customer account information originates in a variety of systems and functions. The originating system may directly provide the information to the billing system or it may pass the information to another system or systems for transmittal to the billing system.

Customer information from the sales process

Customer account information is gathered throughout the sales process and is then forwarded to the billing system. When a customer initially establishes an account with the provider, the information gathered from and about the customer is extensive. Subsequent sales contacts will provide updates of existing account information.

The sales process will obtain the customer's name, address, contact name(s) and telephone number(s), and information about contact availability. Sales may also collect the customer's social security number, employer identification number or Dun and Bradstreet ID, corporate officers or partnership data or, for a residential customer, employment data. Sales may gather bank and/or credit card information for both billing and credit reporting purposes. The sales process will create a customer account identifier (usually called an "account number").

The sales process may also categorize the customer, based on predetermined business classifications or negotiations with the customer. Common categories are based on whether the customer is residential or business, single or multi-line or on revenue potential.

Sales may also identify some special customer classification for specific disabilities or for elderly assistance. There may be provisions for third-party notification if the account becomes delinquent or is otherwise at risk. Tax exemption information will be entered as well, usually after sales has received confirmation that a customer meets appropriate requirements.

Sales may also gather general billing preferences including billing medium (e.g., paper, email, Web, CD) and third-party payment options (e.g., online banking, e-checks, credit card). It may also identify customer account security information such as personal identification number (PIN) and "security word."

Whenever subsequent sales activity takes place, the customer account information normally is reviewed and updated as appropriate.

Service/product information from the sales process

Each sales order generates information about specific services and products. This information is forwarded to the billing system from the sales process. There may be a direct link from the sales process support systems to the billing environment or the information may flow from sales through the provisioning support systems to the billing environment.

Service or product information from the sales process will include an identification of the product or service, generally in the form of a formatted product code for that type of product or service. The product code will indicate that the customer has ordered a residential telephone line, for instance. That product code may be associated with a number of feature codes for such things as call waiting, call forwarding and conference calling. Generally, product codes and feature codes are forwarded to billing simply as "billable elements" or codes.

Sales also identifies the service location or locations, contacts for the service location(s) and contacts for service billing. Sales may identify the applicable regulatory and taxing authorities.

Sales will identify the account to which the service is to be billed, provide the account number and any subaccount configuration.

Credit information

At some point in the sales process, the customer's credit information is requested and reviewed. This usually consists of an automated inquiry to one or more credit reporting services. In some cases, a manual review of credit history takes place after credit reports are returned. In the case of very large custom networks, more extensive credit background or other surety may be required. The customer account will be assigned a credit indicator—either a dollar amount credit limit or a rating related to predefined credit standards. Naturally, if the new account is for a customer who has previous or other existing service, the established track record of payment is usually the most significant indicator.

Service/product information from the provisioning process

"Provisioning" is the process during which a service or product is engineered, if required, and installed or activated. The provisioning process is the source of service identifiers (e.g., telephone numbers, circuit numbers), inventory information and "in-service" dates. This information may be provided directly from provisioning to billing or it may be fed back from provisioning to sales and thence to billing.

Typically, provisioning will supply sales with groups of telephone numbers for assignment at the time of ordering. However, the service engineer usually assigns circuit numbers individually.

Product model and serial number generally are provided by the service provisioning process for installed services. Model and serial number for shipped products are supplied by the shipping and fulfillment function.

Some installation activities are charged on a time and materials basis. These charges are captured in the provisioning systems and forwarded to the billing system. Charge information is linked to a service identifier or to a service order identifier for assignment to a bill.

The "in-service" date is when a service is completely installed and working. This may entail confirmation of service ("acceptance") by the customer. This is an important date for billing. Non–usage-sensitive services start accumulating charges on the in-service date. Usage-sensitive services are authorized to accumulate usage dated no earlier than the in-service date. For stand-alone products without installation, the in-service date may be related to the shipping or to the confirmed receipt date. Shipped products that are required for the operation of an end-to-end service are usually linked to the confirmation that the end-to-end service is working.

Provisioning will provide the date that the service is discontinued or a rental/lease item has been returned. The customer requested date for billing to be stopped (as shown on the sales order) will be the one used, even if the date of discontinuance (or removal) may be different from the date requested by the customer. Billing will usually maintain both dates to facilitate possible adjustments or write-off of unbillable charges.

Payment history/account status from payment and reconciliation process

Customer payments are accepted and credited to the customer account. The billing system payment process collects information on the amount of the payment, the timeliness of the payment and actions required to collect the payment. Payments are recorded in the billing system and subtracted from the customer's amount due. If payments are late or have required additional collection activity, this is noted in the billing system.

Payment history and account status are provided from the billing system to the collections process (this may be part of sales) and are combined with external credit report information in determining if credit limits or classifications need to be changed.

Account inquiry and adjustment from the account inquiry process

If you have ever noted a strange charge on your telephone bill and called to ask that the charge be removed from your bill, you have participated in the account inquiry and adjustment process. Billing systems provide account representatives with access to all the information on charges to the account, payments made to the account, collection efforts on the account, as well as a history of account adjustments.

An "adjustment" is when a change is made to charges, either because they were erroneously charged to the account or because someone criminally used your account, as in calling card fraud, or because the service was in some way unsatisfactory—as when you have a "bad" connection. Adjustments may be made after the customer receives the bill or before the customer receives the bill—often at the time of service or when a calling card is stolen. Adjustments are noted in the billing system and credit is generally applied at the next billing cycle.

Many adjustments originate in customer care systems or in the customer care and account inquiry module of the billing system. These are generally the result of customer-initiated contacts. However, in some implementations—especially for business services—adjustments may be generated by network or service maintenance systems. For instance, if a major network outage resulted in loss of service on a private line circuit or a server outage resulted in a loss of service for an Internet host, the maintenance system might generate an automatic credit into the billing system.

These maintenance interfaces tend to be highly customized, so we do not discuss them in detail. However, they are becoming more common in environments with little tolerance for downtime.

Debit adjustments may also be required. These are less welcome to the customer, but they are necessary if a customer's check is returned unpaid or when unusual situations occur, such as when notice of completion of service order installation is delayed or a rebilling of a charge from one customer to another.

Summary

In this chapter, we examined where billing gets customer account information and how billing uses that information. In Chapters 5 through 12, we'll get a closer look at the many ways in which the billing environment uses customer account information. First, however, we need to review some information about products in Chapter 3 and about business rules in Chapter 4.

Chapter

3

Product Descriptions

Product descriptions are the basis of all billing. In the simplest terms, they define what is being billed and how much to charge for it. Product management defines individual products and product relationships and then develops product descriptions based on overall business strategy, market, sales and cost information.

Billing Elements

Any service or product for which there is a charge or which is shown on a bill is a "billing element" or "rate element." Frequently, companies want to detail all services provided, even if there is no charge at present. Displaying services for which there is no charge may be a strategic decision in order to remind the customer that they are "getting something for nothing" or it may be a regulatory requirement. Also, retaining a complete record of services provided can ensure appropriate revenue if at some point a charge is instituted for those services. Additionally, the billing system may be the company's official inventory of services for customers.

Product name

The product name is used to describe a product in understandable language on an itemized statement. This product name is used in a computer-generated bill format and is usually limited in length. Choosing a product name that will make sense to the customer in the context of a bill can minimize informational customer care queries.

Product code

In complex product environments, alphanumeric product codes are used to identify a product and link that product to customer accounts, to pricing tables, to availability tables and any other product-related information in the sales, provisioning or billing systems.

Relationships

Often a particular service or feature is only available if the end customer also has another service or feature. An obvious example would be the dependency of "call waiting" upon having a telephone service. This type of dependent availability is generally screened in the sales and/or provisioning systems. However, to protect against improperly entered data or against inadequate screening in upstream systems, billing system service descriptions may also contain this type of information.

More closely aligned to the billing process are relationships of services and products which affect service pricing. The price of a stand-alone service may be discounted when the subscriber also has a related service. Many cable companies discount the price of cable Internet access if the subscriber is also a cable television customer. A more complicated type of relationship would be determining the pricing for specific services based on the amount of the previous month's bill or on the aggregate of all undiscounted services.

Pricing

Pricing information is at the heart of all billing functions. Pricing may be simple—"one size fits all"—or subject to temporary or permanent discounts or completely unique.

Pricing basis

In its simplest form, pricing basis is the unit for which a price is assigned. For products, that unit is often a count—most often "1" but consider typical retail promotions of "2 for the price of 1" or "3 for the price of 2." In those cases, the unit might be 2 or 3 respectively. Unit pricing may also be appropriate for services such as email accounts or features such as "call waiting" or "call forwarding."

Pricing for some services may be based on units of time or "usage." We look at usage-based services in more detail in Chapter 5, but a simple example is the much-advertised "ten cents a minute" for all domestic U.S. telephone calls. In that case, the pricing basis or unit may be one minute or fraction thereof.

Often services involving dispatched personnel, such as installation or repair service not covered by contract, involve time-related charges based on each quarter hour or fraction thereof.

Pricing plans

The concept of pricing plans covers a lot of territory. However, the basic idea is "semi-customized" pricing. Pricing plans fall somewhere between fixed pricing (e.g., "widget *x* always costs 10 cents") and negotiating the price of every service and product with every customer.

One simple form of pricing plan establishes a baseline price structure for a product or service (usually the price charged for someone with a single service) and then identifies certain "standard discount levels" against that baseline structure. In that scenario, the baseline price for widget x might be 10 cents (pricing plan A) but there might be a 10 percent discount plan (pricing plan B), a 20 percent discount plan (pricing plan C), etc., etc. Sales and product management can then agree upon the rules for invoking a particular pricing plan.

Another form of pricing plan is based upon volume of purchases. In telecommunications, this is often invoked for usage-based services. In this scenario, the first 100 minutes of usage might be charged at 10 cents a minute, while minutes 101 through 5000 might be charged at 8 cents a minute and minutes 5001 through 20,000 might be charged at 5 cents a minute. Alternatively, all usage minutes might be added up and if the total is less than 100 minutes, everything will be charged at 10 cents a minute. If the total is between 100 and 4999, everything will be charged at 8 cents a minute. If the total is greater than 5000, everything will be charged at 5 cents a minute.

A commonly used form of pricing plan, especially in consumer markets, is "service combining" or bundling. This is where customers are encouraged to buy additional services (such as Internet access) by a discount on a different service (such as local or long distance telephone service). A typical example of bundling in residential telephone service is a plan giving a customer call forwarding, call waiting and third-number calling for one price that is less than the aggregate of the individual services.

Billing systems may or may not "understand" pricing plans. If the billing system does not contain pricing plan intelligence, then sales or provisioning systems must translate pricing plans into unique service and price definitions for billing. For instance, a stand-alone consumer service may be "POTS 1" priced at 10 cents a minute. That same service when sold in conjunction with some intelligent network features may be "POTS 2" priced at 8 cents a minute.

Promotions

Companies often wish to offer promotional discounts for a specific period or for some interval from service installation. Generally, this requires specific logic in both the sales systems and the billing systems for the graceful management of such options. It is desirable for such promotional information to be administered (initiated) in sales or product management systems, rather than directly in the billing systems. These promotions may have to be administered manually when billing systems are too antiquated to handle them or when sales/marketing and billing systems cannot support the automatic coordination of promotional parameters and pricing. The company is safeguarded in a better way when the promise to agreement with the customer is automated.

The variety of promotional discounts in established telecommunications enterprises can be staggering. In some cases these promotional pricing schemes are even formally tariffed (i.e., rates are formally filed with the appropriate regulatory jurisdiction). For instance, one company offers a special discount rate or free calling to residence customers on their birthdays. Another provides for discounts on five bank holidays (as defined by individual states!) following the initiation of service.

Custom pricing

If your company deals with large business customers—or if your company is relatively small and the sales strategy is to negotiate every customer "deal"—then billing solutions will have to support unique pricing for a specific customer. These unique prices may be for a single service or for an entire class of service whenever the customer may order it. Of course, custom pricing schemes may be as complex as any predetermined pricing scheme, including promotional discounts and volume discounts.

Market variable pricing

One of the major variables in pricing some services is regulatory authority. A call from downtown Manhattan to uptown might be priced significantly differently—both in rate and in the rules controlling that rate—than a call provided by the same company at the same local time of day of the same duration and distance across Los Angeles. In part, this may be because the New York regulatory board has different rules and priorities than its California counterpart.

Apart from regulatory influence on pricing, both the cost of delivering service and normal marketing considerations such as stimulating usage where there is spare capacity, as well as other business strategies, can drive a company to vary service pricing by geographic market. "Market" may be defined by the initiating end of the service (the "caller"), by the terminating end of the service (the "callee") or by both.

In addition to geographic market definitions, telecommunications and networking companies often define market and market rating by some characteristic of the customer. The traditional "business" and "residence" designations are one example of this type of market definition. Some companies distinguish between "urban" and "rural" markets. Others might include the industry of the customer, the customer's annual telecommunications expenditures or whether the customer will deal exclusively with the provider.

Usage-Based Services

"Usage-based services" are services where the charges are based on the time the customer is actually using the service. Everyone is familiar with traditional long distance calling rates where charges are determined by the length of a particular call. Some Internet service providers (ISPs) charge based on total monthly connection time.

Units

Chargeable units for usage can be hours, minutes, tenths of minutes or any other unit of time. Internet charges are typically computed using hours and parts of hours as the chargeable unit. Many traditional telephone services (and legacy tariffs for those services) use minutes and parts of minutes as the chargeable unit. Services designed for very short duration calls, such as credit card authorization calls, may be chargeable in units of tenths of a minute or in seconds.

For some services, the units are a volume of something other than time. Data services may be measured in units of data packets transported. Some telephone services are charged on the basis of the number of calls. An example of this is a local calling plan that gives the customer 30 calls a month for a specified dollar amount.

Variables

The traditional variables for usage-based billing are the time the call is placed and the originating and terminating location. Other variables may include operator involvement, emergency conditions and quality of service.

Time. Often services are priced differently based on time of day and day of week. Some of these pricing scenarios focus on reduced rates at times when the network infrastructure is not fully utilized or "non-peak" hours. The classic evening and weekend discounts for consumer telephone calls are an example of this type of pricing. This means that a call placed at 9 A.M. on a Monday from Manhattan to Los Angeles will be charged at a different rate than a call of the same length between the same two phones placed at midnight on a Saturday. Some wireless carriers offer a block of free minutes for calls placed in certain non-peak hours as an incentive for subscribers.

Typically, calls are "rated" or priced based on the location of the originating caller (the person who places the call). Hence, that 9 A.M. call between Manhattan and Los Angeles might be priced differently if the person on the Los Angeles end placed the call, since the call time in Los Angeles would be 6 A.M. even if both callers were on exactly the same rate plan! Time zones, then, are one geographic variable in pricing.

To/from locations. Traditional telecommunications tariffs had some component of "distance," that is, the idea that pricing should reflect the number of miles between the originating and terminating telephones. Today, that concept has largely been replaced by the idea of rates unique to some relationship between the originating and terminating code. This is usually area code (NPA) to area code for long distance calling; country code to country code for international calling; and area code and exchange (NPA/NXX) to area code and exchange for local and regional calling. For the systems-oriented reader, this should evoke the idea of very large relational tables of originating and terminating numbers. To facilitate maintenance of these tables, some billing applications allow these relationships to be defined as groups for which a common rate can be specified.

Fixed-Price Services

The alternative to usage-based services is "fixed-rate service"—sometimes called "non–traffic-sensitive" charges. Dedicated private line service is one example of a fixed-rate service. A flat monthly fee for Internet access is another. The basic monthly service charge for having a telephone is another.

One-Time Sales and Charges

The most common one-time sales are items of customer premises equipment associated with a specific service. These may range from a conventional telephone through a wireless phone or personal digital assistant (PDA) to a cable modem kit to sophisticated private switches or broadband terminating equipment. These may be sold as part of the original service or as "aftermarket" equipment on an existing account.

Common one-time charges would be associated with installation of services. These may be a flat-rate installation fee or time and materials charges. Sometimes customers will request technical support not covered by normal service charges, such as coordinated testing with other vendors. These charges are generally on a time and materials basis.

Distance and Dedicated Services

We discussed the evolution of distance-based usage charges into fixed relationships of to/from locations. For dedicated services, however, charges are generally computed based on "air miles" ("as the crow flies!") between the ends of the service. In concept, for a simple two-point service, that distance would be the distance between the two ends as if plotted with a ruler on a map. Determining distance for dedicated services is discussed in more detail in Chapter 6.

Regulatory Classification

While not significant for currently unregulated Internet services, most telecommunications services are subject to some—or a great deal of—regulatory definition and oversight.

Definitions

"International" services cross national boundaries. However, political and economic treaty agreements such as the formation of the European Union, may create an environment in which services that geographically cross national borders are regulated or administered as "domestic" services.

In the United States, "interstate" services are services that carry traffic from a location in one state to a location in another state. Some exceptions may be made for traffic within a greater metropolitan area that happens to span a state border.

"Intrastate" services are services that carry traffic from a location in one state to a location in the same state. The determining factor is the location of the endpoints, not the physical routing of the service. A service with endpoints in the same state may be defined as "interstate" if the traffic it handles is largely destined for interstate locations. Access facilities between a customer PBX and a toll switch, for instance, are often classified as "interstate," as are off net extensions of interstate networks and some Internet access services.

The agreement between the U.S. Department of Justice and AT&T, which broke up the Bell System in 1984, created a new kind of geographical definition—the local access transport area or LATA. LATAs are roughly based on standard metropolitan statistical areas and are designed to identify geographical areas with a "local" community of interest. Any geographic area of the United States served by a Bell System company prior to January 1, 1984, is included in one, and only one, LATA.

Companies that offer inter-LATA and/or interstate services are considered "inter-exchange carriers." Companies that offer intra-LATA services are considered "local carriers." Contractual billing arrangement between inter-exchange carriers and local carriers are common throughout the industry. Most frequently, a local carrier will agree to include inter-exchange charges on the customer's monthly bill and to collect payments on those charges as an agent of the inter-exchange carrier.

Jurisdictions

A telecommunications jurisdiction is the geographical area regulated by a board or commission. In the United States, for example, there are interstate, intrastate and local jurisdictions. An average U.S. telecommunications bill will be subject to both interstate regulation from the U.S. Federal Communications Commission and intrastate regulation from the state(s) in which the customer does business. In the United States, intrastate services are regulated by the state or (in Texas) by the county. Interstate services are regulated by the federal government, specifically, the Federal Communications Commission (FCC)—Common Carrier Bureau (CCB).

Summary

In this chapter, we covered the basic components of product descriptions from a billing perspective. We looked at the billing elements and common types of pricing, as well as the determination of regulatory classification.

In Chapters 5 and 6, we examine how pricing relates to the determination of charges. In Chapters 7 and 8, we show product description information used in compiling and rendering a customer bill. In Section II, we will take a look at how the billing perspective on product descriptions relates to a marketing perspective or a technical perspective.

Chapter

4

Business and Operational Management

Where Are Things Defined?

Business and operational management information may be defined at several different levels within the billing architecture. If your business determines that these parameters should be identical for all customers and services, the parameters can be defined at the "system" level—either in universal reference data or in the actual control logic of the system modules. How your business determinations link to setting up the billing system parameters will be discussed in Section II.

If your business determines that these parameters should vary by classification of customer, the parameters can be defined at the "group" level—sometimes identified as "market." If your business wishes these parameters to be individually set, then these parameters will be part of the customer account definition.

Market Segmentation/Customer Classification

Businesses with a large customer base often use market segmentation to simplify pricing structures while allowing some flexibility in pricing as a market incentive. Market segmentation divides up the customer base by some predetermined set of criteria. These criteria may be something inherent about the customer or they may be based upon service volumes or spending levels.

In the traditional world of regulated telecommunications, regulatory concerns divided customers into "business" and "residence"—with significantly different rate structures for each. Much of this was driven by a national policy favoring "universal service" or access to basic telephone service for all residences. This distinction lingers in many legacy tariffs and pricing schemes, although newer tariffs tend to combine single-line businesses with residence service.

Typically, a company's very largest customers are given the most favorable pricing. However, business strategy may determine that being well positioned in a particular growing industry or growing region would be highly beneficial, resulting in a unique class of rates for those targeted markets.

In addition to variable pricing structures, different market segments may be offered different billing and account management options. For instance, small customers may be assigned a monthly billing date, while larger customers may get to select a convenient billing date. Small customers may have no choice in billing media (paper bill or credit card billing), while larger customers may get to choose between paper, CD-ROM, Internet, or EDI billing. Small customers may have only sets of "pre-packaged" service options, while large customers may get to "roll their own."

Billing Cycles

In general terms, a "billing cycle" defines the billing period (usually a month) for the customer. Some companies have the same billing period for all customers. Some assign billing cycle dates by customer location or customer name. Others allow customers to choose their billing dates for greatest convenience to the customer. There are several important dates in the billing cycle, all driven by the "billing date," which is usually the date appearing on the top of the customer's bill and has some relationship to the date the bill is mailed to the customer.

One set of billing cycle dates defines the period for which data are collected in the bill, including the "service period"—the days of service for which charges are to be gathered. Other important dates are the "credit booking date"—the date after which credits will not be included in the bill—and the "payment due date"—after which any interest and service charges may accrue.

Another set of dates governs the actual bill production and mailing. These dates identify when any data collection and mediation, call rating and taxing and billing data processing will take place. They also identify when the billing data or image must be provided to a bill print or other bill delivery facility. Finally, they define when the bill must be mailed or otherwise made available to the customer.

Payment Conditions and Standards

Payment conditions include how long after the billing date a payment is due and what the consequences are of an early or a late payment. Is there a discount for early payment? Is there a penalty for late payment? When is interest charged on payment due? What is the responsibility of the company for processing and posting payment? What is the consequence of a problem in payment such as a bounced check?

Credit Classifications

Credit classifications are used to determine credit limits (how much the company will allow you to be in debt to them), the size of initial service deposits, payment terms and intervals and how soon after billing the account will be eligible for collection activity.

Many companies use commercial credit bureau scorecard ratings to establish initial customer credit classifications. For business accounts, Dun and Bradstreet (D&B) is the most common credit interface, although there are several companies that specialize in combining D&B information with other sources to create proprietary scorecards. For consumer accounts, nationwide companies generally have an interface with one of the major national credit bureaus, while many regional companies rely on regional credit bureaus.

Most companies modify initial customer credit classifications based on ongoing payment history with the customer—similar to credit card companies that automatically increase credit limits for long-term customers with good payment history.

Collection and Treatment Rules

A standard approach to collection and treatment is to show a "pay by" or due date on the customer bill. Accounts unpaid within a reasonable period following that date will cause the billing system to generate a written notice to the customer, stating the consequences of an outstanding account. The customer who continues to procrastinate may receive a second system-generated notice or letter with a more strongly worded statement. The next step generally is a billing system referral to the company's sales or collections group for a personal contact. The non-paying customer is then subject to service termination, which may be followed by referral to an outside collection agency.

At one time, many jurisdictions in the United States dictated telecommunications collection and treatment rules, but for the most part, that is not so today. Company policy and standards are what will determine how the billing system will be set to operate. Most billing systems are capable of automating most steps, although some companies prefer to have the billing system merely list delinquent accounts for a personal contact and payment negotiation.

Billing Formatting

Most billing platforms support several types of billing information formatting. There may be a billing format for usage-based hardcopy bills, one for non–usage-based hardcopy, one for credit card billing, one for Internet account management, one for EDI invoicing, etc. Often external vendors for bill printing or distribution limit flexibility of these formats. There may be regulatory concerns for content that must be addressed, too. For instance, rules adopted by the FCC in 2000 specify that telephone bills include the names of all service providers, highlighting any new service provider that did not appear on a previous bill. These rules also mandate the identification of any charges for which failure to pay will not result in an interruption of service, the use of standardized labels and contact information for inquiry or complaint.

Billing Information Routing

Formatted billing information by account can be routed to printing centers for conventional billing, to Web databases for secure display to customers, to credit card companies, to billing bureaus or to email addressed for email-billed accounts. For business customers, it may also be transferred to tape (quickly becoming obsolete) or to CD media, which is then sent to the customer. Business customers may also have direct electronic invoicing interfaces.

Everyone is familiar with printed telephone bills for single line service. Information is formatted for a specific customer account (a telephone number) for a specific billing period. It is then sent to a "bill print center"—usually a high-volume printing and collating center that prints and assembles bills, puts them in envelopes with applied postage, and sorts and bundles for mailing according to postal regulations. All printed bills follow a similar process.

Some commercial credit card billing routes only a total amount due for the billing period to the credit card account. Supporting bill detail information is then provided via a secured Web interface or with CD or other media, if it is available at all. This requires that the billing environment support two separate streams for billing routing.

If your enterprise supports only one type of billing in one location, then billing routing may be a single parameter at the system level. If your enterprise offers different types of billing by market segment or product/service mix, you may want to have routing information at that level. If your enterprise offers choices in billing delivery to customers within the same market, your billing environment must contain information about the routing of billing information at the customer account level.

Summary

In this chapter, we examined some of the business rules that are implemented as system parameters or as operational guidelines in the billing environment. These rules define when and how billing environment activities happen. We'll see how these rules are used in Chapters 7 through 11.

Chapter

5

Processing Usage

This chapter provides a look at usage information as it is generated and completed prior to entering the billing system. This information is helpful to the billing person in several ways. First, of course, it will help you understand where the information is generated, which is helpful in understanding how it should or must be billed. Second, for those interested in setting up a billing system, it is a guide to the path the transaction must take to get to the billing system. Finally, it provides some insight into billing-related requirements and considerations for the acquisition and upgrading of network equipment and call detail collection and mediation systems. This chapter also examines the relationship between the creation of usage data and the rating and taxation functions.

What Is Usage?

In simplest terms, usage is how much service is *used*, where it is used and when it is used, as well as the conditions under which it is used, tied to who is using it. Everyone is familiar with "plain old telephone service" (POTS) long distance: each call is listed on the monthly telephone bill for a charged telephone number (who) with a time and date (when), the number of call minutes (how much), the called and calling numbers (where) and the applicable rate code (conditions). Obviously, to generate this information on the bill, at least the *when*, *how much*, and *where* must be collected at the time of the call—a usage record.

Similarly, an Internet service provider that charges based on connect time will identify the Internet account username (who), the number of connect minutes (how much), whether access is local or using a toll-free number (conditions). The current convention in the Internet industry is to be location independent, thereby removing the "where" from the equation.

Some pricing schemes involve a flat rate for some amount of usage, with charges for usage exceeding that amount. Examples include wireless services that include "500 free minutes" with the monthly flat rate fee and Internet services that charge a flat rate "up to 5 hours a month" or "up to 100 hours a month." These are still usage-based services, since all usage must be recorded and tracked to determine when and if the flat rate "quotas" are exceeded.

Generating Usage Records

Usage records are initiated within the network elements that set up and/or provide the service. A network element is a piece of equipment—such as a switching system, a router or a computer—that is involved in providing a service. Many modern network elements format and store call records or forward them in real time to call record processing systems. Some network elements with little or no data management capabilities (e.g., old electromechanical switching systems, simple modems and routers) interface with external traffic information systems that gather call information, generate call timing, format call information and store and/or forward call records.

In order to process a call or connection, the network element generally has the information required to identify the originator of the call or connection and the location being called, at a minimum. That information is recorded in a call detail record. When the connection is made (or, in some cases, when it is attempted or within 30 seconds after the attempt is made), the start time is recorded. When the connection is broken and the call ends, the end time is recorded. Usage on this type of call is calculated as the time from the start time to the end time of the call.

In a very limited number of cases, a local telecommunications switch may not have the capability to create usage records. In that case, usage recording is done at the nearest "toll" switch—a switch that does have usage recording capability.

Some network elements have the ability to create a message with usage information, but have no capability to store, reformat or relate the simple usage information to stored data. In that case, collection and call detail formatting and translation ("mediation") systems are usually connected directly to the network equipment usage message port.

Some services are based on units of capacity such as packets, rather than on dedicated connections between two locations. These are characterized as "connectionless" services. Usage in connectionless services is generally some sort of count of transport capacity. Date and time recording may be required in relation to time-of-day billing or for "call" identification for these services, but not as a means of determining usage.

Call Detail Record Data and Formatting

Call detail record (CDR) is used as both the generic term for a data record that includes all the usage parameters for a "call," and as a reference to protocols and formats for specific standard types of such records. Unless otherwise indicated, this text will use "call detail record" in the generic sense.

There are numerous types and formats of call detail records. Some of the differences are driven by actual differences in the type, configuration and price basis of individual services. Some differences are the result of differences in business and standards environments in different parts of the world and in different industries. Some of the differences are based on capabilities of the network elements that deliver service. Others are the result of individual vendors' proprietary designs.

Even the switch and call data recording vendors that conform to one or more industry standard for CDRs may vary in specific implementation since the standards have some flexibility. Older switches may not conform at all.

In North America, call detail records initially were called "automated message accounting" or AMA and produced paper punch cards for each call record. You may also find references to "computerized automated message accounting" (CAMA) if you are dealing with a legacy switching or systems environment. The term survives in the current complex Telcordia standard "billing AMA format" or BAF, which is used as the message accounting exchange format by virtually all legacy—and many other—telecommunications companies in North America.

Standards for "call accounting" information address several different aspects of call information. First, they identify what information about a call or transaction is captured and stored for accounting and billing purposes. Second, they specify the format and meaning of that information. Third, they specify the storage and transmission format(s) of that information (record and database layouts, etc.). Fourth, they specify data exchange protocols.

In addition to supporting customer-billing detail, call accounting standards support intercompany transactions for call authorization and prepaid calling. They also support "settlements" and billing between telecommunications companies. In the simplest terms, settlements are a mechanism in which companies that provide services for each other tally up all those services and determine which company owes the other for a specific period of time. Internationally, settlements are often governed by treaty agreements.

Most call detail record standards today also include fields for "real-time" rating of calls. This is most often used when the caller is not a subscriber to the originating provider (as in cellular roaming), when the service includes charges determined on the terminating end or if the call is to be billed by a third party, such as a commercial credit card company. These fields are also used in communicating information about "prepaid" card calls and other nontraditional services. Most calls within a carrier's network are rated in bulk after the call detail is forwarded to a billing system, primarily to minimize the load on network signaling resources that are involved in "real-time" rating.

Generally, if charges are usage based, customers will expect billing detail to support those usage-based charges. However, if charges are only loosely usage based, as in large lumps of wireless minutes or Internet connect minutes as part of the monthly "base rate," then customers may be less interested in seeing usage detail. Indeed, as recent highly publicized complaints to regulators and legislators have shown, customers may even perceive the collection and availability of usage detail for Internet services to be an invasion of privacy!

Call Detail Collection and Mediation

Network equipment may be limited in its ability to format, store and route call detail information. After all, its primary function is to provide service. There is a class of systems, known generically as "call detail mediation systems," designed specifically for the purpose of formatting, storing and routing call detail information. These systems generally support the most common call detail data standards and a range of manufacturers' network equipment accounting interfaces.

Identifying the Billed Account

In a typical POTS call, the calling number is the number to which the call is billed and the call detail record will show the same number for billed number and for calling number. In a "collect" call, the call detail record will show the same number for billed number and for called number. In a "third-party billed" call or a telephone credit card call, the billed number may be different from both the calling and called number. Generally, calls not billed to the calling number are routed to an operator or an automated charge authorization system that captures the billed number and enters it to the call detail record.

What about all those weird special calling codes—"800" numbers and other "toll-free" numbers? When a toll-free number is dialed, the network element identifies that number as a special case requiring a database lookup. The database lookup produces several pieces of information about calls to that number. It identifies the type of service code for calls to that number. It also identifies a routing telephone number (for all intents and purposes, a POTS number) which tells both network and billing where the call is going. It also provides a billing telephone number for calls to that special number. All this information is then recorded in the call detail record.

There are lots of variations on special calling codes. In some, the billed number is also the calling number, but the special calling code is translated to some routing telephone number.

Proposed "content-based" services will create a new wrinkle in usage billing. The call detail for content services will have to contain information on the content accessed and transported.

Routing Billing Information

What if the company recording the call detail is not the company with whom the subscriber has a billing relationship? This can happen in calling card and third-party calls; it can happen in wireless roaming; it can also happen when one carrier contracts with another for usage recording and in many other service configurations. [A brief historical footnote: The largest contracts of this type were between AT&T and the Regional Bell Operating Companies as part of the "Shared Network Facilities Agreement" negotiated at the time of breakup of the Bell System.]

How does billing information get to the right place? In the telecommunications arena in the United States, every telephone number is assigned to a specific telephone company that has a billing relationship with the customer. The assignment of telephone numbers and the maintenance of the assignment database for the United States are performed by Telcordia.

The telephone company may have a default location where all billing records are received or it may designate a "Revenue Accounting Office" or RAO for a specific telephone number. The full format for call detail records includes fields that identify the carrier and the RAO.

Calculating Duration

Generally, call start time and end time are recorded in a structured date/hour/minutes/second format, often identified as UDT or "Universal Date-Time" format in the local time at the originating switch. The local time for the originating switch is also stored in the call detail information. Generally, to calculate the duration for charges, the UDT formats are converted to a number representing the number of seconds from some reference date and time. Then the call start time is simply subtracted from the call end time. At this point, a lookup is performed to determine the usage increment in seconds (6 seconds, 60 seconds, 3600 seconds, etc.) The usage in seconds is divided by usage increment in seconds. If there is any fraction of an increment remaining, the usage amount is typically incremented by 1 ("rounded up").

Determining Rate

Once the usage quantity is determined, the billing system must look up the rate for the identified service. If the service is based on to-from codes, the lookup will include the originating and terminating routing codes, as well as the service identifier.

If the service has a time-of-day and/or day of week component, the origination time of day and day of week will be a parameter in the lookup as well. The rate found in the lookup will then be multiplied by the usage units to determine the cost of the call. In most telecommunications rates and tariffs, calls are rated based on the origination time *even if the rating period changes during the call*. Billing systems generally are capable of identifying when a call passes into another rating period (e.g., day into evening or night into day) and breaking the call into two charges, so this is driven by tariff or service description.

Determining Taxing Authority

Determining taxing authority can be a complex issue. It is very important to obtain specific legal and regulatory guidance for each area of service and for each service offering, as there are far too many jurisdictions, combinations of jurisdictions and constantly changing tax codes for this book to do more than skim the surface. There are several excellent pieces of tax software on the market that are readily compatible with (or fairly easily adapted to) most billing systems, new and legacy.

Quality of Service

Quality of service (QoS) is a relatively new issue as a variable parameter of service, as a result of dynamic bandwidth assignment as a network capability. Today, QoS is primarily considered a service configuration variable used to specify capacity prioritization. However, the possibility of varying QoS from one transaction to another creates the possibility of per-call or transaction charges based upon QoS. Thus, QoS may be added as one of the basic usage parameters. Most call detail formats include fields for QoS information at this point in time.

Summary

In this chapter, we looked at the development of charges for usage-based services. We defined usage and examined how usage data are generated, formatted and collected. We briefly looked at how to identify the appropriate account for billing and we looked at how charges are calculated. We also discussed the determination of taxing authorities. In Chapter 6, we look at the identification and calculation of one-time and recurring charges for products and services that are initiated by a service order, rather than by a "call."

Chapter 6

Processing One-Time and Recurring Charges

In Chapter 5, we discussed charges based on a customer's actually using a service—usually provided by a "network" or some processing capability. In contrast, some telecommunications charges occur only once or are billed periodically—usually on each scheduled bill.

Types of Charges

"One-time charges" are similar to normal retail or service billing charges in any business. Indeed, one category of one-time charge—the direct purchase of equipment—is exactly the same as retail billing. In that instance, a specific type of equipment has a set price (perhaps with customer discounts applied) and is billed to the customer at that price. One-time services, such as service set-up, may also have a one-time set price and are billed based on when the service takes place.

Some one-time charges involve variable labor and materials costs. These could include customer-site installation charges, coordinated testing services (when one company supports testing with other companies' services and products), maintenance and repair services not provided under contract or tariff and consulting and training services. Labor charges are generally assessed based on a set rate per hour or portion of an hour. Materials generally are assessed based on a unit cost for materials consumed or installed.

Recurring charges are those associated with a product or service on a periodic (usually monthly) basis. Many services have a basic monthly charge just for subscribing to the service. For instance, a residential telephone customer will pay a basic service fee each month to be connected to the network, even if there are no calls made from that telephone number during that service month. Many service providers offer a flat monthly charge that covers any basic service fee and some "lump" of usage. Examples of this include some low-cost Internet services that provide an Internet login identity and five hours of connect time for a minimal charge. Similarly, many wireless voice services provide a telephone number and a large number of "free" minutes for a flat fee per month.

Equipment rental is also a recurring charge. If the customer leases customer premises equipment (CPE) such as telephones, PBXs or data service terminating equipment, a fixed recurring fee will be billed for those items.

In the preceding examples, the communications provider charges a fixed monthly fee for a specific type of service. Another class of recurring charges, however, is based on the unique configuration of a specific service, such as distance, number of terminating points, customer control capability and many other features and options. The recurring charge for that specific configuration is determined at the time the customer order is placed and is thereafter associated with a unique service designator. The service designator, of course, is associated with a specific customer account.

Start of Billing

Generally, billing for one-time charges occurs in the first billing cycle after the services are performed or the product is delivered. In the case of time and materials billing, the time and materials quantity information is generally entered into a provisioning or dispatch system by the organization providing the service and materials. These quantities may be priced in the provisioning or dispatch system or they may be passed as quantities to the billing system. Frequently, services provided to business customers are not billed until customer acceptance is verified.

For products that are shipped for customer installation, billing may be triggered by the actual shipment from a warehouse or it may be triggered by customer acceptance of the shipment registered in a shipment tracking system. However, when customer-purchased products are part of a larger service configuration and/or if the contract for the product includes installation, then billing for the product may be linked to the completion and acceptance of that related work.

The start of billing for recurring service charges is controlled by two things—the service due date agreed upon by the salesperson and the customer at the time of the service order and the successful implementation of the service. If the service is not successfully implemented by the service due date, billing will not commence until successful implementation. If implementation is complete before the service due date, billing will not commence until the service due date unless the service due date is renegotiated with the customer.

Successful implementation will be defined for each service or type of service. In some cases, a test (remote or on-site) by service test personnel will be considered sufficient verification of successful implementation. In many other cases, the customer must formally verify and accept the service as working correctly. In either scenario, successful implementation will be recorded in the provisioning system(s). The provisioning system will then notify the billing system of the successful implementation of the service and the date and time when it occurred.

It is worth noting that some companies bill certain recurring charges in advance (i.e., for the upcoming service period). Generally, this approach is used for recurring local service charges in the residence and low-end business markets.

Determining Taxing Authority

If a service is "intrastate"—with all terminations within a single state—then that state has taxing authority over that service. However, if the traffic on that service is primarily destined to terminate in other states (such as a T1 circuit terminating on a toll switch or providing access to an "on-demand" interstate broadband service), it is considered an "interstate" service and is subject to federal as well as state tax.

Service that originates in one state and terminates in a different state is considered an "interstate" service and is subject to federal tax. In order to determine which state jurisdiction has taxing authority for a service between locations in two different states, the "two out of three rule" states that there are three locations to consider: originating station, destination station, and the location that the bill is sent to. If two out of three are the same, then that state receives the tax.

When a telecommunications company introduces a new service or enters a new territory, it should conduct a strategic legal review for each jurisdiction in which the company operates or offers the service.

Distance

As mentioned earlier, dedicated services often are priced by distance between service endpoints. This is easy to visualize for a two-point service as "miles as the crow flies" from one end of the service to the other. In more complex service configurations, each service segment is generally treated as a simple two-point configuration and then the segment distances are added up for the total service distance.

There are two common ways of describing a service endpoint location. In many older tariffs and legacy services, locations are described by telecommunications-unique parameters known as "V&H coordinates." V&H here stands for "vertical and horizontal." For more information about the V&H algorithm, go to http://www.trainfo.com/products_services/tra/vhpage.html. Often V&H coordinates for a service endpoint are taken as the coordinates for the closest "serving office" (an office that offers the capability to connect to that service), whether or not the service is actually connected to that office.

Many newer providers and service descriptions use generic latitude and longitude map coordinates for computing service distance. Of course, there are a variety of tools for converting one to the other.

Summary

In this chapter, we looked at the processing and pricing of one-time charges such as installation or equipment purchase. We also covered the processing and pricing of fixed recurring charges such as equipment lease payments or dedicated private line charges. In Chapter 5, we examined the processing of usage-based charges. Now we are ready to look at how separate collections of charges for many customers are organized into individual customer bills. In Chapter 7, we take you step-by-step through the compilation of the bill, using the customer and product information reviewed in Chapters 2 and 3, the business and operation information covered in Chapter 4 and the charge information developed in Chapters 5 and 6.

Chapter

7

Compiling the Bill

So far, we have discussed the various pieces of the bill. Now we will talk about when and how to put all those pieces together to form the bill to send to the customer. Billing systems may be scheduled to produce bills on a one-time basis, monthly, quarterly, yearly or some combination of those periods. Such variables as the credit classification, the market classification (e.g., business or residence) and individual customer circumstances will determine the period of time covered by a customer's individual bill.

Regardless of the frequency of billing, the basic steps required to go from accumulated information to an accurate, timely and complete bill are the same. Figure 7-1 shows the process steps, and each is described below.

When Does an Account Become Eligible for Bill Compilation?

A customer bill can be triggered by a regularly scheduled billing cycle date, by a customer request, by a company request or by the closing of an account.

Billing cycle date

As noted earlier, a "billing cycle" defines the period of time covered on a customer bill. Billing cycles are established to support the company strategy, which we will address in Section II. Generally, companies want to bill customers monthly. This facilitates cash flow and, for high-usage customers of as-yet-unproven credit, can limit company revenue loss if the customer defaults on an initial or early bill.

Billing accuracy requires a tremendous amount of coordination and management of data. For regular billing situations, modern billing systems rely on a system calendar to identify when to initiate a billing cycle's various steps. This facilitates the necessary coordination and management of data.

The billing cycle calendar identifies the date to be shown on the bill. It identifies the date through which charges and credits previously received by the

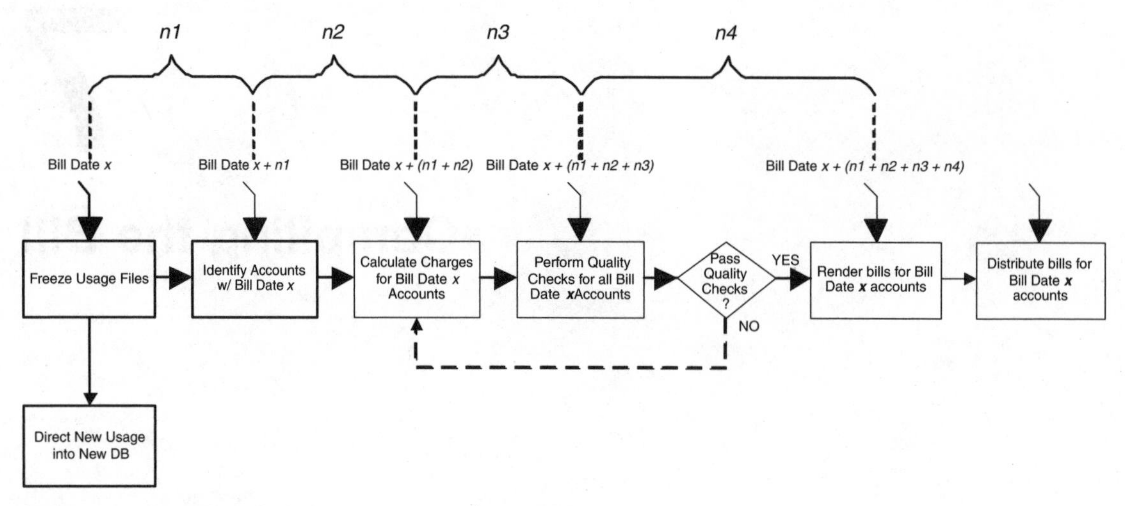

Figure 7-1 Compiling the bill.

billing system will appear on the bill "in production." It initiates a "freeze" date for the usage files—databases holding previously received CDRs—for the accounts "in production" and it initiates the direction of all usage received subsequent to the freeze date to usage files for the next billing cycle.

Management of the charges and credits to appear on the bill in production versus the charges and credits that continue to come into the system is vital. Without these controls, it is difficult—if not impossible—to recompile a billing cycle if an error is detected or some catastrophic situation occurs.

In addition to the dates in the system calendar, there are other dates that are important to the billing cycle preparation, such as the bill due date to be shown on the formatted bill and the date by which a late charge will be applied to unpaid amounts. These dates are found either in a billing calendar or in billing tables accessed during the billing cycle processing.

Final bills

When a customer account is closed, the last, or final, bill may be prepared during the next regular billing cycle date. Alternatively, the company may wish to have it prepared on a schedule driven by the last date the customer is in service, such as 5 days after the date the customer gives for account discontinuance. In either case, the billing system will use the date requested by the customer as the last "in-service" date to prorate any service or product charges that have recurring rates and it will capture all usage activity for the account with dates through the last day of service.

If the customer account has a recurring charge for a service or product that is billed in advance, the proration may produce a credit amount for the final bill. As with a regular monthly bill, any payments, adjustment credits and late charges will be identified and applied to the bill. Several additional steps could be taken as well. The billing system could review the account for any deposit being held and reflect the deposit amount against the total amount due.

Another possible unique final bill-processing step would be a review for an unexpired service contract period. For instance, if the customer has committed to use a service for 12 months for a special rate, understanding that a penalty will be charged if the service is discontinued early, the penalty amount will be calculated and reflected on the bill.

Some billing systems require the manual entry of an address for final bills. This is a good idea, as it forces the customer contact person taking the request for account closure to verify to whom and at what address the company can look for final payment. This is a helpful revenue protection step, especially as many account closure requests are related to a customer's relocating or to a business' reorganizing.

Occasionally final bills will show that the company owes the customer money: a credit final bill. The final bill may be sent to the customer showing that credit balance, generally accompanied by a note to say that a check for the credit balance will be forwarded separately. Alternatively, the final bill distribution may be routed to the financial organization that then will send a single envelope to the customer that contains both the bill and the check.

Special one-time bills

Some products or services may call for a special one-time bill. Usually this is for a customer with whom the company will not have an ongoing relationship. These bills are generally produced in existing billing cycles and show only the non-recurring charge for the product or service.

Customers may request special bills on a one-time basis. These may be requests for duplicate bills or they may be requests for a reformatted bill. Duplicate bills are usually available as a standard system option, but typically, if reformatted bills are to be provided, it is as a chargeable service.

Special one-time bills may be needed because of system changes or errors. For example, a billing cycle date change may require a partial month bill for a whole group of customers. This frequently occurs as a result of converting to a new billing system or structure or as a result of customer account redefinition or market realignment.

Calculation of Charges

The items in each account will be reviewed to determine if a charge (or credit) is applicable for this billing cycle. Charges are applicable if they were incurred prior to the billing cycle date and not previously billed. Items with an "in-service" date starting subsequent to the current billing cycle will be ignored; that is, held for billing in the next billing cycle.

Services

Services may be eligible for a one-time charge or a recurring charge. The service code (received from sales) assigned to the service tells the billing system where to look for the charge, what type of charge it is and provides the service descriptor to go into the bill. Charges usually are obtained from a rate table, but may be an amount that was provided on the service order applicable solely for that action and customer. No calculation may be required. For example, the billing system will go to a table for the code for a one-time, standard charge for connecting a line on a date during the billing cycle, associate that charge with the descriptor of the service and identify this as an input to the bill being prepared. The service code will tell the billing system which revenue accounting classification (including tax jurisdiction) to associate with the charge and, possibly, provide sales codes for compensation and tracking reports.

Calculation is required on a one-time charge when the service code tells the billing system that the charge is unit based. In that event, the service order will have been required to contain the units to be charged, which the system will multiply by the rate associated with the service code. The calculated amount and the service descriptor will be input to the bill being prepared. Also, the service code will tell the billing system which revenue accounting classification to associate with the charge and, possibly, provide sales codes for compensation and tracking reports.

Services eligible for recurring charges may have a unique recurring charge contained in the customer account record, but more normally, the service code contains the chargeable billing cycle amount (e.g., $10 for a full month's service), which is applicable to all customer accounts billed for that service. The first step the billing system must take is to determine for what period the service must be billed. An active service that has been billed to this customer on this account in the previous billing cycle will be identified as chargeable for the period from the first day of the billing cycle to the last day. The billing system will know that the full period charge applies and will input the amount and service descriptor to the next step in the process. Of course, as with the nonrecurring charges, accounting and sales code information will be associated with the charge.

Partial-month or multiple-month recurring charges are a bit more work for the system. Let's look at new services first. It doesn't matter whether the whole account is new to the billing system or if the service is new to an existing customer account: the *charge for the individual service will be recognized as new to the system* and, therefore, potentially eligible for proration. The number of days between the in-service date shown on the service order and the date of this billing cycle determine the length of time. Generally, systems using monthly charges prorate assuming a 30-day standard month. This means that each prorated day is equivalent to 1/30th of the recurring monthly charge. However, different billing system implementations handle prorating in subtly different manners, which becomes significant when comparing one system to another, as in parallel testing during a conversion.

Most new services will have been received by the billing system subsequent to the last billing cycle date for the account, so the prorated amount will be less that a whole month charge. But, because billing systems "know" that the service has never been billed to this account, the amount may be more. This allows billing systems to cope with the occasional (one hopes!) late or correction service order. In the event that the new service is one that is billed in advance, the number of days to be prorated will be those between the in-service date shown on the service order and the date of the next billing cycle, making it possible to have up to two months' charges on this bill. The service descriptor to be inserted on the bill is usually set to show the "from" and "to" dates for the proration, and it is passed along with the amount to the next step in the process. And, as usual, accounting and sales code information will be associated with the charge.

A canceled or discontinued service works in reverse of the new service proration. The billing system recognizes that the service is no longer to be charged to the account and uses the date that the customer requested service be discontinued and the last billing cycle date as the bounds for proration. The result will be a charge, unless the service is billed in advance; advanced-billed services will usually show that the company owes the customer a refund of previously billed charges.

Usage

As usage records (CDRs) have been received into the billing system, during the billing cycle they have been: checked to ensure that there is an associated account and service, rated, and held in a data base waiting for billing cycle processing. We won't cover rating here since previously we have discussed the ways that individual usage is calculated (see Chapter 3 for usage-based service pricing). The beginning of the billing cycle will cause the usage file to be "frozen" and any subsequent usage received for the accounts in this billing cycle will be directed to a new usage file for the next cycle.

The usage records in the frozen file for each service can now be gathered and processed to determine how they will be reflected on the customer bill. The service code in the customer account record will direct the billing system on how to aggregate usage, apply applicable discounts and display usage on the customer bill. The billing system will reject usage records that are received prior to the in-service date on the service—or after the date requested by the customer for service discontinuance.

Most modern billing systems direct all rejected usage records to an "unbillable" usage file. In addition to the types of unbillable records mentioned above, the file may contain records which the billing system received from a switching system or router and was unable to associate with a valid service/account or which contained incomplete or invalid data. The value of the unbillable usage file is usually estimated (if it cannot be specifically identified) and information is passed to the financial system to be booked as an uncollectable amount. It should also be sent to a manual error correction group for investigation and rebilling. It is very useful as a tool to determine where the service order/usage processing/billing processes need repair or strengthening.

Products

Essentially, product charges are handled in the same manner as services. The billing system will handle the sale of a product in the same way that it processes a one-time service charge. The date of sale will be used in the same manner as the provision of service date. A leased product charge will be treated in the same way a recurring service charge is treated. Products are assigned product service codes that contain the same types of information and are used in the same manner as those for the services.

Taxes

What is a tax and what is an "other charge?" It is tempting to designate all charges that are mandated or regulated by a regulatory body as taxes, but technically this won't do. Customer bills *must* show taxes, which are those monies telecommunications companies must collect from their customers on behalf of the government. Additionally, bills *may* show various other charges that regulatory bodies levy and permit the company to pass along to the customer. The number and types of taxes vary from state to state as do the requirements to explicitly state the amount on the customer bill. Some states also mandate the verbiage that must be used in the bill to describe the tax.

Federal excise tax and most state and municipal taxes are assessed on the customer account total charges. Other taxes, such as local jurisdictional 911 charges, the federal lifeline surcharge and the telecommunications relay service charge are applied to individual services in a customer account. The service code and the customer account record contain indicators that tell the billing system which taxes are applicable to specific services, the calculation methodology and the descriptor to use in the next step of the process. As with other billed amounts, accounting classification information is associated with the billed amount to direct the charges into the appropriate accounts and tax returns.

Some few customers qualify for "tax-exempt" status. These are customers who, when establishing the account, could provide documentation that taxation does not apply to them and/or their type of business. The customer account record will be identified as tax exempt, and tax will not be generated for products and services charged to the account.

Other charges

The company may initiate additional charges and, of course, regulatory jurisdictions initiate additional charges. Companies often wish to assess late payment charges and returned check charges to customers. The company's payment and collection procedures will identify when such charges are appropriate and input a transaction to the customer account. The billing system will see the transaction much as it does a one-time service charge: it is a new, never before billed item; it has a date appropriate to the current billing cycle; it has a service code that advises the billing system what amount and descriptor to show on the bill; and it has accounting classification information for the financial systems.

As with the taxes noted above, the number and types of surcharges that regulators levy vary from state to state, as do the requirements to explicitly state the amount on the customer bill. And as with taxes, some states also mandate the verbiage that must be used in the bill to describe the tax or surcharge. Examples of these charges include subscriber line charges, universal service fund surcharges, and local number portability charges.

The FCC subscriber line charge (SLC) allows local phone companies to recover a portion of the costs associated with interstate access to the local phone network. State subscriber line charges, which some state regulatory jurisdictions require, compensate local phone companies for a portion of the cost of providing local telephone lines associated with state and local services. The universal service fund surcharge (USF) is a federal program designed to support affordable basic service for various groups such as poor Americans, schools and rural health care providers. The local number portability (LNP) Charge is a charge that the FCC allows companies to apply in order to recover equipment upgrade costs necessary to make moving from one local exchange carrier (LEC) to another without changing telephone numbers technically possible.

The service code and the customer account record contain indicators that tell the billing system which of these charges are applicable to specific services, the calculation methodology and the descriptor to use in the next step of the process. As with other billed amounts, accounting classification information is associated with the billed amount to direct the charges into the appropriate accounts and tax returns.

Aggregating Charges, Taxes, Payments, and Adjustments for a Customer

Now that all the pieces of the customer account have been identified and calculated, they must be brought together to form a cohesive bill. As with any building project, this is done from the "ground up," and starts with each service within the account.

The billing system will determine that all data required for each service contained in the customer account are present and that charges have been calculated for associated service products, usage, service affiliated taxes, and other charges. It will determine that all products and charges to be billed at the account level are present and that charges have been calculated.

The billing system will identify the need, if any, to sum usage charges for the application of discounts to each *service*. This is generally driven by a unique product or promotional code contained in the service code. The billing system will calculate the usage discount(s), identify the accounting classifications that will be associated with those discounts and associate the discount with the appropriate bill descriptors for the individual service.

The billing system will identify the need, if any, to sum usage charges for the application of discounts on an *account* level (versus a service-by-service level). This is generally driven by a unique product or promotional code associated with the account—not the individual services. The billing system will identify which services are included in the discount program (this may be all or a selected group), calculate the usage discount(s), identify the accounting classifications that will be associated with those discounts and associate the discount with the appropriate bill descriptors at the account level.

The billing system will summarize all charges for both federal and state taxing jurisdictions. This may entail adjusting tax amounts calculated on usage (CDRs), depending on how discount plans have been structured. The strategic planning of such offers will determine if the discount is to be applied before or after tax is calculated.

The billing system will identify all payments received since the last bill was created. If this is the first bill for the account, it is possible that an advance payment may need to be applied. It will identify any adjustments that have been applied to the account. It will determine if these should be used as post-billing items or, if the identified transactions were not billed in a previous billing cycle, used to suppress the charge from this bill so that it is never seen by the customer.

Finally, the billing system will determine the due-by date to be shown on the customer bill. This may be standard for the billing cycle or determined by the credit classification in the customer account record. It will obtain from the customer account the customer name, address and any special directional information needed to get the bill to the customer.

After all the customer account pieces have been brought together, there may be one more item to be added. The sales organization may have the billing system insert messages on bills that are unique to customers using particular services or who are in a certain regulatory jurisdiction. This allows these types of messages to be more directly tailored to a segment of customers than those that are inserted in the bill format stage of bill creation.

Figure 7-2 illustrates a sample structure for a simple account with two services: a service for local telephone access and a long distance service. A typical billing system would take the following steps to compile the bill for this customer account:

- Determine the recurring charges and taxes for the local service and the three features associated with the service. Apply any discounts that may be appropriate for having combined features.

- Determine the recurring charge, if any, for the long distance service. Retrieve all usage accumulated during the billing period for the service, rerating it if called for. Determine appropriate tax treatment.
- Determine the recurring charge and tax treatment for the monthly report consolidation service.
- Determine the recurring charge and tax treatment for the CD-ROM additional bill feature.
- Determine what, if any, account level discount may be applicable due to having the combination of services (local and long distance) and additional products (monthly report consolidation service and CD-ROM additional bill feature).
- Aggregate the taxes, with appropriate consideration for discounts resulting from account-level consolidation.
- Obtain the billing name, billing address, delivery medium and any unique formatting requirements from the customer account record.
- Send to the bill rendering process.

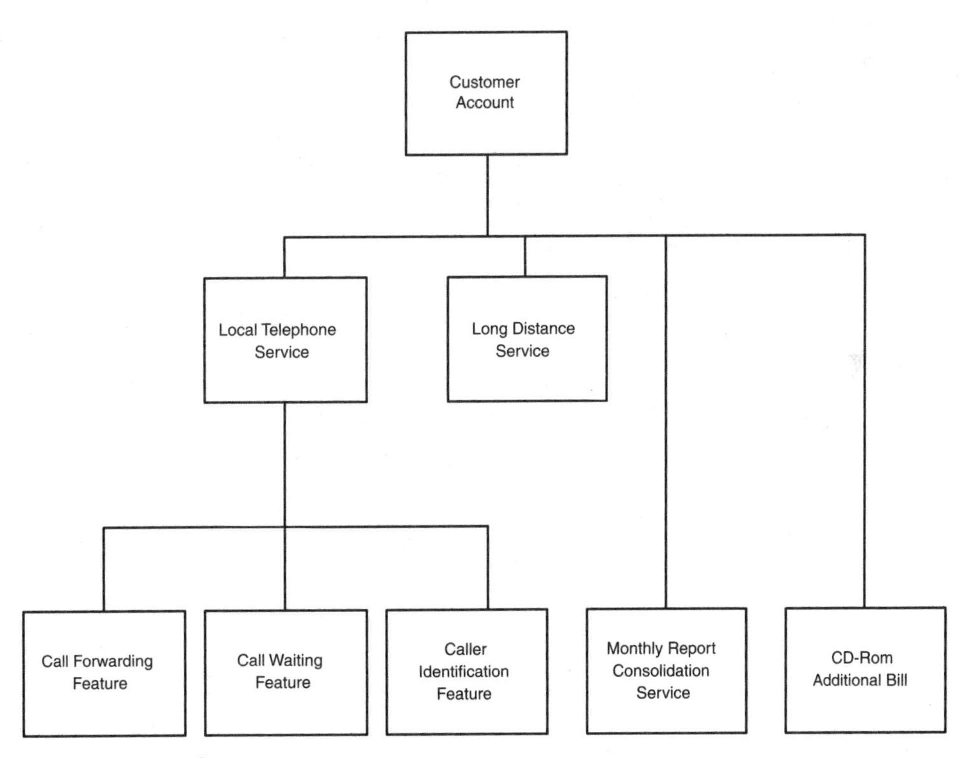

Figure 7-2 Sample account structure.

Most modern billing systems allow customers to establish account hierarchies, where accounts may be designated as subaccounts of a "master" account. The subaccounts may be billed with the master or billed separately, at the customer's direction. Figure 7-3 illustrates a sample structure for this type of more complex arrangement. As with the simple account compilation we just reviewed, the bill(s) will be compiled from the "ground up." The typical billing system compilation steps for the Figure 7-3 would be:

- Compile subaccount I, which is separately billed, using the steps shown in Figure 7-2. Maintain all information to be able to display data on the master account consolidated report.
- Compile subaccount II, using the steps shown in Figure 7-2 through the aggregation of taxes, for inclusion in the master customer account bill as a separate section.
- Compile master customer account, using the steps shown in Figure 7-2 through the aggregation of taxes, for the master account services.
- Obtain the billing name, billing address and delivery medium from the customer account record.

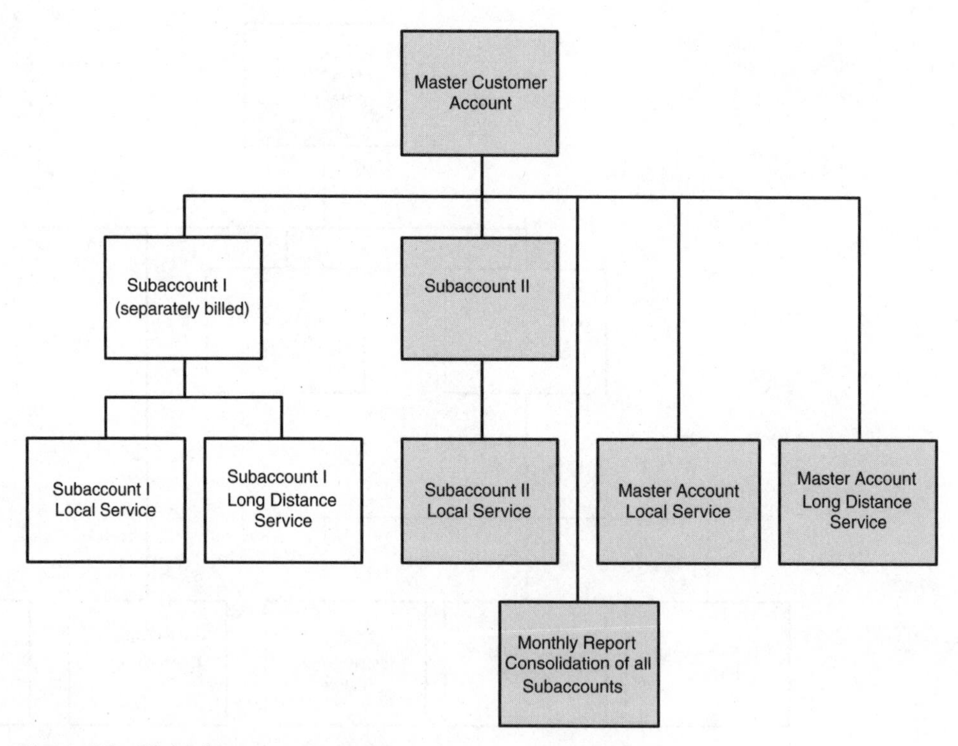

Figure 7-3 Multiple account structure.

- Obtain the formatting requirements for showing the master customer account and subaccount II in the same bill. All data may be combined or separate sections may be required, based on agreements or formal contracts with the customer.
- Compile information to prepare the monthly report consolidation of all subaccounts. Data will include master customer account, subaccount II plus the information that is separately billed to subaccount I.
- Send to the bill rendering process.

Note that all items that are shaded gray in Figure 7-3 will be included in the master account bill. Items that are shown in white in Figure 7-3 will appear on the separate bill for subaccount I. The service address on either sub-account may be the same as or different from the service address on the master account. It is entirely possible that subaccount I could be a large department or subsidiary at the same physical location as the master account, while subaccount II might be a small remote sales office. Account and billing structure does not necessarily reflect physical locations or service structure.

Quality Control of Compiled Information

One last activity should be done before the bills are prepared for delivery: a check of the validity and accuracy of the calculated and compiled data. Generally, this is done by an offline group, which reviews a representative sample of the accounts. The sample should include accounts from all jurisdictions served by the company as well as accounts that use all types of products and services. Many times the sample will include "sensitive" accounts that "must" be 100 percent accurate. For very large accounts, where item-by-item usage verification is impractical, aggregate bill amounts may be compared to typical monthly-billed amounts to identify unusual or suspect high or low charge amounts.

The billing cycle calendar should include a "window" of time to complete this review—with the understanding that the cycle processing cannot go forward until the billing media are confirmed to be accurate and complete. If errors are found, the next step is to analyze the impact rapidly. Is the problem affecting all accounts, some accounts or localized to a small minority? Is it a calculation or compilation problem versus one that is cosmetic? The problem may require restarting the entire billing cycle, which would be the worst case, as it could negatively impact the bill rendering and distribution functions, cash flow and customer satisfaction.

Figure 7-4 shows the possible impact on the billing calendar of a problem detected during the quality control review.

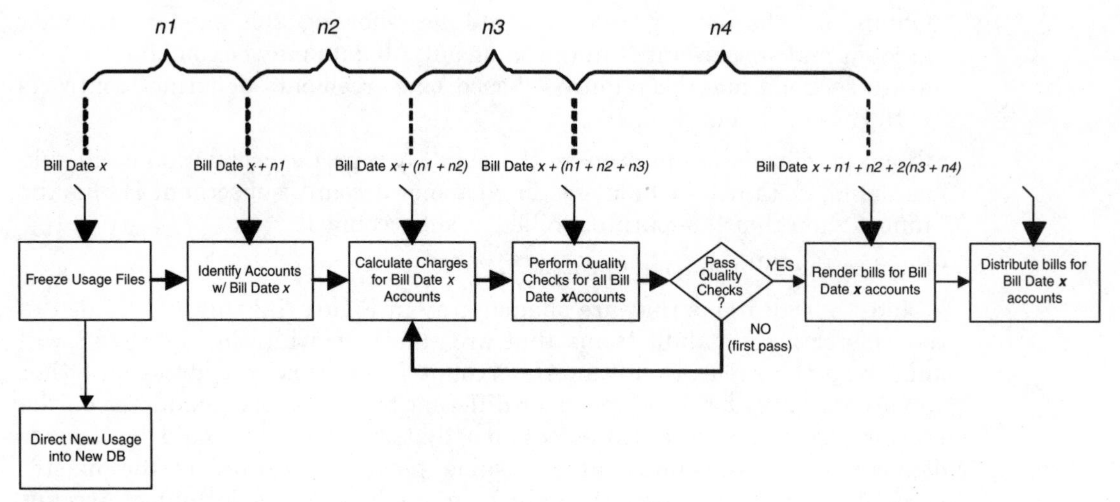

Figure 7-4 Impact of errors.

Summary

In this chapter, we discussed the frequency with which bills are prepared and how the various chargeable items of the bill are brought together. We examined different types of customer account structures and illustrated how they are compiled. And last, but certainly not least, we talked of the need for a quality control review before sending the information to bill rendering. In the next chapter, we look at the process of "rendering" or formatting this information for delivery to the customer.

Chapter

8

Bill Rendering

The appearance of the customer billing media is a key strategic company decision. The bill is often the only regular communication the company has with the customer, and formatting it to provide clear, easily read entries is a primary goal. Regulators and consumer advocacy groups may request (and/or require) entries with specific language, placement and even type fonts!

The content of bills sent to the customer in one medium will be essentially the same as that of bills delivered via other media. Company strategy usually demands that there be logo standardization and that other major appearance factors (such as name and address placement or service information displays) be consistently presented regardless of the medium in order to promote brand awareness.

Format Identification

The building blocks for the formatted bill were prepared during the bill calculation and aggregation steps. The product and service amounts, taxes, totals, payments, adjustments and associated descriptors from those steps are the items that will be inserted into the formatted bill and usually will be the same in each billing medium. However, formats may differ by market segment and the delivery medium.

Market specific

Bill messages often are tailored specifically for the customers in a market segment. These messages may be basic information placed in every billing cycle's bills on how to reach the company (for billing questions or dispute resolution), one-time messages providing sales promotional announcements or a combination of both.

Format specific

Delivery medium-specific format requirements may or may not be necessary, depending on the type of arrangements the company has with the entity responsible for bill distribution. The billing system may contract with a delivery medium to provide explicit direction on the appearance and placement of every item on the bill as well as the content to be placed in the bill. This would include such items as logos, fonts, spacing and placement direction. Alternatively, the billing system may contract to identify the type of item being passed to the delivery medium (e.g., billing name or service charge) and the delivery medium will apply agreed-upon formatting conventions to create the bill for the customer.

Bills in billing cycles that encompass large numbers of accounts are often resequenced to the print delivery process to facilitate using the U.S. mail's most economical rates. For example, accounts that will be in the same weight classes for a particular postal zone are grouped, allowing the bills to be printed, sent through automatic bill enclosing machines, supplied with the lowest rate-appropriate first-class postage and bundled for post office delivery with a minimum of human intervention.

Regulatory Requirements

Regulators in federal, state and local jurisdictions can and do dictate billing formats. As discussed in Section II, the requirements for the specific jurisdictions where the company plans to do business need to be understood and recognized when developing market, product and service strategies.

The direction given by regulators may be very specific, but often it provides information on a goal to be achieved and leaves the specific approach up to the company. For example, look at an excerpt from the Federal Communications Commission regulation Truth-in-Billing Requirements § 64.2001 dated June 25, 1999:

(a) **Bill organization.** Telephone bills shall be clearly organized, and must comply with the following requirements:

(1) The name of the service provider associated with each charge must be clearly identified on the telephone bill.

(2) Where charges for two or more carriers appear on the same telephone bill, the charges must be separated by service provider, and the telephone bill must provide clear and conspicuous notification of any change in service provider, including notification to the customer that a new provider has begun providing service.

(i) "Clear and conspicuous notification" means notice that would be apparent to a reasonable consumer.

(ii) "New service provider" is any provider that did not bill for services on the previous billing statement. The notification should describe the nature of the relationship with the customer, including a description of whether the new service provider is the presubscribed local exchange or interexchange carrier.

(b) **Descriptions of billed charges.** Charges contained on telephone bills must be accompanied by a brief, clear, non-misleading, plain language description of the service or services rendered. The description must be sufficiently clear in presentation and specific enough in content so that customers can accurately assess that the services for which they are billed correspond to those that they have requested and received, and that the costs assessed for those services conform to their understanding of the price charged.

These directions are similar to those that are promulgated by state and local jurisdictions.

Summary

In this chapter, we discussed the factors affecting rendering the bill. Customer needs, delivery medium requirements, company marketing strategy and regulatory requirements all play a role in the way that bills are formatted. In the next chapter, we look at the many options for bill delivery now available to telecommunications providers and their customers.

Chapter

9

Bill Distribution

Traditionally, telephone bills have been standard format paper bills containing multiple pages of call-by-call detail. In the 1970s, very large companies began exchanging billing information electronically, but these customized interfaces were not available to the small business or residential customer. Even 10 years ago, a book like this one would be focused primarily on issues of printing and mailing. While paper bills continue to occupy a large place in the world of telecommunications billing, the rapid expansion of Internet connectivity and World Wide Web access has added many new options for bill distribution.

Print

The print medium is still the most widely used method of getting billed information to the customer, although a number of alternative media are being provided to customers now in addition to (and, frequently, in lieu of) paper bills. A printed bill offers several advantages. Among them are: the customer gets a tangible record of the activity on the account supporting the billed charges and credits; a document is provided to return to the company with payment; and regulatory jurisdictions will accept the fact that the bill was mailed as evidence the customer was officially notified of charges due.

The printed bill may be as simple as a summary of the services and product charges. Generally, the bill includes an explicit statement of underlying component charges and usage, such as individual call data.

Internet

As more and more individuals and businesses have access to the Internet and become accustomed to doing business on the Internet, Internet billing becomes both feasible and desirable. Standards for XML "Electronic Bill Presentment and Payment" (EBPP) are the latest "hot topic" in the billing world—for telecommunications and many other industries, implementations of EBPP are becoming generally available.

In current practice, Internet invoicing is usually combined with some other form of billing notification or actual billing. The two most common scenarios for consumer and small-business billing are the combining of an email notification with Internet invoicing (and payment) or actual billing by commercial credit card of the total billed amount with bill detail provided on the Internet.

In either scenario, call detail and other itemized charges are formatted as a Web page (e.g., HTML or XML formats) and made available to the customer on a secured Web page. The customer-specific Web page will generally be secured by the use of a customer ID and password or PIN and may also require recognition of the customer's Internet login and domain using ID capture utilities. If the invoice information is formatted in a Web-standard format, this requires only that the customer have a compatible Web browser.

Fully Web-formatted invoices can also be readily structured to support online bill inquiry and challenge. They can also be integrated easily with Web-based bill payment capabilities.

A possible transition strategy from conventional print-format billing to Web billing involves spooling "print image" files and processing them through document imaging software (e.g., Adobe Acrobat®) for Web-based display. This requires the customer to download a utility such as the free Acrobat Reader® for display. This requires the implementation of the same type of user security environment employed in a native Web format implementation, but may serve to extend the viability of conventional bill formatting implementations.

Email

Email billing may involve sending a complete detailed invoice to the customer via email or it may simply be a notice of new invoice availability on a Web site as discussed earlier. A common "user-friendly" combination of email and Web technologies utilizes an email that automatically invokes access to the user's secure invoice Web page or provides a "hotlink" URL for the secure invoice Web page, which the user can "click" to access the bill. Pure email billing raises significant privacy and security issues, due to the notoriously poor security on email systems and transactions. Since formatting for email may depend upon the user's configuration of a mail reader, readable formatting of complex bills for pure email billing may be an almost insurmountable problem.

CD Media

Compact disc (CD) storage media have replaced magnetic tape for conveying large quantities of billing detail information to business customers. Many business customers who have large quantities of billing detail information and who analyze that detail information for departmental accounting, usage patterns, cost control, network design and other purposes prefer to receive billing detail in a bulk electronic format, which can then be loaded into their internal accounting and analysis software. Telecommunications companies may offer fixed formats for this information or they may negotiate customized information formats, generally for very large customers.

The production of CD media requires coordinated labeling of the media, as well as production of addressed mailing containers for the media. Once the media have been produced, labeled, packaged and addressed for shipping, they may be processed for mailing or for other types of delivery service. Using tracked shipping or mailing services may be especially important when a customer contract or service description includes a provision for timely delivery of billing detail information.

EDI

Business customers often prefer invoices that feed directly into their accounts payable processing systems. Standards for business-to-business EDI have been around for decades and many companies have implemented EDI for receipt of invoices.

Older EDI implementations require specific agreements regarding business rules and detailed formatting between the "trading partners" (i.e., customer and service provider). Generally, these are negotiated contractual agreements that are then implemented in EDI utilities and customized for each trading partner for periodic (monthly) billing. This type of legacy EDI interface is somewhat labor intensive and has generally been invoked only for the largest customer accounts. In these implementations, the billing environment forwards an invoice (billing) message to the EDI handling utility. The EDI utility transmits the invoice message to the designated IP address of the customer company's accounts payable system or (most likely) their EDI handling utility.

Recent work on XML-based EDI standards for EBPP combines the benefits of EDI with widely available Internet capabilities. This supports EDI capabilities for telecommunications billing for smaller business customers and even for consumer accounts. Specific EDI standards and standards organizations will be discussed in Section III.

Commercial Credit Card

An initial customer service agreement may include authorization to bill account charges to a commercial credit card. This is quite common for Internet services, but is becoming more widely used for a variety of communications and information services. The service provider must establish a "merchant account" with the credit card company or with a credit card clearinghouse. When the customer account is established, the customer will provide—and the provider will verify—relevant credit card information including card number, name on the credit card, card billing address and card expiration date.

Most commonly, the service provider will calculate the entire monthly bill amount for a customer and will forward that as a single transaction for payment by the credit card company and it will appear as a single line item on the customer's credit card bill. Bill detail is generally provided in another format (e.g., Web page, email, paper, CD). Occasionally, customer service agreements or regulatory requirements will specify that items above recurring monthly charges must be separate transactions to the credit card company.

Summary

In this chapter, we outlined the various methods that are used to provide customers with billing information. The ubiquitous paper bill of yesterday is giving way to electronic delivery media for a steadily increasing number of customers.

Chapter

10

Account Maintenance

Account maintenance covers the regular day-to-day work that a billing system must do to ensure that accounts reflect all payment and adjustment activity. Another important aspect of account maintenance is keeping the customer name and billing address and credit information current.

Security Deposits

Typically, security deposits are requested from prospective customers who have not yet established a credit record—or from those whose credit history is known to be poor. When such a deposit is required, service provisioning will not be completed until the deposit has been received by the billing system. Occasionally a security deposit will be requested from an existing customer from whom collecting money is a difficult and risky job.

The deposit amount is shown in the customer account record, indicating the date it was received, the length of time for which it is to be held (this may be indicated as "indefinitely") and when, if at all, any interest may be paid to the customer. The billing system provides the financial system with the amount of the deposit and accounting classification information to ensure the deposit is accounted for appropriately.

The best time to return a deposit is when the customer has established credit to the company's satisfaction. The amount of the deposit, with interest if warranted, usually will be applied to the customer bill as an offset to billed charges. In other cases where the customer defaults on the current bill and the account's services are suspended or discontinued, the held deposit is used to offset final bill charges.

Payments

Payments may be received at any time in the life cycle of an account: as advance payments prior to service provisioning; as payments for accounts with current product and service charges; or to apply to bills on discontinued accounts.

When a payment is received, it will be identified as belonging to a specific customer account and credited to that account on the date it enters the billing system.

- If this is an advance payment (i.e., the account has never been issued a bill), the account balance will now be a credit, which will be brought forward to the first customer bill.
- Payments on previously billed accounts will be deducted from the amount of the previous bill's balance, and the new balance will be calculated.

Most billing systems track balance due amounts on bills to understand how long the period is between the customer's receipt of the billed amount and the company's receipt of payment. The information is often referred to as "payables aging data" and is used for collection activity and financial reporting.

Adjustments

Requests for customer account adjustments may be initiated by either the customer or company personnel. Most adjustments are for items that have appeared on a bill, but adjustments may also be made for items that are being held by the billing system for future billing.

The most efficient auditor of billed charges may be the customer. The billing system and the manual processes supporting the quality of the system are designed to prevent billing errors, but there will inevitably be some that slip through and will require removal.

Customer requested adjustments

Customers request adjustments for charges appearing on the bill that they do not recognize or that they feel are incorrect. The customer may accept disputed charges after the company investigates and provides additional information. But when additional information confirms the charge is incorrect, or the customer is still dissatisfied, an adjustment transaction is created and input to the billing system.

The adjustment input should contain a reversal of data that billed the charge plus a code showing the reason for the adjustment. For example, if the charge to be adjusted was a call from Chicago to Atlanta of seven minutes' duration billed at a weekend rate, the adjustment transaction will contain those same pieces of information plus a code indicating that the customer denies making the call. This, in essence, not only removes the charge from the customer's account, but also alters the information used to reflect this charge in the company records. The customer will see a credit for the previously billed charge—and any associated taxes—on the next bill. In this example, the company found no error in billing during the investigation of the charges, so it must treat the adjusted amount as a properly billed amount that is "uncollectable" and adjusted for "customer satisfaction."

When an investigation results in the identification of a charge truly billed in error, the customer account appearance after the adjustment will be affected in the same manner as for the situation described above—all charges for the transaction will be removed from the account. The company books will not be affected in the same way; instead, the original transaction amount will be taken out of the original usage revenue category and directed to an "unbillable" account. This type of accounting classification change may impact company taxes. It may also cause the company to investigate what caused the incorrect billing and work to some long-term process correction.

The investigation into an incorrectly billed amount may show that there are similar transactions being held in the billing system for insertion on the next bill. These can be adjusted before they are shown on the bill. The adjustment input is prepared the same way as for removing a billed adjustment and the system can suppress showing either the charge or its reversal on the customer bill.

Company initiated adjustments

The company may initiate either pre-billing or post-billing adjustments to a customer account if error conditions are identified. The company may determine that adjustments are required as a result of quality assurance reviews or service order error correction activity. The adjustment inputs to the system are prepared in the same manner as those for customer initiated requests.

Automatically generated adjustments from the maintenance process will be calculated based on the duration or estimated customer impact of an outage or maintenance activity.

Fraud correction adjustments

A common situation that causes the company to initiate adjustments is the identification of fraudulent usage directed to a customer account. Most billing systems review customer account incoming usage volumes and compare them to previous months' billing or to preset levels for the customer account credit classification. A sudden precipitous rise in the amount of usage entering the system for a customer account will cause the billing system to flag the account for manual review and possible action. The review may show that the usage does belong the customer and no further steps are needed—other than a possible credit classification update. If the review shows that fraudulent usage charges are being received, the company will initiate adjustments and take appropriate steps to identify the source of the problem and correct it.

Billing Information Changes

The customer account record stores information on the person—or company—responsible for paying the bill, the billing address and credit information. The sales force initially obtains this information when an account is established and it is important that it be kept current when opportunities arise to have contact with the customer.

Names

The billing name should be the individual or corporate entity responsible for the bill. Identifying the proper person or name for the bill is vital should the customer default on charges, thereby requiring the company to take legal action for correction. Residential accounts or small companies owned by a sole proprietor should have that individual named as the person responsible for billing. Accounts for companies that are partnerships should show the primary partner as the billing name. Accounts for corporations will be billed to the corporate name, but the customer account record should contain a list of the principal officers of the corporation, the state of incorporation and the name of the individual to contact for billing questions or concerns.

A customer may request that a billing name be changed in an existing customer account "just for the direction of the bill." This type of request should be carefully considered and only done if it will not result in a change in billing responsibility. Changes in billing responsibility (e.g., the sole proprietor of a business asking to have someone else's name entered as the billing name in the account) should be directed to sales and handled as a new service.

Some billing systems allow name changes to be entered directly into the billing system, but many require that all changes be entered via a service order initiated in sales.

Addresses

The billing address in the customer account record may or may not be the same as the address of services. When discussing billing issues with existing customers, and certainly with final bill customers, it is prudent to verify that the billing address is the correct and most current one.

Some billing systems allow billing address changes to be entered directly into the billing system, but many require that all changes be entered via a service order initiated in sales.

Credit information

Credit information may be reviewed as part of collection activity or during discussions with the customer who is requesting an adjustment. Any credit information changes identified during such contacts with the customer should be entered into the billing system.

Summary

In this chapter, we covered the vital and ongoing job of maintaining customer account records. We described the types of payments and adjustments that are applied to customer accounts. We also discussed the name, address and credit information maintenance that is required to provide accurate and timely billing. In the next chapter, we look at how payments are made and what happens when they are not made.

Chapter

11

Payments, Collection, and Treatment

This chapter provides an overview of how payments get to the service provider and what happens when payment is not made in a timely manner.

Receipt of Payments

Payments may get to the company in a number of different ways. It is important to establish payment methods that customers can readily use and that can expeditiously move the payments into the company accounts.

Direct to the company

Payments may be mailed or delivered directly to the company. Several processing steps are needed to properly reflect the payments in the billing system and the financial system concurrent with putting the money into a bank account. Most companies that receive payments in this manner have specialized mail receiving units that are trained to control the receipt of payments and provide timely notification to billing and financial systems.

In the event that a customer's check for payment is returned unpaid, the customer account will be debited—or charged back—the amount of the returned check and collection action will be initiated. Most companies have made the strategic decision to charge customers a fee for each check returned unpaid. This transaction is entered into the billing system in a manner similar to an adjustment on the customer account. The charge will be billed to the customer on the next billing cycle.

Via lock-box bank arrangements

The company may contract to have a bank receive all customer payments. The customer is asked to mail payments to the company at a postal box number (the "lock-box") address, which belongs to the bank. The bank identifies the payment amounts for each account and electronically notifies the billing system of the transactions. The bank also prepares the input to the company's financial organization and moves the money into the company's bank account.

Lock-box bank arrangements are able to shorten the interval for getting the payment into a bank account where it is available for company use. They also provide timely customer payment information to the billing system.

Lock-box bank arrangements include notification to the company that a customer check has been returned unpaid. In that event, the account will be referred back to the company to take collection action and initiate a returned check fee, if appropriate.

Via debit or credit card

Many companies now offer customers the option of bill payment by credit or debit card. When the customer account is established, the customer provides the card number and authorizes the company to automatically collect the amount billed within a specified period of time after each bill. The company regularly delivers a bill for review to the customer, who may dispute charges and withhold payment if desired.

Via collection agencies

Customer accounts may be referred to an outside collection agency after all company efforts to collect overdue amounts have been exhausted. Typically, the collection agency agrees to take the accounts for a percentage of the amounts collected—sometimes as much as 50 percent or more. The customer is told to direct all payments to the collection agency and the collection agency then remits the payment amount less the agency percentage to the company.

Electronic bill payment

Options for electronic bill payment are becoming more common. Electronic payments may be made directly to the company or, more commonly, to an electronic payment agency or bank that provides billing system updates and cash accounting updates similar to those provided by lock-box banks.

Identification of Overdue Accounts

The due date for payments is shown on the customer bill. Bills not paid by that due date are overdue. Many companies now assess customers a "late payment charge" if the payment is not received by the due date. The billing system can identify the accounts to which a late payment charge is applicable and generate the charge to appear on the next customer bill.

Collection Steps/Treatment Schedules

Not all overdue accounts will be flagged for collection action. The credit classification will determine the amount of time granted to a customer account before it is eligible for action. Customer accounts that have established good credit, and carry the most highly rated credit classification, may be allowed to be 45 to 60 days overdue without being contacted by the company for payment.

A standard approach to collection is to start with gentle reminders, followed by increasingly stronger actions that could result in termination of service. Most billing systems will set up collection schedules for customer accounts based on the credit classification, automatically produce standard letters and notify collection personnel to make collection calls or initiate account discontinuance.

- **Letters**. The initial letter may actually be a reminder message on the next customer bill. A follow-up letter may be more strongly worded, mentioning the possibility of service termination and possibly giving a date for account termination. Some overdue customers who have special relationships with the company (e.g., company board members or local political figures) may never be sent collection letters. Instead, a designated employee usually contacts such customers for individual handling.
- **Calls**. A call to the customer may be made to ensure that the customer recognizes how serious the company is about the need to collect the amount due. The customer needs to be told the consequences of failure to pay the amount due.
- **Discontinuance of service**. Discontinuance of a customer account should be a last resort step, and taken to protect company revenue.
- **Referral to outside collection agencies**. An account should only be referred to an outside collection agency when it is deemed to be unlikely that additional reasonable efforts will produce payment. Many companies do not want to devote the time and personnel effort to extensive final bill collection effort, and use of the collection agency is preferable to not making any additional collection effort.

Table 11-1 provides a sample of a collection approach for a single credit classification. Note that bill calculation, compilation and invoice generation all occur on the Bill Date. This is desirable for process controls and management. In this example, accounts will be due and payable when bills are rendered. A late payment charge will be assessed if payment is not received within 20 days of the bill mailing date. Customers who have not paid a monthly bill in full by 45 days after the bill mailing date will receive a call asking for payment in full. If payment is not received, nor any arrangement made, one letter will be sent stating the service will be subject to suspension of service and, ultimately, total denial of service.

TABLE 11-1 Sample Billing and Collection Activity Cycle

Date	Event	Limitations/Requirements
Bill Date (BD)	Billing process is initiated	
BD	Invoice generation process is initiated	
BD + 1-4	Bill verification and approval	
BD + 5 (=BM)	Bill mailing date (BM)	
BM + 20	Payment due date	Bills not paid in full will be subject to a Late Payment Charge (LPC) of *x* percent on the total amount outstanding. This will appear on the next monthly bill even if the payment is received before the next bill is generated.
BM + 45	Full or partial amounts still outstanding will be eligible for a telephone call requesting payment or a payment arrangement.	If the amount still outstanding is less than $50 and a balance that has been carried forward equal less than 3 months of charges, no call will be made.
BM + 60	Full or partial amounts still outstanding will be eligible for a generated treatment letter giving 10 days for receipt of payment to avoid a temporary interruption of service.	If the amount still outstanding is less than $50 and a balance has been carried forward at least 2 months, a call will be made.
BM + 70	Accounts with full or partial amounts still outstanding (where there has been no negotiated payment arrangement) will be output on the "temporary interruption for non-payment" report.	The report will be directed to sales for temporary interruption service orders to be written.
Temporary Interruption Date (TID)	The date that the provisioning organization completes the work to place the customer's service on temporary interruption.	The TID will be the effective date on the in-effect report associated with the service order.
TID + 1	Generate a letter giving 10 days for receipt of payment to avoid complete discontinuance of service.	
TID + 10	Accounts still outstanding (where there has been no negotiated payment arrangement) will be output on the "permanent interruption for non-payment" report.	The report will be directed to sales for total disconnect service orders to be written.
Final Bill (FB)	Produce a bill showing all charges and events through the last day the account had items in service. This will be produced at the next regularly scheduled bill cycle date. (Note: The FB may be the result of customer requested disconnection or non-payment disconnection.)	If the final bill shows a credit balance (i.e., the customer is owed a refund by the company), generate a report for the finance department, where a check will be generated and sent separately to the customer.

continued on next page

TABLE 11-1 Sample Billing and Collection Activity Cycle (continued)

Date	Event	Limitations/Requirements
FB + 10	Generate a letter for accounts disconnected for non-payment. The letter will advise that payment must be made within 10 days to avoid referral to a collection agency and/or legal action.	
FB + 20	Generate a report for managerial action on accounts disconnected for non-payment.	
FB + 30	Generate treatment letter reminding of overdue payment of full or partial amounts still outstanding for accounts disconnected at the customer's request.	
FB + 70	Generate treatment letter giving 20 days for receipt of payment to avoid referral to a collection agency and/or legal action for full or partial amounts still outstanding on accounts disconnected at the customer's request.	
FB + 90	Generate a report for managerial action on accounts disconnected at the customer's request.	

After billing is calculated and charges are formatted for printing and reviewed for accuracy, an invoice will be directed for output to the customer (either electronically or to a printer for print and mailing). In our example, there will be approximately five days between the bill date on the bill and the bill mailing date. The "payment due date" is determined by adding 20 days to the bill mailing date.

Maintenance of Treatment History Files

Billing systems will maintain a record of all programmed treatment for a customer account and will allow collections personnel to enter notes about customer promises to pay or other aspects of customer contact. The treatment history file is useful in determining how, if at all, the credit classification for a customer account should be updated.

Aggregation and Use of the Payables Aging Report

As discussed in Chapter 10, the billing system tracks all outstanding balances due on customer accounts to maintain information on the number of days amounts have been unpaid. The billing system can produce reports displaying the data in various ways. For example, the finance organization can obtain a breakdown of the total of balances due for a specific billing cycle and how much of that total has been outstanding for 30 days, 45 days, or 60 days.

Payable aging reports may be available by market, customer or jurisdiction. The collection organization should review the reports by customer on a regular basis to ensure that overdue accounts are getting appropriate attention. An analysis of the payables aging report allows detection of accounts that have been assigned an incorrect credit classification or where no follow-up on collection action has been done.

Summary

In this chapter, we saw the avenues available to customers to get payments into the company. We also discussed what to do when the customer does not pay. The steps most often used to affect payment were described and the tools available for monitoring customer payments and treatments were detailed.

In Chapter 12, we turn to various reports that can be generated from billing information. Some of these reports may support the billing and collections process, including tracking customer payment status and uncollectables. Others are used to summarize billing information for other business processes.

Chapter

12

Reports, Data Feeds, and Query Databases

The billing environment creates, processes and aggregates critical information required by many other business functions. In legacy billing environments, this information may be contained in hard-copy reports specifically programmed to meet an ongoing business need. Many financial transactions are structured as direct data feeds from the billing environment to standard accounting system formats. Other information needs are best served by databases specifically created to support information queries or by direct queries against the billing databases.

Corporate accounting requires advice of receivables, payments and adjustments, uncollectables and payables from the billing environment. Fraud management (which may be a financial or security or operations function) requires advice of suspected or confirmed fraudulent activity. Product management requires feedback on the financial performance of products and services, markets and promotions. Billing data also may drive sales compensation.

This chapter looks at the most common requirements by other processes for information from the billing environment. In previous chapters, we have mentioned the collection and maintenance of some of the data that support this type of reporting such as accounting classification, jurisdiction, product and service codes and market designation.

Posting to Accounting Systems

The most critical and sensitive of the billing environment's external interfaces are the direct feeds of financial advice to the corporate accounting systems. Controls on these interfaces are discussed in Chapter 13. Generally, these interfaces are fully electronic, but often not fully automated. Accounting systems frequently require manual intervention before updating ledgers. Frequently, automated notification processes, such as email to the manager responsible for the ledger, are implemented in the billing process following creation and availability or transmission of the advice file.

Receivables

Each billing cycle will create a receivables advice file, identifying to the accounts receivable system all new receivables identified in that billing cycle. Note that this is not the same as total billed amount in a billing cycle, since total billed will include previously billed but unpaid amounts.

Payments and credits

As payments and credits are processed, the billing system creates payment advice records for accounting. A payment advice log is generally provided to the accounts receivable environment on every business day.

Adjustments

Chapter 10 discussed the different types of adjustments and their accounting classification. When adjustments are made, the billing system creates an adjustment advice record and enters it either into the "unbillable" adjustment file or to the "uncollectable" file. It is good practice to post adjustments daily. However, if the volume is small, the corporation may decide to post less frequently.

Uncollectables

Uncollectables are billed amounts that cannot be collected at all or that cannot be collected without expenditure and risk not commensurate with the amount of the uncollectable. When an account is deemed uncollectable, an uncollectable advice record is created and entered into the uncollectable file. The uncollectable file will be forwarded to accounting as agreed between billing and accounting.

Payables

For each billing cycle, the billing environment aggregates tax information by taxing authority and type of tax. This information may be further aggregated for a month or some other reporting period mandated by the taxing authority. This information is then used to create reports that go to the taxing authorities and to generate payable advice records to the accounts payable system, which then generates payments to the taxing authorities.

Fraud Management Flags and Queries

Controlling "theft of services" in the telecommunications arena is a critical business function. Indicators of potential fraud occur at many points in the call accounting and billing flow. These indicators must be forwarded to fraud investigation in a timely manner. While fraud investigation activities may lead to apprehension and prosecution of the perpetrator, the corporate driver of fraud investigation is limiting potential losses.

Loss analysis for network advice

Fraud investigators use analysis of billing adjustments, unbillable calls and uncollectables to identify geographic areas, specific telephone numbers and accounts that are particularly susceptible to toll fraud. When patterns of fraud are identified, network flags may be set to create real-time notification to fraud investigators of activity matching those patterns. For instance, analysis may show that calling cards used at a public telephone in an urban bus terminal such as New York's Port Authority terminal are often subsequently used for fraudulent international calls. Any calling card number used at that telephone may be monitored for several days for subsequent "high-ticket" calls. If such a call is placed, the call information may be routed immediately to fraud investigations who can verify the caller's identity and who may check legitimate usage patterns in the customer's billing history for similar calls. This protects the company and the customer from loss, while minimizing inconvenience to the customer.

Billing fraud

Customers often request credit for "high-ticket" individual calls. While such calls may be indications of fraud by a third party, a review of the customer's billing history may show a history of calls—and even credits for calls!—to the same number. This may be detected by the customer care person dealing with the customer or by the fraud investigation group when looking at uncollectables. Once such behavior has been identified, it should be flagged in the customer account record.

"Bad accounts"

Sometimes customers open accounts and use services with no intent to pay. While credit screening and service deposits can limit the risk of this type of fraud, if initial usage is not limited to the amount of a service deposit, losses can be significant. Some companies do usage screening—monitored by fraud investigators—on new accounts to limit this type of risk. Wireless providers are particularly vulnerable to this type of fraud, since their service terminals are not fixed to a specific customer premises.

Product Management Reporting

Product management and strategy need to know how much revenue is being generated by specific products and services, promotions and campaigns by account type and market. Generally, modern billing systems carry indicators of numerous parameters of interest for marketing analysis in each customer and billing record. Comprehensive data on billing and payment detail are generally provided to product management query systems on a monthly basis—or more frequently if required.

Sales Reporting

The sales organization requires billing information for many reasons—ranging from calculation of sales commissions to identifying additional sales opportunities.

Sales compensation

Some companies compensate their sales forces on the basis of initial orders. However, companies interested in long-term relationships with customers, which are critical to telecommunications profitability, compensate based on actual revenues generated by a sale. This generally requires that sales compensation systems receive advice of revenues received according to the original service order number. This is a large amount of detailed data and may be a monthly or quarterly feed to sales compensation. Since this is the basis of payments, it should be subject to the same controls as other financial data interfaces.

Customer account analysis

Often, information about a customer's usage and payment history is useful in targeting ongoing sales activities. Billing history is often provided to sales and account management query databases.

Summary

In this chapter, we looked at an overview of the most common data reports billing systems provide to other company processes, systems and departments. These reports contain information from the billing system that allows other parts of the company to operate appropriately in areas such as corporate accounting, customer care, personnel compensation, sales/marketing and network management.

In Chapter 13, we survey some important account and systems controls that are used to maintain the integrity of billing information and of the customer bills. Some of the reports identified in Chapter 12 play a role in those controls.

Chapter 13

Account and System Controls

The billing system is very dependent on other company systems for input and support. The sales system provides customer names, addresses and credit classification; the provisioning system supplies service and product data; the network organization captures, rates and supplies usage (CDR) transactions.

Similarly, other systems and business functions depend on information collected and/or generated within the billing environment. As discussed in Chapter 12, corporate accounting, fraud management, product management and sales all require billing information.

In this chapter, we discuss some of the process controls—both automated and manual—that help keep system and process interfaces functioning effectively. We also discuss some of the tools that ensure the systems can communicate well and that data are not lost between one system and another.

Process Controls

The service order processing flow between the sales, provisioning and billing organizations must be well coordinated to get the customer order from start to finish. The initial request for service is processed through to a working service with established billing records available to all the company's customer contact personnel. Each group must be clear on its responsibilities and the system handoffs of information have to be coordinated.

Service order receipt and error correction

With three major systems involved in service order processing, it is imperative that there be some common designation for the order as it moves from system to system and back again. Billing systems expect the service order number and the customer account number to provide that linkage.

Each of the systems will perform data validation checks as the service order goes through processing. The processing systems must develop processes to ensure that the data from one system meet the expectations of the next. Some of this coordination may be accomplished by middleware in modern provisioning and billing environments, but not all. When a service order transaction is sent from one system and is unacceptable in the receiving system, the company must have an intersystem process in place for problem notification, resolution and error correction.

Billing systems track both service orders in progress and service orders in error. Both reports assist billing personnel to work with sales and provisioning on error correction and problem resolution.

Product catalog coordination

The product catalog, or list of valid company products and services, may be a billing database shared with sales and provisioning. Most frequently, it is not. The individual systems create and maintain individual databases that contain information about the same items. Processes for synchronization of these databases must be developed and implemented.

Customer care

The billing organization can have significant impact—both positive and negative—on the customer care organization and vice versa. Processes need to be in place to keep the customer care personnel aware of such things as bill date schedules, bill appearance changes, and rate change implementations that must be explained to customers. The customer care personnel must be accurate in the verification, preparation and entry of billing information. Billing should track and provide reporting to customer care management concerning any billing process failures resulting from inaccurate customer care transactions, such as mailed bills returned because of erroneous customer address information.

CDR error processing

Call detail records received by the billing system from the network may not be acceptable for billing. This is another example of the need for an interdepartmental process approach for problem identification and resolution. The billing system needs either to bill each CDR to a customer account or estimate the financial impact (the unbillable/uncollectable value) so that the company books of account are correctly stated.

The billing system will produce reports detailing the unbillable transactions to use in both the problem identification and resolution. Network personnel and possibly salespeople often will be important members of the resolution team. If the problem involves CDRs from another company, the resolution may include regulatory and/or strategic relations personnel, as well as representatives of the other company.

Revenue reporting

Product management analyzes and reports revenues in different "buckets" than do billing and accounting. However, total reported revenues for a specific period should be compared from product management, billing and accounting for "reasonableness."

System Controls

System controls and reports that are designed to ensure accurate transfer of data between systems may also be used to validate billing media. Here are some major examples of the additional use of these types of reports.

Usage

During the course of a month, the billing system usage databases will receive CDR files from switches and routers. A report will be produced for each transmission, showing the date, time, and number of records transmitted. Billing system personnel use these reports to identify patterns and anomalies. Problems such as missing files or duplicated files can often be identified before critical dates in the billing cycle are reached, thereby avoiding adverse customer impact.

Customer payments

Customer payments received and processed by a lock-box bank on behalf of the company are transmitted to the billing system electronically. Billing system personnel use these reports to track and analyze customer payments in much the same manner as the usage file.

Accounting advisements

Most interfaces between billing and corporate accounting are highly controlled. Log files are created for each feed, which contain version and synchronization information, counts of records and bytes transmitted, financial totals and other business and technical validations of the feed. In a well-controlled process, these log files are sent separately from the actual data feed and are compared to the receiving system's log of received data, as well as to results from posting the transactions. Any fallout from posting is sent back to billing for resolution. A well-controlled environment will track the resolution of fallout. This may be fully automated, partially automated or completely manual.

Reconciliation

Reconciliation processes may be highly automated or entirely manual. Reconciliation is required when two separate data streams should represent the same physical or financial conditions. In general, any time data streams diverge, it is good practice to institute some form of reconciliation process. Reconciliation is simply the comparison of the output of separate data streams and the resolution of any discrepancies between them.

The frequency of and the appropriate investment in reconciliation activities may be driven by the attendant risk of a discrepancy. If the data streams concern low-volume and low-value transactions, then reconciliation may be relatively infrequent and not very elaborate. If the data streams concern high-volume and/or high-value transactions, then reconciliation should be frequent, highly controlled and well supported with tools and automation.

Payments

In Chapter 11, we discussed the payment and collection process. It was briefly noted that payments are recorded to the customer account in the billing system. The record of the actual deposit of funds is forwarded to the "Cash" process in the finance organization for booking. In Chapter 12, we reviewed the feed of payment advice records to accounting. Comparing the "cash receipts" to the payment records is a critical financial control that should be given the highest level of attention and support.

Inventory

The billing system often needs to know inventory information such as product serial number, as well as the disposition of that inventory for billing-related purposes. As a result, some billing system implementations act as the "database of record" for equipment and capacity inventory. This requires the periodic reconciliation of actual inventory with inventory records in the billing system. This may be an automated reconciliation with warehouse automation systems or some form of physical inventory.

Summary: Where We Have Been and Where We Are Going?

In Chapter 13, with the discussion of account and system controls, we concluded a tour of typical billing systems. We have gone from defining sources of billing system information and describing inputs and outputs through discussions of how all the billing system functions work and how they work together. In Section II we will discuss how company strategy affects the choice of a billing system and how that billing system is configured. Section II is where we see what strategic and operational questions have to be asked—and answered—in order to select the best billing system solution for your company.

Section

Billing Systems and Business Strategy

There is no "perfect" billing solution, despite some vendor claims to the contrary. The right solution for your business depends on your business. To make a good decision on a billing system, you must understand clearly what your business is, what the business will be in the expected life of this billing implementation and what the role of billing is within the business strategy. It is also important to understand your current and targeted customers.

In Section I, we looked at the day-to-day work of billing systems and the billing system data flow. We introduced the business processes and organizations that are part of that day-to-day work. Sales sets up a customer account and takes orders from the customer. Provisioning and fulfillment literally "provides" the customer with the products and services they order. Network supplies usage detail. Billing operations actually bills the customer. Customer care handles billing inquiries, complaints and other customer interactions.

In Section II, we will discuss many of the critical business and strategic considerations in determining the right billing solution for your enterprise. We will introduce the organizations and processes involved in formulating, supporting and implementing business and market strategies. Here the cast of characters begins with corporate senior management who oversee company strategy and direction. A major player in this arena is the marketing department. While some companies organizationally combine marketing and sales, the functions of marketing are quite different from those of sales. Marketing determines product strategy and tactics, ranging from what products to introduce or eliminate, through which customers to target, to pricing strategy, promotion strategy and a host of other details.

Throughout this section, we refer to the "business analyst" or the "billing business analyst." By this we mean whoever is gathering and analyzing business information in order to formulate a good set of business requirements for a new billing environment. The billing business analyst is the principal target audience for this section, although this is a valuable guide for senior management and other decision makers, as well. In most cases, there will be a business analysis team comprised of some senior management resources to review and interpret high-level strategy and business objectives and some less-senior resources to gather and collate huge amounts of detailed information.

This section may be used as a guide to questions that should be asked and answered before choosing or entering into any contracts for billing software or services. This is also where the authors reveal all the trade secrets of business requirements consultants—so you can do this work yourself or monitor the effectiveness of those you might hire to do it!

Chapter

14

What Does Your Business Need?

Determining what your business needs in a billing environment requires asking and answering many important questions—at a high level as well as in detail. What is your business size, product mix, and customer base? Is billing a key differentiator in your business strategy? Is your business truly unique? Do you need to integrate with legacy/custom provisioning, bill print or account management systems?

What Business Are You In?

This may seem to be an unnecessary question—you know what business you are in! The point here, however, is to articulate your overall direction (mission) and the ways and time frames (objectives) in which you hope to achieve your mission in terms that are meaningful to billing system vendors.

Mission

Your official mission statement may be quite global—along the lines of "we will end world hunger"—or very detailed. The aspect of your mission that is important for billing system selection is the *scope* of telecommunications products and services that you envision providing and the *areas/jurisdictions* that are to be served. Documenting it at an early point in project planning can improve requirements for the acquisition and implementation of billing system capabilities and can minimize costly mismatches between needs and "solutions."

A typical telecommunications company mission is of the global variety. Let's examine a couple of representative examples of published mission statements.

Company X. Company X provides a brief and high-level mission statement:

> *Company X will be our customers' choice for high-value, high-quality data, voice and video communications and services.*

This statement contains several important cues for the billing business analyst. First, it indicates a high-end customer base. That type of customer base is more likely to be interested in "high-tech" billing options and customized billing statements. From this statement, it does not appear that this company has much interest in making a strategic commitment to residential POTS, for instance.

Second, it indicates strategic product lines that include "high-value, high-quality data, voice and video." This should trigger some key questions about quality of service (QoS) guarantees in service definition and about capturing QoS information in transaction detail records.

Company Y. Company Y, like Company X, has crafted a high-level mission statement:

> *To build shareholder value by becoming the customer-focused market leader for worldwide broadband communications and applications services.*

This statement indicates the intention to provide "worldwide" services. The impact of this intent is not clear without further investigation. Does Company Y intend to provide worldwide connectivity for U.S. customers? Do they intend to provide services to companies elsewhere in the world? Do they intend to bring this about by reselling services (in some fashion) provided by foreign partners? The billing analyst must seek answers to these questions, as well as focus on follow-up issues of currency and language support, international standards support and lots of questions detailed in Chapter 15.

Additionally, Company Y has indicated a strategic interest in "applications services." This will lead the business analyst to assume a need for supporting leading-edge content and processing services in the billing environment. This may surface a requirement for a whole new type of transaction detail recording, rating and reporting within the expected life of the billing environment.

Company Y has gone a step further in its public documents by adding the following supplement to its high-level statement:

> *We now are closer to making our vision a reality with:*
> *Digital subscriber line (DSL) services for video and data*
> *Wireless communications services*
> *Advanced broadband communications services*
> *Local and long distance services*
> *Leadership in hosting, applications services and Internet protocol*

This statement provides a clear list of the categories of services considered strategic by Company Y. For the billing business analyst, this triggers preliminary criteria for prioritizing billing system features and functionality. The billing system should gracefully handle both business and residential DSL configurations, a broad range of wireless services, broadband services, traditional local and long distance services and Internet-based applications services and IP-based voice and data communications.

In order to select the best billing system vendor for your company, it is most helpful to identify the types of telecommunications services and products, in a manner similar to the statement above, that you expect to provide now and in the future.

Objectives

This is where you need to identify what your company provides now, how long you plan to keep providing it and when or if you plan to begin providing additional services. Your company's plans for both the short term—12 to 18 months—and longer term—four to five years—will be very important for selecting the billing system that both accommodates today's book of business and will accommodate the future. The effort to state the marketing and corporate planning directions in this fashion greatly speeds the data gathering processes.

Objectives should be stated in terms of fairly specific services or products, markets and jurisdictions. The time element is another important factor in this set of statements; usually an identification of the quarter of the year targeted is sufficient. Lastly, the anticipated volumes of customers served and instances of service (e.g., "lines in service") initially and projected growth rates should be estimated.

Why a New Billing System?

One event or several may occasion the need for a new billing system. It is helpful—and again it sounds elementary—to keep the specific need for change in mind as you explore billing system options. For example, the data gathering and analysis process may show that your business needs can be met without a move to a new system at this time if anticipated growth in one service/jurisdiction will be offset by a planned reduction elsewhere.

New billing systems will be needed for a new business, but also to support growth, a need for new billing functionality or because the existing billing platform is outdated.

New business

Obviously, a new billing system will be needed for an entirely new business start-up. However, billing system acquisitions are also frequently triggered when an existing company enters a new market. For instance, in recent years interexchange carriers have been entering the local exchange business and vice versa. For both regulatory reasons and because these businesses have enough critical differences, these new market entries have triggered the acquisition of part or all of a new billing environment. When a "facilities-based" broadband provider enters the switched services market, they will immediately need a billing system that can support usage-based services.

Growth

Many billing solutions do not scale up gracefully after a certain point. If your company has been wildly successful—or extremely aggressive in growing through acquisition—the current billing system may simply not be capable of handling the growth. Billing processing is time bounded, as discussed in Section I, so processing cannot just be extended indefinitely.

Another scenario for growth is entering geographic markets not supported by your current billing system. This could include the need to replace "home grown" support for a single regulatory and taxing authority with support for multiple regulatory and taxing jurisdictions or the need to support additional currencies or languages.

New capabilities

Perhaps your company has a business strategy of migrating from paper billing to electronic billing and your current system is fundamentally paper oriented. Perhaps your company needs greater flexibility in defining price-based promotions. A strategic or operational need for a capability not available in the current billing system is a good trigger for surveying available options in billing systems.

Platform

Hardware and software are evolving at a rapid pace. At some point, either the hardware platform or the software platform—or both—of your current billing environment will become obsolete. At that point, continuing support for the platform—especially if regular updates are required to support new versions of standards and interfacing technologies—may become difficult and uneconomical at best and impossible at worst. For instance, many older computer platforms were never upgraded to support Y2K gracefully. Systems based on these platforms are essentially unmaintainable.

Gathering Data

Different types of products and services require quite different billing capabilities. In this section, we examine a variety of issues that should be fully explored before making any decisions about a (new) billing environment. This information is required for any billing implementation. Documenting it at an early point in project planning can improve requirements for the acquisition and implementation of billing system capabilities and can minimize costly mismatches between needs and "solutions."

Much of the information we discuss in this section is necessary for the implementation and configuration of a working billing environment. In order to produce a bill, the billing environment must know all about services and rate plans, categories of customers and markets, regulatory jurisdictions, credit policies, options for bill delivery and payment and a host of other information. Information on the size of the customer base, volume of billable items or transactions and account turnover determine the appropriate size of the billing platform.

Documenting this information early in the procurement cycle can also significantly shorten vendor time and expense for integration and implementation of billing capabilities.

Other information is used to determine what capabilities will be needed during the projected life of the billing environment. So, in addition to looking at the current customer/service/product mix and the plans for adding services in the very near future, a look at the company's longer-term plans is necessary to select the best billing system solution.

Asking questions

It is really important to put someone in charge of asking the questions we identify in the rest of this section and in charge of *getting answers to those questions*. This may seem pretty fundamental. However, an extensive if highly unscientific poll of billing solutions vendors and experienced billing systems implementation managers showed that every one of them had encountered serious project crises and delays because of unanswered or unasked questions. The effort to profile your business's current business and strategic direction at a very detailed level will take some time in the beginning, but will produce the big benefits of identification of the best billing solution implemented in the shortest time.

Often, getting complete answers requires gathering information from more than one source. For instance, a complete description of a service may require talking to the product marketing manager to determine the product name and related codes, product pricing and dates of market introduction. If your product management process is pretty mature, this information may be contained in a product marketing description or a formal product catalog entry. Other product questions can be answered by the product engineering manager or in a product technical description. Determining the life expectancy of a product may require input on strategic direction from executive management in either marketing or engineering or both. The accounting classification information for the product may require discussions with the finance department personnel and the legal organization may be the source for taxation or jurisdictional requirements.

Your company may have a more ideal situation: a product management process team that includes members from all "stakeholder" groups—including billing—that gathers, documents and maintains information on the various aspects of a product and has it available for coordinated uses throughout the company systems.

Once questions have been asked and answered, it is important to make sure that all related answers are in agreement. If there appear to be discrepancies in the actual information supplied or in the underlying assumptions, point this out to the sources that appear not to be completely congruent and ask them to explain any apparent contradictions or ambiguities. Many companies vest the product manager with the responsibility of resolving such issues; this approach often facilitates appropriate changes in the systems or processes not directly interfacing with billing.

Documenting answers

Gathered information is most valuable when that information is organized and written down in a way that allows careful review, formal agreement and use in downstream decision-making and configuration activities.

In order for review and acceptance of this documentation to be meaningful, the documentation must provide careful cross-reference of information and its impact on very different disciplines. Marketing managers frequently do not share much vocabulary—or focus—with engineering managers or IT managers. For instance, a marketing manager may indicate that a particular service may be offered in the United States and in Europe. To a regulatory manager, that may mean dealing with the European Union (EU) plus individual country regulators in Europe, as well as with the FCC. To an engineering manager, it may mean dealing with a backbone network containing both SONET and SDH facilities. To a billing manager, it may mean receiving usage records in BAF and some European Q.825 formats. The billing manager will be concerned with currency conversion issues, too: will billing be done in both euros and dollars? To the unwary, these might not appear to be addressing the same issue. This tower of Babel becomes even more messy when you are dealing with "voice" and "data" folks or mobile and "wired" folks—inevitable in various "convergence" applications and services.

One of the reasons for this section is to act as a "Rosetta Stone" among the various disciplines. Even with this as a guideline, however, it is important that everyone who reviews and signs off on the description of a particular service or market or strategy actually understands the implications and language of the document.

Change control

If your company does not have formal processes for documentation and change management in the areas we examine in the rest of Section II, this is a good time to put those processes in place, since gathering the baseline data is a wonderful first step. Changes will occur between initial data gathering and the implementation and configuration of a new billing environment or function. When formatting this baseline information, keep in mind that each piece and group of information should be clearly and uniquely identified in order to easily identify any changes.

Once the baseline documentation is created, a process for managing, controlling and communicating changes should be implemented. Typically, a mature change management process includes a way to publicize proposed changes for review, a formal review process, an approval process, followed by amending the baseline and the publication of change notes.

This sounds rather complicated, but in a small company it can be as simple as generating an email to a designated mailing list to "publish" the proposed change, holding a quick meeting or teleconference for review and approval, editing a version-controlled document and emailing the change notes to the same or an expanded email list. In a large enterprise, it can get a bit more formal, but it need not be particularly difficult or painful.

It is far more painful to implement a multi-million dollar new billing environment—and discover at conversion that you are missing or misbilling a whole class of customers or services that just happen to have been added or changed in some way since the initial data gathering!

Evaluating Data

Once you have collected the relevant information, analyzed it and summarized it, you can use it to create formal business requirements for your new billing environment. However, there is more work to be done!

Setting priorities

Setting priorities for the new billing system implementation is critical to making good choices for solutions, as well as in developing a good implementation plan. The first step in setting priorities is to identify the absolutely non-negotiable requirements. Generally, these are requirements that are mandated by legal or regulatory direction.

If your company is billing customers currently, continuing to bill those existing customers for their existing services is generally a top priority. Simply put, that means that the new billing environment at a minimum has to be able to accomplish everything the old billing environment is doing. It does not mean, however, that the new environment must accomplish those things in the same way the old one does.

Requirements that have some flexibility but also have an outer boundary should be expressed that way. Typically, this would be quantifiable items such as cost or time to implement or personnel time or customers impacted. For instance, your company may wish to have the new billing system fully operational in six months from contract signing. However, the current contracted billing service will no longer be available when the current contract ends 10 months from now at the end of the next calendar year. Having the system fully operational and all data converted by the termination of the contract is the required outer boundary of implementation and conversion timing, while business strategy may make the earlier date highly desirable.

The rest of the business requirements should be ordered by relative criticality to the business. It is frequently useful to rate requirements in several dimensions. Some common evaluation dimensions include potential for revenue generation, potential to reduce operational costs, importance to strategic direction and improvement in cash management. It is often valuable to consider these dimensions for the first year after implementation and for the fifth year after implementation, in order to capture lifecycle value.

Evaluating tradeoffs

When choosing a billing solution—or making any other business decision—you may be faced with tradeoffs. Typical tradeoffs might be cost versus a better fit to your requirements, system complexity versus manual process complexity or better support for the requirements of one product line over another. Other tradeoffs might include sizing for growth—at a cost!—versus sizing for current needs and planning for the disruption of scaling up at a later time.

Some of these tradeoffs can be evaluated on the basis of pure economic cost and benefit and we will discuss economic analysis factors. Some tradeoffs, however, must be evaluated on more strategic grounds. For instance, one vendor may offer flexible and easily manageable rate configuration but have no support for electronic billing (EBPP), while another may have a complex and rigid structure for rate management but have extensive support for EBPP. Of course, if this were a simple two-factor equation, you would evaluate these parameters on the basis of previously identified priorities and choose the solution that best meets the higher priority.

Even in complex decision making, weighted priorities can be used to assist in choosing a product and vendor. Many government procurements are chosen exactly that way. In order for a weighted priority system to result in a quality decision, the parameters must be fully understood and evaluated in the context of your business strategy and operations.

Let's go back to the rate management versus EBPP problem for a moment. We will assume that there are two companies that have determined that their business will require lots of rate management activity and their markets will desire EBPP. Company A has a large and low-cost rate management staff located in a small town in North Dakota (just to make the "low cost" believable!) Company B has one rate manager located in New York City (as an example of "expensive labor"). With similar customer bases and strategies, Company B might choose a vendor for rate management features while Company A might choose a vendor for EBPP.

In the real world, however, Company B may have the option of outsourcing rate management to a contractor in Nepal at a very low cost, making the EBPP solution more desirable for them as well. Opening up that kind of additional option is what makes a qualitative factor analysis a critical complement to any quantitative evaluation.

Economic analysis factors

When looking at the costs and benefits of any new billing solution, it is important to identify all related costs and potential benefits. While costs may be easier to capture than benefits, many indirect costs are often overlooked or underestimated.

New system implementation costs. First, there are the costs of implementing a new system. The obvious costs are the purchasing of new software and the hardware to support it. Somewhat less obvious are the costs of integrating the new billing solution to your company's existing systems, which may require systems support resources from the interfacing systems, as well as from the vendor or the billing systems support staff. Other costs involve the initial configuration of the system—building information on products and services and rate tables, on customer classification, on billing options and other business and operational parameters. There also may be considerable cost involved in customizing reports and controls to your company's business processes and needs. Lastly, there may be employee training required for users and support staff to use and support the new system most effectively.

Conversion costs. There are unique costs involved in converting from an old system to a new one. We look at this process in detail in Section IV, but all that effort represents substantial costs. First, there is the need to extract, validate, format and load information from the current billing systems. Unless your company has been unusually rigorous in data quality management and the formats for information in the new systems are nearly identical to the formats in the older system, there will be significant fallout from any automated conversion and that fallout will require manual examination, resolution and data entry. In our experience, the range of fallout can be from 10 to 100 percent of total data! The amount of fallout increases with the age of the book of business and of the old billing systems and with any complex data mapping decisions.

Once the conversion data are actually loaded, there are costs incurred in keeping the data synchronized with the old production system throughout testing and parallel billing cycles. Any new business, payments, disconnects or other account changes must be captured in both the old and the new environment. This may require additional software development, periodic incremental conversions and/or additional manual effort.

Lifecycle costs. Lifecycle costs for the new environment include system maintenance and support, both internally and from the vendor. More important, however, are the impacts of the new environment on process costs. Identify the time and associated costs involved in characteristic user transactions such as configuring a new service, changing a rate schedule, implementing a new discount structure, implementing a promotion, researching an account query and crediting an account. Then, you should determine how often those activities will occur in your business over the life of the new system—or some reasonable future horizon. Examining the product of transaction cost and frequency of transaction over 1, 5, and 10 years will provide a real picture of the costs of a particular solution—and begin to establish a handle on the benefits of that solution.

Other life cycle costs include the cost of ad hoc reports and analysis. A good user query system can save lots of programmer hours over the life of the system. Having mature and well-specified programming interfaces can minimize ongoing custom development. If your company plans to buy business as a strategic method of growth, the availability of good tools to bulk-load customer accounts can be a major advantage over the system life.

Benefits. Quantifying benefits in economic terms is often more difficult than estimating costs. Operational benefits are the flip side of operational costs. We've already looked at estimating user operational costs and benefits. On the systems side, identify the costs of continuing to maintain and support an aging platform. Often vendors stop providing regular maintenance for technical and standards conformance, etc., as a platform ages. They begin to insist on payment for customized support, where your company pays directly for all development costs, when development costs no longer can be spread over many customers. Additionally, there are often business costs associated with poor data quality or with the manual effort to maintain good data quality on older platforms.

Revenue and market advantages from billing system implementation are always somewhat speculative. If an existing customer states, "If you don't provide EBPP by the end of my current contract, I'll take my business elsewhere..." you may be able to estimate the benefit of EBPP capability to include the margin on that customer's business. This is not usually the scenario for market and revenue forecasts, however. The best an analyst can do on that side of the ledger is to make sure that senior management understands, buys into and supports the forecasts.

Stakeholders

Throughout this section, we identify "stakeholders" in specific kinds of information. A stakeholder is an individual or an organization that uses, originates, manages or is impacted by the information. Stakeholders are people who need to know what the information is and/or how it is formatted, how it is used and when it changes in order to do their jobs.

Sales systems

Sales systems management are stakeholders in data that originate in, pass through or are destined for the sales and customer order systems. Sales systems management includes the managers responsible for business requirements for sales systems, they may be functional managers from sales or product management personnel. It will also include the IT management team responsible for sales support systems. The IT team may include both development and support managers for both applications and platform.

Provisioning systems

Provisioning systems management are stakeholders in all data that originate in, pass through or are destined for the provisioning systems or which must match the output of provisioning systems at some point downstream. Provisioning systems management include the managers responsible for business requirements for provisioning systems—often functional managers from the provisioning organization—as well as the IT management team responsible for provisioning support systems. The IT team may include both development and support managers for both applications and platform.

"Middleware" systems

"Middleware" systems are systems that are designed to "translate" messages between one existing or preconfigured system and another. They manage message timing and handling, reformatting when necessary, correlation and synchronization. In this context, middleware is often employed between the provisioning and billing environments and/or between the customer order and billing environments.

Product management

Product management is the portion of the marketing department responsible for product development, introduction and performance. Product management is the ultimate decision maker on most information concerning a product or service. However, unless you are lucky enough to have a mature product management environment with multifunctional teams, product management may only be able to provide the outline or market-facing view of a product or service. Product management may not necessarily represent the interests, capabilities or concerns of all stakeholders.

Carrier relations/strategic partner relations

Carrier relations are stakeholders whenever data passes from or to another carrier. Carrier relations become involved in negotiations concerning such things as the content of optional data fields in the OBF (Operations and Billing Forum of the Alliance for Telecommunications Industry Solutions) standards for service orders. They also are key to establishing control processes between carriers.

Strategic partner relations are stakeholders whenever data pass from or to noncarrier strategic partners. Essentially, these folks negotiate all interfaces, management and control processes and reporting responsibilities, among other duties. Examples of strategic billing partners are financial institutions, collection agencies and printing vendors. Other strategic partners may be the originators of call detail records for calls originated outside your company's network.

Regulatory

The regulatory organization is a stakeholder in anything that might be specified by the various regulatory bodies. This ranges from the format and appearance of information on the bill through classifying the regulatory jurisdiction for individual types of service to the format and production requirements of reports.

Legal

The legal organization is a stakeholder in any considerations or data relating to contracts or other legal commitments. The legal organization is also a stakeholder in such legally defined arenas as regulatory requirements, tax applicability, data retention and archiving.

Billing operations

Billing operations—the folks who actually run the billing process—are a stakeholder in everything having to do with the specification, implementation and day-to-day operation of a billing system. They are especially concerned with architectural trade-offs that could affect their workload and ease of system use and management.

Fraud systems

Fraud systems management includes the managers responsible for business requirements for fraud control systems, as well as the IT management team responsible for fraud control systems. The IT team may include both development and support managers for both applications and platform.

Network equipment engineering

Network equipment engineers are stakeholders in any data that are produced in the network, especially call detail records, as well as in any provisioning data that are shared with network equipment and operations systems.

Server resources management

Server resources management are stakeholders in anything that impacts the capacity or configuration of the billing systems servers and related data network servers and services.

Desktop/workstation support

Desktop and workstation support staff are responsible for the desktop and workstation computing environment within the corporation. They generally determine the standard platform, as well as manage installation, upgrade and support of that platform. In most companies, they are held accountable for desktop and workstation computing costs. If the new billing environment requires a new desktop platform or an upgrade of the current platform—or simply additional workstations—this group is a key stakeholder.

Accounting systems

Accounting systems management includes the managers responsible for business requirements for accounting systems, as well as the IT management team responsible for accounting systems. The IT team may include both development and support managers for both applications and platform.

Accounting standards

Larger companies often have a group responsible for accounting standards. In a smaller company, this may be the office of the CFO or the CFO personally. Accounting standards is a stakeholder in decisions impacting accounting classification or auditability.

Tax accounting

Tax accounting is just what it sounds like—the folks responsible for all the accounting interfaces to taxing authorities. Tax accounting is a stakeholder in any information or decision impacting tax application, calculation, tax reporting, tax payments or tax auditability.

Customer care

Customer care addresses the needs of customers post-sales. Customer care functions include account inquiry, problem resolution, adjustments, collections and account maintenance. Customer care may be a stand-alone organization or the functions may be part of sales or billing operations. Customer care transactions may be directly entered into sales or billing systems or there may be a customer care support system that interfaces with sales and billing systems.

Network OS

Network operations systems (OS) management is responsible for systems that manage network technology and service. These systems may be highly specialized to "talk" to specific pieces of network equipment (often known as element management systems or EMS) or they may be configured to manage services across the network or to manage the performance of a network.

Standards

The standards function in a telecommunications company may be in a specific standards organization, often reporting to the chief technology officer, or it may be distributed among network engineering, systems and possibly research and development. However the function is organized, at a minimum it tracks standards activity relevant to the company's business and technology, identifies potential impacts and advises on compliance.

Standards groups may also represent their company's views and requirements to various standards bodies. For an overview of potentially relevant standards and standards organizations, refer to Chapter 26.

Summary

In this chapter, we looked at determining what business your company is in and what business it will be in. We have discussed the important characteristics of developing business requirements for a new billing environment and we have identified and described major stakeholders in various aspects of billing requirements and data. In the next few chapters, we look in some detail at the information gathering and analysis process that will result in quality business requirements and that will support the system implementation process.

Chapter

15

Defining Products and Services

In this chapter, we look in detail at the information you will need about each product or service in order to select billing solutions, implement those solutions and convert any existing business to the new billing solutions. We also provide guidance about who the key stakeholders may be (with some variation due to differences in corporate structure), where critical information may be found and what some of the most critical questions may be.

If your company has a mature product development process, product management may be able to provide formal documentation that will provide a head start on gathering information about products and services. A *marketing service description* will include information on the offering from the market perspective, including: how the customer will see the service, billing basis, price and discount structure, market structure and market sensitivities, as well as projected revenues, sales compensation structure and marketing and advertising costs. A *technical service description* will contain information on the technical structure of the service—including call detail formats and recording—as well as information on the cost of delivering the service. The technical service description may also contain information on the formatting of information in and out of the provisioning and sales systems.

While current marketing service descriptions and technical service descriptions are very useful in understanding products and services, this information must be confirmed with the current stakeholders and, in the case of existing products and services, against current billing practices.

Time Horizons

New billing system implementations should accommodate any existing products and services, any planned products and services and, at a minimum, have a plan for accommodating foreseeable strategic directions in service and product offerings. A planning period of five years is recommended.

If you are selecting a billing system for a new enterprise that is not billing any active customers, you will not have any "current products and services" to research. In that case, you will probably retain your hair and its youthful color through the initial months of your implementation project. You can also skip over the material specific to "Current Products and Services."

In the unlikely event that your company is not actively planning any new services, you can skim information only relevant to "Planned Services" since it will come in handy at some point in the future! Material in this chapter is keyed to the data gathering forms in Appendix A and on the included CD.

Planned services

Most companies have products and services "in the pipeline" for some time prior to actually selling and billing them. Significant lead time may be required for acquiring network capability, gaining regulatory approval, notifying sales and training all involved personnel, advertising the service, establishing accounting classifications, determining taxation jurisdictions and other preparatory activities. These products and services will certainly require billing capabilities in the life of any new billing environment or function and should be considered in acquisition planning.

Current products and services

Since any new billing system implementation should be capable of billing products and services currently offered, capturing the current "product catalog" is critical to selection of a billing environment. A new billing solution should be capable of supporting all the current billing configurations. If it cannot, then the solution may not be appropriate for your business. Alternatively, if the billing configurations that are not accommodated are being phased out, it may be possible to continue to bill those configurations in the old environment until customers can be migrated to current service structures.

Sometimes a comprehensive documentation of all products and services offered identifies unnecessary overlaps and complexity in offerings. This is particularly true in the wake of merger or acquisition activity—whether the acquisition is of another company or of a "book of business"—that is, a selected group of the existing customers of another company with their existing services.

Identifying these overlaps and complexities early in the billing environment acquisition cycle can provide an opportunity for product management to design a more manageable catalog of services and to work with sales to migrate customers to the new product structure in advance of conversion. This can immensely simplify configuration and conversion activities.

Currently offered versus "grandfathered." Usually the products and services currently being billed are also currently being offered to new customers for sale. Since sales support systems and materials often contain a lot of the information required to understand how to configure that service in billing—and since sales support systems and materials are often designed in a way to make viewing that information possible and even somewhat "user-friendly"—these systems and materials can be a very helpful source of product and service information.

However, sometimes a product or service currently being billed is no longer available for new sales. This is particularly true in highly regulated areas of conventional telecommunications where "old tariffs never die" and you can only hope that they fade away quickly! These "grandfathered" services often do not appear in product catalogs designed to support new sales. They may or may not properly appear in provisioning and maintenance inventories. Data gathering for these items is more challenging, but just as necessary.

Describing a service. The usual authority for products and services that are candidates for conversion to the new billing environment is the old billing environment. Examination of that old billing environment is often excruciatingly difficult, especially if the billing information lives in systems that belong to another company from which your company acquired the business. Also, many older billing systems have proprietary data structures that may or may not be viewable or understandable even when viewable.

If you are very, very lucky, a data catalog will be available for the current billing environment—or for the entire corporation. In that case, finding relevant data elements, interpreting them and identifying where they came from and how they are used is fairly straightforward.

As you gather requirements information for a new billing environment, you are creating the basis for a new or updated data catalog. On behalf of future billing business analyst and conversion managers, the authors enter a heartfelt plea to capitalize on that opportunity and leave those who follow you a good data catalog!

Even if you have full access to viewable and understandable data in the old billing environment, other sources of information may be needed to understand and use those data. For each type of information, we will look at appropriate sources of information, what that information means and how it can be used, as well as some potential issues and caveats gleaned from real-world experience.

Service and Product-Specific Data

For each item of data concerning a specific product or service, we look at a definition of the data and its use. We also identify the best sources for such information for planned products and services and for current products and services. We describe the impact of gathering and baselining this information on the process of new product development and implementation. We also review potential issues to look out for when documenting existing products and services for conversion into the new billing environment. Finally, we identify the key stakeholders for the data.

Product/service code

The product or service code is usually a short alphanumeric code used to uniquely describe a product or service even if it has a plain language descriptor. Generally, this code is used in all systems that are involved with the product or service.

Where to find product/service codes for planned services. Product management should maintain an inventory of these codes and should be the initial source of this information. If the product management team has not assigned a code yet, they may be willing to commit to format standards at a minimum.

Where to find product/service codes for current services. Identifying this code in the old billing system and verifying it with the sales and provisioning environment is appropriate for current services. The code should appear in the product/service definition data and also in individual customer account records for customers subscribing to the service or product. The codes in both places should be in agreement.

What information. Note the format of the code, including field length, any embedded intelligence (such as state or market codes) and character set.

Product development process impact. Identifying the code format and/or the actual code for this effort can improve communications between stakeholder systems. Information collected on codes for other services can assist in preventing non-unique code selection.

Potential issues for current products and services. Remember that this is a unique code. If you are converting data from multiple old billing systems, they may be using the same code for different services. This is particularly true if you are converting traditional telecommunications services from RBOC databases.

Since these codes are usually linked to systems in other environments, it would be advantageous if the new billing environment could accommodate the existing formats. However, if the sales and provisioning environment will also be replaced in a coordinated effort with the new billing system implementation, it is critical to verify this information with the new sales and provisioning systems management, as well as with the legacy sales and provisioning systems management if they are different groups.

Stakeholders

- Sales system management

- Provisioning system management
- Middleware system management
- Customer care management
- Customer care systems management
- Product management
- Carrier relations/strategic partner relations
- Accounting systems
- Tax accounting
- Billing operations

Product/service name/description

The product or service name or description should be what the customer will see on the formatted bill to explain the charge or group of charges. It should also be the name that sales personnel will use with the customer when describing and selling the service or product.

Where to find product and service names or descriptions for planned services. Again, product management, with input from sales and customer care, should be the principal source of information here.

Where to find product and service names or descriptions for current services. Product management should be the principal source of information. Information gathered from product management should be verified against information in the old billing environment and with customer care. Any differences between the old billing environment and the terminology to be used in the new billing system must be flagged to explain expected conversion testing differences and, possibly, generate notification to current customers on the first bill produced by the new system.

What information. When asking product management about this information, be very specific that you want the exact wording, *as it should appear on the customer's bill*. Other (often more expansive) wording may be used in sales and promotional literature. The more expansive information should be captured as "product/service detail."

Product development process impact. Business analysts working with product management and the product development team can develop standards for product and service names and descriptions that would minimize potential systems issues such as misinterpretation of embedded special characters.

Potential issues for current products and services. Often this description is not unique to a product or service. If it is not unique, then it is important that the new billing environment accommodate a unique product/service code and that the product or service have or be assigned a unique product/service code. For example, suppose three communities are each sold local service for which the monthly rate and number of calls allowed without charge are different. All are billed with the descriptive of "local monthly service" but each has to have a unique product service code. Note the format of the existing descriptions, including field length, character set (e.g., does it use any special characters or numbers?) and any fixed formatting. Special characters that can cause lots of problems include the use of double quotes to indicate inches or single quotes to indicate feet, slashes to separate related words such as "product/service," asterisks, exclamation points and question marks. These special characters are often interpreted strangely by programs used to bulk-oad old service and billing records.

Stakeholders

- Customer care management
- Customer care systems management
- Product management
- Sales management
- Sales system management
- Billing operations
- Billing systems management

Product/service detail

The product or service detail consists of a "plain language" description of what the product or service is or what products or services are bundled under this description. This description should be sufficiently detailed to identify the service to sales, network, billing, other infrastructure technical and operations personnel and to make clear what the customer perceives the product or service to be. This plain language description is not entered into the billing system; it is used as a reference by all the humans to ensure that the systems are properly populated.

Where to find product/service detail for planned services. Product management is the principal source of information on the details of a product or service.

Where to find product/service detail for current services. Sales literature and product brochures are useful sources of product and service detail. Provisioning and fulfillment systems often contain implementation detail on complex products or services that can support product or service detail. For regulated services, the relevant tariffs are definitive and useful sources of this information.

What information. This information is primarily descriptive of the product or service. It should be sufficient to distinguish this product or service from other offerings. If the product or service is actually a combination of other products and services, these should be enumerated individually.

Product development process impact. Gathering this information facilitates communication between stakeholder functions. It can also assist in identifying any product overlaps, as well as clarifying differences between similarly named offerings.

Potential issues for current products and services. When acquiring a book of business, it is important to get full definitions of services being provided. Offerings with billing descriptions of "call waiting package," for instance, could bundle the call waiting feature with any number of other items, ranging from basic local service with call waiting to just the call waiting feature bundled with call forwarding, to local service bundled with call waiting, call forwarding, call conferencing and calling number identification.

Stakeholders

- Provisioning systems management
- Customer care management
- Customer care systems management
- Sales management
- Sales system management
- Product management
- Regulatory
- Billing operations
- Accounting standards
- Tax accounting

Product/service identifier format

Usually, some unique identifier identifies individual instances of products and services. In the world of conventional telecommunications, these are known as "circuit number designators" or "telephone number designators"—terms that imply specific formats. Traditionally, circuit numbering formats were used to identify dedicated services ("private lines") and telephone number formats were used to identify switched services and/or access to switched services, although these boundaries are less rigid than they once were.

In the IP or Internet world, a service may be identified by a fixed IP address, by a domain name or by a user access ID. Essentially, these are the "telephone numbers" of the Internet world.

IP addresses and domain names are coordinated for uniqueness and standardized for format internationally by the Internet Assigned Numbers Authority (IANA). Telephone numbers (TN) when used with country codes are unique worldwide. Without country codes, they are unique to the numbering region or country. TN format is always all numeric. The numbering scheme and length is specified by the numbering region or country, while the country code is assigned by the International Telecommunications Union (ITU). IP addresses and TNs represent terminating locations for dynamic transactions and are widely standardized to allow global connectivity. Email and Internet service providers (ISPs) coordinate user IDs within their domain(s) for uniqueness. The maximum length and character set for user IDs is specified by the Internet Engineering Task Force (IETF).

Dedicated service designations are not internationally standardized. Within the United States, virtually all legacy carriers use some version of the Telcordia common language circuit identifier® (CLCI) for dedicated circuit identifiers. Since the Ordering and Billing Forum of the Alliance for Telecommunication Industry Solutions (OBF) standards for access service orders specify the use of CLCI format, this is generally used for external communications between carriers. However, that does not mean that these companies use CLCI formats internally or with their customers.

Many companies identify individual products by unique serial number. This is desirable for tracking inventory on leased equipment and for verifying duration of warranty and support obligations on purchased equipment.

Where to find product/service identifier format for planned services. Product management should provide information on service and product designators. For some third-party products, the manufacturer may be able to provide serial number information, unless your company plans to reserialize the products.

Where to find product/service identifier format for current services. For regulated services, tariffs may provide definitive information on service identifiers. Provisioning systems—and in some cases, sales systems—often do the actual assignment of identifiers for services and can be a valuable source of information on formats, number ranges, source of numbers and other identifier information.

Cross-verify documented formats with those actually in use within the old billing system and those currently being assigned in the sales and provisioning environment. Often documentation will fail to include the use of suffixes to designate parts of complex services or other variations from the stated standard.

What information. Identify the length, format and character set of the identifier. Does the identifier conform to any industry standard? Does the identifier contain subfields? Do those subfields contain intelligence? (CLCI, TN, and IP formats do have fielded intelligence.) Are there regulatory or legal constraints on these formats? (Telcordia owns common language standards, for instance.

Companies that do not subscribe to Telcordia's numbering services are constrained from using Common Language standards for internal designations, although they may communicate externally with a Common Language subscriber using Common Language formats.) Is there a controlling organization to guarantee uniqueness of designations?

Product development process impact. Determining numbering format early in product development can ensure the readiness of systems and processes in multiple areas for product introduction. It can also notify appropriate organizations of the need for new blocks of numbers or for a new numbering format.

Potential issues for current products and services. It is important to understand which identifiers have been assigned or allocated by external organizations and which are generated internally by your company. Often companies use formats that appear to be consistent with external coordination, such as Common Language formats, but are not actually registered with or allocated by the external coordinating organization.

Stakeholders

- Provisioning systems management
- Customer care management
- Customer care systems management
- Regulatory
- Standards
- Legal
- Billing operations

Billing basis

Billing basis, as discussed in Chapters 5 and 6 of Section I, includes whether the service charge is usage based, one time, or recurring; the units for calculating charges; and the formula for calculating charges. Chapter 7 contains important details on how billing basis information is used.

Where to find billing basis information for planned services. Product management should be able to identify the billing basis for calculating charges for planned services. If the service will be regulated, regulatory personnel may also have relevant input. In all instances, the planned billing basis should be reviewed for tax implications.

Where to find billing basis information for current services. Product management is always the final arbiter of the basis for calculating charges for any product or service. However, the high turnover in telecommunications-related businesses and service ownership may require digging this information out of the old billing system(s). If the old systems were custom developments or highly customized for the products or services you are investigating, this information may be embedded not in the databases, but in unique processing codes for determining charges. This could be in the context of conventional call rating or in unique calculations of distance and grade of service for dedicated services. These charges may be based on individual customer contracts, which may be needed to verify the billing basis if all else fails.

In the arena of regulated services, published tariffs are an excellent and definitive source of information on the billing basis for specific services. Since tariff language is not always a model of clarity to the uninitiated, ask the fegulatory gurus to explain the application of the tariff for a range of service examples. If everyone understands how a specific charge amount is obtained for each example, you should be able to document the information in a way that will be understandable to both product management and systems folks.

What information. Documenting the chargeable units, how they are expressed (e.g., tenths of minutes versus seconds) and what to do with partial units (e.g., truncate, round up to the next whole unit, round up if greater than .5 otherwise truncate or prorate) is critical in evaluating billing solutions capabilities as well as in planning conversions. If more than one parameter is involved in calculating charges (e.g., time and QoS), capture the relationship of those parameters in the format of a mathematical formula. This both lends clarity to the relationship and actually makes reviewers think through the relationship.

Use the formula to test the identified billing basis against that used for the existing customer bill. This is an appropriate point in the process at which to do this, because it will provide assurance that all components of the formula are identified. For example, obtain a CDR from a previous billing period and manually reprice it based on the formula that you have documented. Make sure that you have included the tax impact. Compare the result obtained with the billed amount. Any differences must be ones that you anticipated—or the formula needs to be reviewed and amended. We know that this sounds very elementary, but this step has been missed often enough in the past to make us feel that it needs to be emphasized here.

For all recurring charges, identify whether they are charged in advance or afterwards. Also, determine what should be done with partial billing periods (e.g., prorate or not; formula for proration), as discussed in Chapter 7.

For recurring charges based on some quantity parameter such as distance, identify whether that charge will be provided precalculated from the sales/provisioning environment or if the billing system will be calculating the charge. Typically, this will be precalculated, but it is a good idea to verify that. In either event, the billing system may need to receive all the parameter information if product management or regulators require it to be shown on the customer bill or in the customer billing record.

Identify the format and precision of any chargeable units or of data from which chargeable units are to be calculated. In some cases, the billing system may receive start and end timestamps in Universal Date Time (UDT) format. In other cases, the billing system may receive the number of units (e.g., seconds, minutes) of call duration.

Product development process impact. Defining the billing basis for any product or service is an essential part of any product development process. Identifying that basis and expressing it in a formal calculation may clarify issues with the billing basis early in the process. Requiring sign-off by all stakeholders will ensure that all platforms are prepared for and capable of supporting the specified billing basis. For instance, if the network call detail collectors only capture whole minutes, a billing basis of 6 seconds has a significant impact on the network folks.

Potential issues. Usage-based services may have some fixed rate recurring charge for some number of calls or some number of units. Those units may be the same as or different from the chargeable units for usage beyond the initial block of service. For instance, the first 30 calls may be covered by the recurring charges, but succeeding calls may be billable by the minute.

Stakeholders

- Product management
- Regulatory
- Legal
- Sales system management
- Network equipment engineering
- Network OS
- Billing operations
- Accounting standards
- Tax accounting

Billing information sources

The billing environment bases charges on information from a variety of sources. Recurring and one-time charges usually have a fixed source and format. Usage-based charges originate within service-providing equipment, but may pass through various storage and formatting functions—mediation—before reaching the billing environment.

Where to find billing information sources for current services. Billing information for recurring and one-time charges is generally provided by sales and provisioning systems internal to your company, as well as the billing system itself for items such as payments, returned check charges and late fees. Call detail for usage-based services may come from many sources in differing formats.

Recurring charges. In Section I, we reviewed the logical relationships between customer order and provisioning systems and the billing environment. However, to adequately research the billing information sources for recurring charges, it is necessary first to understand the architecture of your sales and provisioning environment and its relationship to the billing environment.

Recurring charges may be identified by the billing system based on the product/service code or calculated by customer order systems and supplied to billing as a calculated amount with the initial service order information. In that architecture, customer order system management is the primary resource for information about the content and format of this information and the protocols involved in providing this information to the billing environment. Figure 15-1 illustrates the direct feed of recurring charges information from the customer order system to the billing environment.

In some implementations, customer orders are fed through the provisioning system to the billing environment. These orders may originate in a separate customer order system or be entered directly into the provisioning system. Figure 15-2 illustrates a data flow architecture in which the immediate source of information on recurring charges is the provisioning system. In this architecture, the primary source of information on the source of recurring charges will be provisioning system management. The content and proposed use of recurring charge information should be verified with customer order system management and with product management.

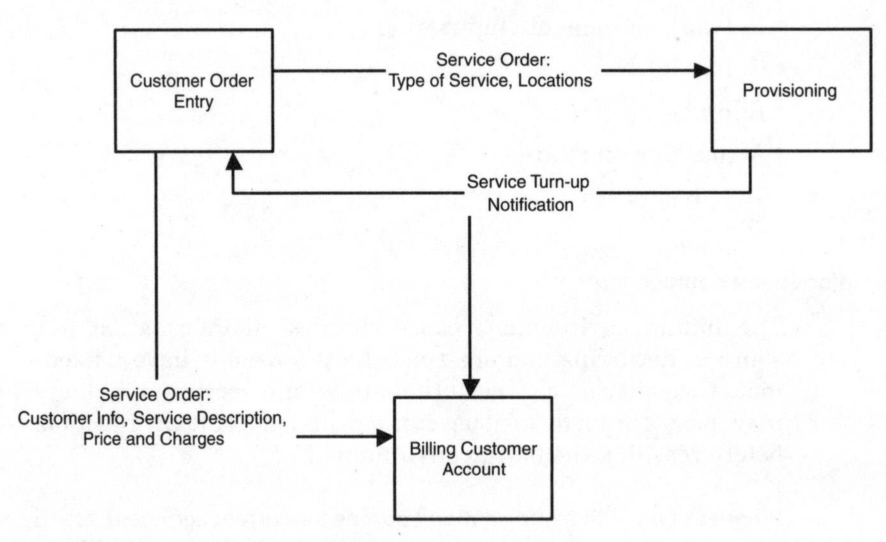

Figure 15-1 Customer order to billing.

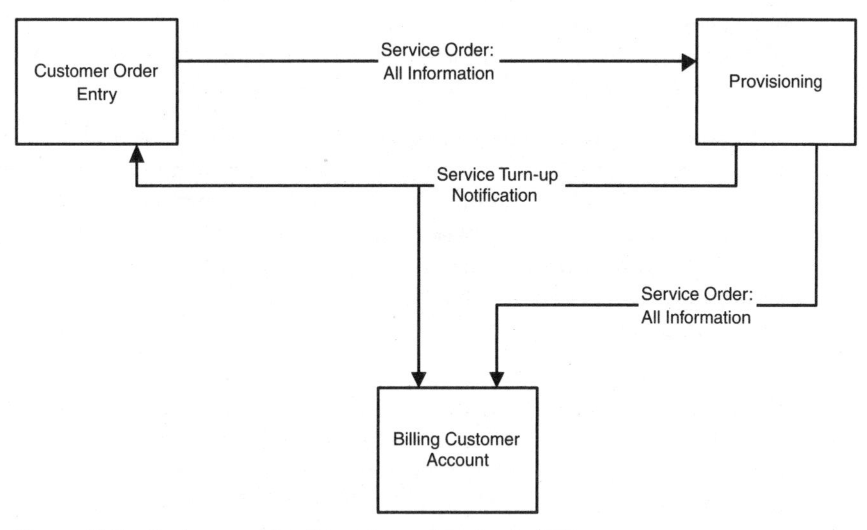

Figure 15-2 Customer order through provisioning to billing.

Some (fairly recent) architectures for interfacing sales and provisioning systems with billing systems employ stand-alone "middleware"—software designed to translate formats and protocols inherent in the sales and provisioning systems into the formats and protocols understood by the billing system and vice versa. Figure 15-3 illustrates this type of architecture. If the

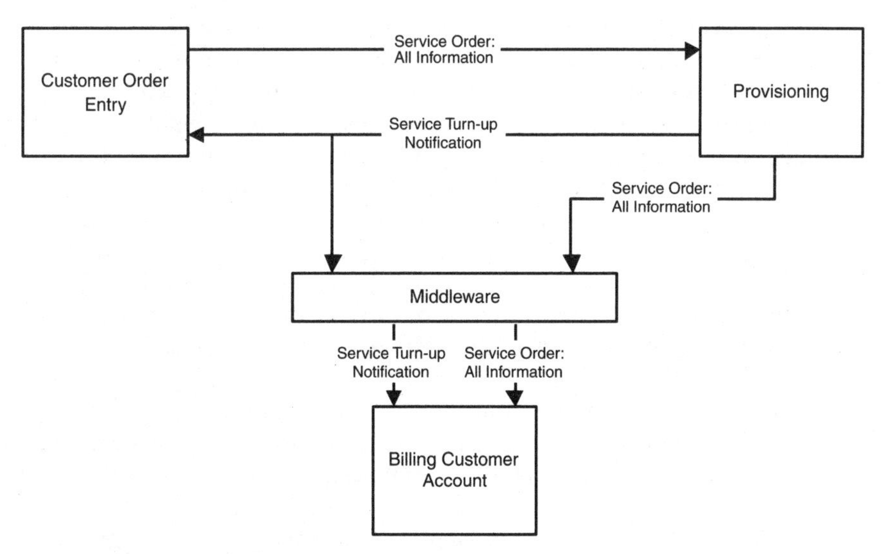

Figure 15-3 Service order middleware.

service order data flow in your company resembles this architecture, the middleware management staff will be your primary resource for billing information sources for recurring charges.

One-time charges for labor and materials. One-time time and materials charges may be supplied by provisioning, maintenance, or dispatch systems. Typically, the system capturing the labor and materials information will directly feed the billing system. In this case, the capturing system will be the primary resource for billing information sources for labor and materials charges. This information should be verified with customer order and customer care systems and functional management.

One-time charges for product sales. One-time sales charges may be supplied by the sales system or identified by the billing system based on the product/service code, but shipping costs may be supplied by the warehousing and fulfillment system. Typically, the shipping costs will be routed to the sales order system with a notification of order fulfillment. However, some implementations route the shipping costs and fulfillment notification directly to billing. (This is most likely when the billing database is the inventory database of record.)

One-time non–labor-based set-up fee. Non–labor-based set-up fees or installation charges are generally provided by the customer order system to billing, either as a product code associated with a charge amount in the billing system or as an actual charge amount in the service order record. If service orders are routed through the provisioning system and/or middleware, then the primary source will be the system that actually "talks" to billing. This information should be verified with the customer order system and any other systems in the data flow.

Usage detail records. Frequently, both the network equipment and the mediation equipment are under the control of your company's Network organization. The organization responsible for service-providing equipment and the organization responsible for data collection and formatting systems are key resources in understanding billing source information. However, some usage information may originate outside your network, as in wireless "roaming" or when your customer makes a collect or third party call charged to their account with your company.

Where to find billing information sources for planned services and products. Product management is the ultimate resource on billing information sources for planned products and services. However, carrier relations personnel are good resources for billing information originating outside your company. Network equipment engineers and network operations systems management are good resources for billing information originating within your company.

Where to find billing information sources for existing services and products. Managers of the "old" billing environment should have some perspective on where call detail data originates. However, if your company is buying a book of business from another company, those sources may change.

It is valuable to find the managers responsible for the data collection and formatting systems and those responsible for the service-providing equipment that triggers the call detail recording. This allows an accurate definition of the interface as it exists and also supports the exploration of additional options for formatting, timing, protocols and other interface configuration specifications.

What information. What is the source of billing information? Identify the owner and responsible manager of the source of billing information. Keep in mind that the billing information source may be outside your company. If so, you may also need to identify the person or organization within your company who is responsible for interfacing with the billing information source. Identify the vendor/manufacturer, version, standards supported, commitment to updates and hardware, software and network platform. Determine if there are any plans for upgrade or replacement. What content and formats of billing information does the source support? If this is expressed as supporting specific standards, identify what version of the standard and what use is made of optional data fields.

What timing and protocol does the source use to transfer billing information to the billing environment? Does the source platform support other transfer timing and protocol options? Are transfers done periodically of bulk files? Are the transfers initiated by the collection equipment or the billing environment ("push" or "pull")? Are transfers based on a schedule, on volume or whenever there is activity? Are transactions rated when they occur and then held for a batch feed to processing? Where is the billing information source—geographically and logically? Does the billing system provide feedback to the source? Are there other data flowing from the billing system to the data source?

Product development process impact. Considering all potential data sources and interface formats early in the product management process eliminates "surprises" resulting from failure to understand the nature or capabilities of an interface. Also, early identification of external interfaces provides adequate lead-time for contract or tariff activity to guarantee a specific implementation of a standard or a specific capability.

Potential issues for current products and services. It is a harrowing but worthwhile activity to sample a significant percentage of billed items in several bill runs and verify that you have captured the billing information source in your documentation fully and accurately.

Stakeholders

- Product management
- Network equipment engineering
- Network OS

- Carrier relations/strategic partner relations
- Customer care
- Customer care systems
- Standards
- Regulatory
- Legal
- Billing operations
- Provisioning
- Sales
- Accounting standards

Geographic market

Where will this product or service be available? In what country or countries? If available in the United States, in what states and local jurisdictions?

In some instances, this is limited by physical availability (e.g., does your company own or lease connectivity in the geographic market?) or a strategic decision not to do business in specific areas. In other instances, the limiting factor is regulatory approval by the relevant jurisdictions.

Where to find geographic market for planned services. Product management should supply a list of geographic markets, along with a date for introduction of the service or product in each market. Regulatory, standards, tax and legal personnel may be able to provide important information about the geographic market.

Where to find geographic market for current services. There presently may be a geographic code indicator; if not, the "service address" carried in the customer account structure at the product/service level is a good indicator of geographic market. If you have a modern database, it is fairly straightforward to request a query on "unique" or "distinct" (depending on your version of SQL) states, state-country pairs or countries—or all of the above if you've got a really cooperative systems support staff!

This information should be checked with product management and regulatory staff.

What information. In what states, countries and regions is this service offered? If this is a new service or an expanding deployment, what is the initial availability date for each market? What regulatory bodies have jurisdiction in this geographic market? What taxing authorities have jurisdiction? What standards are in common use in this market?

In what currencies will this service be billed for which geographic markets? If all markets do not use the same currency, what is the basis for conversion, if needed, to/from your primary currency?

Product development process impact. It is important for the entire product development and delivery team to understand geographic deployment plans as early as possible in the product planning cycle. Network must ensure adequate capacity in that locale, regulatory must file tariffs if regulated services are involved and customer care must have capacity covering appropriate time zones with appropriate language skills. Bill formats may need to be changed if the bill will be produced in a different language. The tax and legal departments may have to understand a whole new set of taxation and contract laws. The accounting department may have international revenue settlement changes to implement.

Potential issues for current products and services. Often expanding deployment of services is overlooked in planning a new billing system implementation and conversion. Especially when examining recently introduced or high growth services, be certain to capture plans to enter new markets.

Stakeholders

- Product management
- Network equipment engineering
- Network OS
- Carrier relations/strategic partner relations
- Customer care
- Customer care systems
- Billing operations
- Accounting systems
- Standards
- Regulatory
- Legal
- Tax accounting

Customer market segment

What kind of customer is the target market for this product or service? If this is a regulated telecommunications service, it will be defined as "residence" (sometimes referred to as consumer) or "business" for regulatory purposes. However, your company may choose to segment customers in many different ways. There may be offerings that are made available only to the largest customers, others for the mid-size business market and others that target only consumers.

Before starting any product and service data gathering, it is helpful to understand how your company segments customer markets.

Where to find customer market segment for planned services. Product management should always define the target market segment. How is that target market segment defined? A specific expression such as "business customers spending more than $50,000 a month with our company" is more useful information than "big businesses." The exception to this is if your company has a list of the customers they consider to be "big business" or have so designated some customers in customer account records.

In larger companies, customer markets may be formally segmented by size and industry and customer accounts are so designated. For example, providers might designate their multilocation customers who are billed more than a specified monthly amount as "national accounts," with smaller business customers in specific industries designated as "major," "select," and "general business."

Where to find customer market segment for current services. Product management input on market segment should be verified with either a sample or a complete analysis of existing customers for this product or service. The market segment is usually included in the existing billing system records for the customer.

What information. Identify the customer market segments your company currently uses or which your company plans to use in the near future. These may include "business" and "residence," along with categorizations based on number of lines or equivalent voice-grade lines, levels of telecommunications spending—either overall or with your company—numbers of locations, ordering and billing arrangements, subscription to some strategically important service or set of services or any other parameter that is important to your business.

Product development process impact. Creating clearly defined target market specifications can be useful for product positioning and also for prioritizing feature development and availability. It also supports a focused sales strategy and is useful in the sales process for categorizing pricing plans and other market customizations.

Potential issues for current products and services. If your company has not segmented its current market formally, it may require extensive analysis of current customer accounts to understand the customer market for current products and services. If your company's customer base is small or very uniform, this may not be worth the effort. However, if your customer base and service mix is complex, it is worth negotiating some support from product management to do this analysis.

Stakeholders

- Product management
- Sales
- Sales systems

- Customer care
- Customer care systems
- Billing operations

Price structure

Price structure is the information required to actually calculate charges once you have the billing basis information. This can be very simply a price per unit (e.g., 10 cents a minute). It can be some minimum price for an initial bulk amount and then a unit price (e.g., $39 a month with 400 "free" minutes and 10 cents a minute thereafter).

Where to find price structure for planned services. Product management is the definitive source of pricing information for planned services. If there is regulatory control of pricing via tariffs or international treaty or other structure, then regulatory and/or legal should provide the appropriate documentation. It is always wise for the billing organization to review material from all these company sources to verify that they are consistent. Any identified anomalies need to be resolved before further billing development can proceed.

Where to find price structure for current services. The old billing environment should contain price structure information. This information can be verified with customer order systems data and with customer care systems data.

What information. The basic price structure information is a currency (dollar, euro, etc.) amount for some unit. If the billing basis is more complicated, there may be a currency amount and discount percentage for each specified value in the billing basis formula. You must capture the values of all parameters in the billing basis formula (quantity discount points, "bulk" quantities included with minimum charge, etc.) and confirm the formula needed for the billing calculation functions.

What is (are) the currency(ies) for this pricing structure? Are currency conversions required for billing displays?

Product development process impact. Clearly specifying pricing structure early in the product development process helps to ensure that sales and billing systems can support that pricing structure at first service availability without extensive manual interventions or "workarounds."

Potential issues for current products and services. Sometimes pricing structures are built into custom software, rather than being stored as data elements in older billing systems. This can make identifying and interpreting these structures particularly difficult. Even uglier are systems in which part of the pricing structure is held as a data element and other portions are contained in custom software. The current system formula applications, when restated for the proposed billing system approach, may produce subtly different results (usually rounding differences in discounts or tax calculations) that require special reviews during conversion and parallel testing.

Stakeholders

- Product management
- Regulatory
- Legal
- Accounting standards
- Sales
- Customer order systems
- Customer care systems
- Billing operations
- Tax accounting

Discounts

Discounts are often applied to an aggregate of a service or of multiple services. Individual services and usage are calculated without the discount and the data are available for aggregation and discount application. Discounts may apply to all services in a bill cycle, selected services in a bill cycle or even selected services over the course of a contracted period (e.g., six months or a year). Discounts may apply to an account based on the existence of other accounts for the customer or affiliation with some third party group.

Where to find discounts for planned services. Product management is, of course, the primary source for this information. However, determining a discount structure is often done after product introduction in reaction to customer responses. Discount structures for similar products or services or for products and services with the same target customer market are a useful indicator of future discount requirements. Product management may require that each discount arrangement for a product or service be given a unique product/service code. Some companies allow the sales force to enter a discount percentage amount on the service order to be used as an "override" to the regular price of the item.

Where to find discounts for current services. Discounts for current services may be identified in designated "discount" databases, in "override pricing" in individual account records or they may be applied in some unique "postprocessing" code. If you are very lucky, someone in product management or sales or in billing system development documented the discount methodologies approved for each type of product and service and the documents are still around. If the basis for the discount is an individual contract, either the billing operations staff or the sales department is usually responsible for keeping the documents that contain the information.

What information. For what discounts is this product or service eligible? What is the discount structure? Are other products and services also aggregated for this discount? What is the structure of each applicable discount? Do discount structures interact? Can one be applied "on top" of another? Does one take precedence over another? What are the specific formulae applicable to the discount? Is there a tax impact?

Product development process impact. Identifying all potentially applicable discounts can help refine revenue and margin estimates—or cause certain discount exclusions. Discounts that can be determined by the sales force at the point of sale may need to be limited or additional billing system edits may need to be initiated to protect company revenue.

Potential issues for current products and services. Identifying all applicable discounts and the order in which they are applied is critical. Unfortunately, that may require some code tracing and "chasing" by folks who really know the old billing system. And, as with price structure, the current system formula applications, when restated for the proposed billing system approach, may produce subtly different results (usually rounding differences in discounts or tax calculations) that require special reviews during conversion and parallel testing.

Tax treatment may be impacted by the application of discounts. This is a good time for reviewing the points at which taxes are applied to discounted amounts to ensure compliance with current tax law.

Stakeholders

- Product management
- Regulatory
- Legal
- Account standards
- Sales
- Sales systems
- Customer care systems
- Billing operations
- Accounting standards
- Tax accounting

Promotions

Promotions are the unique pricing of products and services to support market strategies. Promotions generally have a time-bounded characteristic, but may have long-lasting impact on billing systems. Promotions often have the same characteristics as discounts, with the added aspect of a time limit for the price reduction.

Where to find promotions for planned services. Product management, in conjunction with sales, should be able to project which near-term initial introductions of products or services will be eligible for promotional introductions. Asking about the impact of subsequent promotions on projected revenue figures may elicit additional information with a longer horizon.

Where to find promotions for current services. Promotions may be captured as override pricing or as unique services. While some systems are configured with a "promotions" rules base, this is not very common in legacy systems.

What information. What are the terms of the promotion? Promotions may impact billing basis, price structure and discount structure of products or services in individual accounts. They also may impact the billing basis, price structure and discount structure of another service or even other accounts for the same customer.

How long will the promotion be offered (initial offering date through termination of offering)? How long will the promotional conditions be reflected in billing? For instance, a promotion offered from January 1 through April 1, 2003, could apply in perpetuity to services sold under that promotion or it could apply for "the next three months of service"—meaning it would no longer be applicable to any customer after the bill cycle of July 1, 2003. As a third option, it could only apply for "any usage between January 1 and April 1, 2003," in which case it would no longer apply to current billing after the bill cycle containing April 1 usage.

Allowing for the late receipt of CDRs means that the promotion logic must be retained at least through the June bill cycle and be selectively applied based on call date.

Product development process impact. The method by which the promotion is to be managed must be identified. If the billing system is not able to automatically stop a time-bounded promotion, a process for ending the special condition must be determined. The work force that will be responsible for promotion management must be identified and prepared to do the needed tasks. Additionally, identifying all potentially applicable promotions can help refine revenue and margin estimates.

Potential issues for current products and services. Often system databases continue to hold promotional information in the customer account record for the life of the customer account record—far past the applicability of the promotion or legal retention requirements. This can add complexity to any data conversion activities.

The expiration dates in the current promotions may need to be documented if the life-cycle management of these accounts must be done manually—normally this type of task would be given to billing operations. Customer care may need to take action if it is determined that the current system is continuing to apply promotional rate reductions beyond the agreed to promotional expiration date.

Stakeholders

- Product management
- Regulatory
- Legal
- Accounting standards
- Sales
- Sales systems
- Middleware systems
- Customer care systems
- Billing operations

Special contracts

Services to the largest (business) customers are often sold based on a unique negotiated contract with the customer, rather than based upon any standard rate structure and billing basis. While these services may be and generally are identical to services available to other customers, the rate structure and billing basis may be entirely different from standard pricing.

Where to find special contract information for planned services. Product management is the primary source of information concerning likely special contract applicability to any planned service. Regulatory and legal may supply information on constraints to special contracts.

Where to find special contracts for current services. Information about special contract pricing may show up in the old billing system in one of several ways. It may be seen as "override pricing" on standard products and services. It may appear as special bundled services or, for very large customer accounts, as customer-specific unique products and services.

If the old billing system is not the reference database for managing special contracts, it may be necessary to identify and document the reference database—which may be hardcopy files. The responsibility for maintenance of the original contracts may reside in sales or customer care, but frequently it is assigned to the billing operations organization.

What information. How many special contracts are there for this service or product? What percent of all instances of this service have been sold under special contracts? What customers or types of customers have special contracts for this product or service?

How is contract information captured and retained? How are contract effective dates (start and end date, renewal date) managed? Is there a reference database for special contracts? Is there a management process and system for special contracts? How are special contracts controlled? What is the process for special contract billing audits? Are these special contract procedures to be provided by the new billing system? Should the new billing system provide automated special contract management capabilities not currently available in the old billing system?

Product development process impact. Identifying the potential for special contract sales early in the product development cycle can facilitate the development of good controls and management processes for what is essentially an "exception" process.

If the stakeholders identify a product or service as a likely candidate for special contract pricing, product management may want to specify a highly flexible price structure or an applicable discount structure that could preclude the need for special contract pricing.

Potential issues for current products and services. The special contract management process may not be well documented or controlled. Special contracts are frequently the result of personal relationships between account managers and customers that may be disrupted by corporate reorganizations or merger and acquisition activities. All the products or services shown in the contract may not be slated to be part of the new billing system; this may necessitate a short-term "workaround" in the proposed billing system until the contract can be renegotiated or it runs its course.

Stakeholders

- Sales
- Sales systems
- Product management
- Customer care
- Customer care systems
- Regulatory
- Legal
- Billing operations

Related products, services, and features

There are lots of ways in which two or more products, services and/or features can be related. A service or feature may require the existence of another service or feature. As an example, residential DSL requires the existence of at least one voice-grade "local loop" for POTS provided by the same carrier.

Where to find related products, services, and features for planned services. Product management should specify dependencies between services or between products and services or services and features. Product management should also specify likely marketing relationships between products, services, and features. This should be documented in the "Marketing Service Plan" for a new product or service. Technical dependencies or mutual exclusions should be identified in the "Technical Service Plan" and also reflected in the marketing service plan.

Where to find related products, services, and features for current services. Product management is, of course, the primary resource in identifying related products, services, and features. Existing marketing service plans and technical service plans are useful sources of information. The current billing system may contain information that documents these relationships, such as service bundling codes or other indicators that edit for various relational conditions.

This information can be supplemented by an analysis of account-level and product-level discounts and other pricing accommodations for combinations of products and services.

What information. Identify each product, service, or feature for the possibility of reliance on or combination with another item. For each, determine the nature of the relationship. Does this product depend on another product? Does another product depend on this product? Is the relationship pricing only or other relationship? For items that have a combined price that is less than the aggregate price of individual products/services, how should accounting allocate monies on the corporate books? Do the relationships apply in all jurisdictions or only in selected areas? Do the relationships apply in all markets?

Product development process impact. Identifying potential "related products" can involve the product managers of the related products in the overall planning for this product or service.

Potential issues for current products and services. If there are discrepancies between product management information and the current billing system related products structure, these must be referred to the product manager for resolution. If the current billing system is not billing related items as intended and the new implementation is different, parallel bill runs will indicate discrepancies.

Stakeholders

- Product management
- Product management for the related products and services
- Sales
- Sales systems
- Customer care
- Customer care systems
- Billing operations

Warranty and support information

Product sales often include a warranty that covers some or all support, repairs or replacement during a period following the sale or registration of the product. This information has significant billing implications. Work or replacement performed under the warranty may have a zero price or some proportionately reduced price.

Where to find warranty and support information for planned services. Product management should supply information about warranty and support options. There may be related billing elements.

Where to find warranty and support information for current services. If the old billing system is the database of record for warranty and support information, then that system should be the primary source of information. If the old billing system is not the database of record for warranty and support information, customer care is an excellent resource for determining how warranty and support information is stored, handled and referenced—since it is the primary users of such information.

What information. Determine if any warranty or support is included in the sale of the product. How many warranty and support options are available for this product? Will any instance of the service have more than one? Is there a fee for various levels of warranty or support? Is that a related billing element? What are the conditions of any warranty or support option?

Product development process impact. Identifying warranty and support conditions can impact the economics of an offering. Gathering this information early in product development supports incorporating these considerations in the business plan for a new offering.

Potential issues for current products and services. This information may be supported in manual processes and unevenly applied. If there is to be a change from manual support in the old system to processing in the new billing system, special conversion steps may need to be identified.

Often, legacy systems do not support serial numbering of individual products, making the application of warranty periods in larger installations difficult at best.

Stakeholders

- Customer care
- Customer care systems
- Product management
- Sales
- Sales systems
- Billing operations

Accounting classification

An accounting classification must be assigned to each product and service that generates a charge. The accounting classification directs billed charges to the corporate accounting systems, which allows charges to be appropriately booked. Accounting classifications are derived from the company's chart of accounts, which is used by the corporate accounting systems to categorize all monetary activity.

Where to find accounting classification for planned services. Product management should supply the accounting classification information for each planned service or product. Product management may need to consult with the accounting standards group to identify the appropriate coding if the planned service/product is a new line of service or a unique item that the company wishes to track in a special manner.

Where to find accounting classification for current services. The old billing system should contain the accounting classification. Product management, working with accounting standards, may need to confirm the accounting classification for services and products in accounts that are being purchased from another company. This may be an ideal time for product management to confirm that all product and service monies are assigned to appropriate accounting classifications.

What information. Identify the specific code for input to the product or service record.

Product development process impact. Gathering this information early in product development will give the product management and accounting organizations time to determine what new reports or processes may be needed to most effectively book and display the classified monies.

Potential issues for current products and services. In the event that a full review of accounting classification codes assigned to current products and services identifies the need for corrections or changes, it is suggested this be done prior to implementing the new billing system.

Stakeholders

- Product management
- Accounting standards
- Accounting systems
- Billing operations

Quantity

How many instances of this service are being billed currently? How many will be billed over the life of the new billing environment?

Where to find quantity for planned services. Product management should be able to provide market forecasts for planned products and services. While market forecasts are one indicator of potential quantity of service, remember that they are notoriously optimistic—especially in the early days of product deployment, so reviewing actual market performance of similar offerings may provide a reality check when sizing the new billing platform.

There is one caveat in the other direction when looking at market forecasts. Sometimes companies with very large and loyal customer bases will be overwhelmed with orders when they enter a new market—far outstripping normal market forecasts.

Where to find quantity for current services. The old billing system should provide a count of currently installed services. Product management should be able to provide deployment and growth history that can be extrapolated for a perspective on growth rates.

What information. How many instances are there of this product or service? How many are forecast for the first year of the new billing system production? How many are forecast for the life cycle of the new billing system?

Product development process impact. Forecasts of individual services or products in services are fundamental to business cases for a new offering.

Potential issues for current products and services. There should be agreement between product management and billing operations on a consistent methodology for counting instances of the product or service.

Stakeholders

- Server resources management
- Desktop/workstation support
- Product management
- Sales
- Billing operations

Order volume

How many orders does this product or service generate? For both products and services, orders include new orders, changes and disconnection or removal of the product/service.

Where to find order volume for planned products and services. Product management should provide order volume estimates of new services, changes, and disconnects.

Where to find order volume for current products and services. The current billing system, the sales system, and the provisioning system all should have counts for order volumes. If there is a discrepancy between these counts, it should be resolved in consultation with the product manager and the systems managers.

What information. Identify the quantity of orders and the type of orders. If there are multiple order sources, provide counts by source. Identify how many disconnect orders for this service resulted in closing an account.

Projected new customers. Identifying the number of new customers will help in estimating how many service instance records will have to be established in any specified period.

Projected turnover. Projected disconnect orders will help in estimating the volume of account closing activities and final bills for this service.

Product development process impact. This information is fundamental to an offering's business case.

Stakeholders

- Sales systems
- Provisioning systems
- Product management
- Server resources management
- Desktop/workstation support
- Customer care
- Billing operations

Product Strategic Direction

Billing systems are expensive and difficult to convert—as we discuss in Section IV. The choice of a billing environment should include consideration of the corporate strategic product and service direction. Any billing solution selected should be able to support likely future directions for products, services and markets—or have a clear migration plan to be able to support those directions.

Types of anticipated products and services

Does your company's strategic plan include moving into significantly different types of products and services during the life of the new billing environment? If your business today is all broadband dedicated services and the company's strategy calls for moving into the voice switched services market, that strategic direction will have significant impact on the billing environment requirements. Dedicated broadband services are billed on a fixed recurring basis. They are generally sold on longer-term contracts. This results in low volatility, comparatively low order volume and little or no need to process and rate transaction-based call detail. Voice switched services, on the other hand, even in the large business market, tend to be more volatile—resulting in higher order volume. They require processing, rating and making available to the customer large quantities of call-related information.

If your company is a consumer-oriented Internet service provider (ISP) and you plan to move into the business-to-consumer Internet presence provider (IPP) business, that new line of business will significantly impact business requirements for the billing environment. Consumer ISP billing is overwhelmingly credit card billing, whereas most businesses prefer monthly invoices to support accounting controls.

Plans to add any significantly new line of business or to merge billing or offering structures for existing lines of business can have major impact on the billing environment. In the next decade, most telecommunications and Internet providers will be at least contemplating "content-based services" in conjunction with their existing infrastructure offerings. This may modify the pricing basis of the infrastructure offerings and create a whole new class of transaction-based services that would require data gathering and processing not unlike current call detail collection and rating.

In a related arena, cable companies that are tentatively entering the data services access market are likely to be moving from pre-onfigured "pay per view" to a wide variety of on-demand entertainment services. This may require a more flexible billing environment with more complex data gathering and transaction rating capabilities.

Despite legal and regulatory barriers in the United States, local carriers are entering the interexchange business and interexchange carriers are entering local carrier markets. As many telecommunications behemoths have discovered to their detriment, billing requirements for these markets can be quite perplexingly different.

Traditional telecommunications companies have entered the ISP fray at a breakneck pace and are discovering that online customers present a whole new set of billing opportunities and challenges—ranging from integrated electronic bill presentation and payment to electronic bill inquiry and other customer care functions. Ideally, those online functions would be integrated with billing for traditional services, but that presents significant integration issues.

Whatever wonderful products and services your company is contemplating for the life of your new billing system, it is clear that telecommunications and related businesses will remain hotly competitive arenas for many companies. For those companies, billing systems will need to be flexible and to support a variety of market-driven and promotional requirements.

The big picture. Longer-term product and service strategy begins with the "big picture"—corporate strategic planning. If possible, interview your company's senior executives, separately or together. Explain the anticipated practical life of the new billing environment. If they are not already involved in the detailed financial impact of deploying the new system, review the total costs of implementation and—especially—of conversion with them to underscore how important it is not to do this very often!

Big-picture questions. Are there any strategies or plans to establish or acquire new lines of business? Are there any strategies or plans to move into different market segments? If so, with what type of services? Are there any plans to move into new geographic markets? If so, which markets and with what types of service? Are there any strategies or plans to abandon existing lines of business or market segments? If so, which markets and specifically what services?

Where do your company's senior executives see the company and the industry going during the life of the new billing system?

Are there any strategies or plans to divest any lines of business, market segments or geographic markets?

Using big-picture answers. The information about the big picture can be used to identify longer-term requirements for the new billing environment. It can also be used to validate billing solution procurement priorities. Examples of this type of data gathering are provided in Chapter 14. Big picture information feeds product and market strategy.

The next level—product and market strategy

Line of business managers can provide a more tactical view of things to come. They can address directions within their line of business and in their portion of the industry in general. What are they asking market researchers, R&D staffs and vendors to address? What are the dominant industry trends in pricing plans, billing and payment options, markets, and standards?

Anticipated "combinations" of products and services. Often, companies offer stand-alone products and services and then determine that there are market synergies in "bundling" existing products and services. Are there identified corporate or industry trends that would drive combining or bundling some types of products and services with other types of products and services during the expected life of the new billing system? For example, your company might currently offer Internet connectivity through one business unit and "content" through a different business unit. The product line managers should be able to indicate whether bundling of these complementary services is likely in the life of the new system.

Are all products and services for a customer to be billed together on the same invoice? If not, what criteria are to be used to determine which services/products, if any, may be billed in combination?

Geographic markets for anticipated products and services. If your company is primarily active in North American markets, but anticipates expanding into Europe or the Pacific Rim or elsewhere, it is important to know this when choosing a billing system. The billing system may need to handle a variety of currencies and currency conversion, as well as being able to present a bill in multiple languages and character sets (alphabets, et al.) for the expanded markets. Of course, this will also require expanded information on tax rates and conditions as well as expanded regulatory requirements for billing. If your taxation vendor is United States or North American based, can they support international markets?

If your company currently operates in a single state and intends to expand to other states, you may require more complex support for tax information and application, as well as needing to conform to different regulatory standards for billing formats. If you choose a vendor for tax computation and processing, this may be as simple as buying additional modules of tax information. If you are maintaining your own tax tables (because you are currently only dealing with one or two jurisdictions and it was easy!), this may require rethinking the entire architecture of your taxation processing.

Pricing basis for anticipated products and services. Is the strategic direction of your company toward usage-based services? Do your strategic customers want transaction detail (e.g., call, connection, content delivery, QoS)? Is your strategic direction toward flat rate pricing and less billing detail? Does your strategic direction point to complex quantity discount structures—at the level of the individual service or at the account level?

These directions should influence the weight given to support for specifying pricing basis for individual services and for accounts.

Using Data Gathering Forms

We have provided some sample data gathering forms in Appendix A. The included CD provides a full set of data gathering forms—as Microsoft Word® forms. For each type of information, there is a form corresponding to the data gathering processes described earlier in this chapter. Each form provides a structure for gathering the appropriate data, identifying the information source and the appropriate stakeholder review structure. These forms can be used individually or "bound" together—in hard copy or electronically—to collect all information from a particular source for a specific product or service or to collect a specific piece of information for multiple products and services.

Summary

In this chapter, we have looked at the process of gathering data about specific products and services—both existing and planned. We have also looked at expanding the scope and time horizon of that information by understanding the implications of corporate and product strategy. In the next chapter, we look at combining, correlating and analyzing the information across all products and services.

Chapter

16

Consolidating and Analyzing Product Information

Once you have gathered all the necessary information on individual products and services—as described in Chapter 15—it is time to look at the information *across* all the products and services. This consolidation and analysis of product and service information will provide a solid foundation of business requirements for a new billing environment.

Product Mix

Compile all the service and product detail information for all current and planned services. Categorize by type of product or service into major product lines. Cross-reference this information by market segment. Then, sort the products and services within each product line and market by whether the billing is usage based, fixed, recurring, or a one-time fee. This should provide a comprehensive view of the product mix.

Does your company primarily provide traditional usage-based telecommunications services? Does it primarily provide Internet-based consumer-oriented services or business-oriented services? How much of your billing is usage based?

Is most of your billing fixed rate? Do your customers expect to see call detail for usage-based billing? If the majority of your billing is fixed rate and customers do not expect a detailed bill for usage-based charges, this may determine the most desirable system features for your company. You may want to structure your primary billing system to handle a single billing line for each service in an account. You can then use a separate system processing usage for any usage-based services that would supply a single line item to the main billing system for those services. Even if call or transaction detail is not required on the bill, that detail may be required to support account inquiry functions and should be available to customer care. The major impact in not providing call detail is on the bill print room, which can whip through a billing cycle output like magic!

On the other hand, if your customers or regulatory agencies require detailed billing, your architecture should incorporate detailed billing as a primary requirement for the new billing environment.

Uniqueness

Are there lots of other companies out there doing what your company does in the same or different markets? If your company is a small independent local exchange carrier or a new competitive local exchange carrier and has no other lines of business, there are a lot of "off-the-shelf" billing systems which target your market—because there are lots of other players with the same needs.

If you company distinguishes itself by providing a unique combination of services not offered by other companies or in a context not provided by other companies, you may have a need for significant flexibility or customization in the billing environment. These capabilities may have a higher initial cost and require a higher level of ongoing support. This is justified to support a critical strategic direction or initiative.

Sometimes "uniqueness" in product lines actually consists of novel combinations of nonunique offerings. In this case, it may be possible to utilize the "a la carte menu" approach to systems architecture—combining separate off-the-shelf solutions into a unique overall environment. One common version of that approach is seen when conventional land-based carriers enter the wireless market. Wireless call rating may be significantly different from legacy land-based call rating. Many of these companies have chosen to use a system specifically designed to rate wireless calls for wireless usage processing only and then feed the rated wireless calls into the billing stream after other items are rated by the main billing system. This approach may require some customization at the point where multiple data streams converge, but is far less expensive and more maintainable than full custom development.

Complexity and Packaging

How many of your company's products and services have complex billing formulas? How many have complicated promotional or discount structures? How many products or services indicate that they have "related" products or services?

If the answer to these questions is "a significant portion," then evaluating the capabilities of a mechanized solution to support this type of complexity and costs of managing such complexity should be an important part of your solution assessment. These costs must be compared to the costs of manually maintaining the existing complex arrangements (and manually handling any such structures proposed for the future) or establishing procedures that require a mixture of manual and mechanized steps.

Pricing Variability

If your data gathering indicates frequent changes in rate structures, lots of discounts, promotions and special contracts, this may be symptomatic of a business climate requiring very flexible price responses to market pressure. That would indicate a need for systems that support such flexibility in user-friendly, labor-minimizing ways.

However, sometimes these are indicators of the inflexibility of the current systems environment, so verify all your assumptions with the sales and product management folks. For instance, a system that doesn't support fixed categories of discounts may be overrun with "override" (unique) pricing or special contracts, which wouldn't be necessary if the system simply supported the entry of a discount level or percent!

Volumes

The total number of customers, total number of products and services, total usage records per billing cycle plus total order volume all drive the sizing of a billing solution. In some cases, sizing is simply a matter of specifying adequate hardware. However, many solutions do not "scale up" gracefully and many vendors are not prepared to support extremely high volume operations. On the other end of the spectrum, if your company is a small start-up, you may not have the capital to size your system for all of the targeted growth during its effective system life. You may want to place a high value on the ability of a platform to scale up without disrupting billing operations unduly.

Volatility

How many new products and services does your company plan to introduce per year? How many special promotions or temporary discount offers are anticipated? If the answer is "not many," then an "ugly" user interface for configuring new products, services and special offers may be of limited concern. On the other hand, if you are targeting a leading-edge market segment or a very competitive market segment and plan frequent service and product introductions, as well as lots of changes to service "packages," then the ease of adding or changing product and service configuration may be critical to your evaluation of possible new billing environments. A need to get products to market in speedy time frames, even if you do not foresee a need for a high volume of new products/services, may be another reason to look for expedited configuration capability.

Geographic Markets

What geographic markets does your company serve or plan to serve? Does the platform you are evaluating handle the currencies of those markets? The languages? The standards? The legal and regulatory structures?

Data Flow Analysis

It is often useful to develop detailed data flows for specific information in the old billing environment. This helps identify stakeholders for that information. It also indicates other systems in which data should agree with the data in the current billing system. If data validation is a goal and criterion for this project, then data from the old billing system should be verified with data from other systems in the data flow for that data item.

Data Review and Consolidation

Now that we have surveyed some of the considerations that apply across various categories of product and service data, let's take a look at each category of information across the entire product and service catalog. Keep in mind that conditions that are characteristic of a large portion of your company's business and hence of your billing should have significant weight in evaluating potential billing solutions. Therefore, in reviewing and consolidating data, we often look at "how much," "how many," or "what percent" of your business has certain characteristics.

Product/service code

It is important to verify the uniqueness of product and service codes. Summarizing formats, including subfields and embedded intelligence, will provide the basis for detailed system requirements for these codes.

If some products and services use unique product/service codes distinct from the product/service description and others do not—which can happen if you are planning to consolidate multiple billers—work with product management and the interfacing systems management (usually customer order and provisioning) to determine what will be used in this field for the products and services without distinct codes. This may require mapping to a new product/service code from a product/service name or description. that was unique to the old system.

Any changes made as a result of consolidation that alter the description shown on customer bills need to be flagged for two reasons: to prepare for parallel testing differences and to identify the need for and the cost of customer notification and support.

Product/service name/description

Review product and service name/description for duplications. Review these duplications with product management, sales and customer care to verify the acceptability of duplication. If changes are required, the parallel testing and customer notification issues need to be recognized.

Product/service detail

Product and service detail information can identify overlapping offerings. This is especially useful when combining books of business from two or more sources. Although it may not be easy to eliminate a specific offering, especially if the offering is a regulated service, identifying overlaps early allows product management to determine which offerings should continue to be sold. It may also give the sales organization the information and impetus to migrate customers from offerings targeted to be phased out to ongoing offerings. In the interest of eliminating "product clutter," some companies offer incentives for sales to affect these migrations. This could be in the form of eliminating commissions on the services phasing out or of an actual one-time bonus for effecting migrations.

Product/service identifier format

It is useful to determine how many services and instances of services utilize specific standard formats and what organization coordinates those formats and assignment of identifiers. This can be the basis for negotiating support from the coordinating organization—ranging from the fee structure for registering Internet domain names to contracts with Telcordia for Common Language identifiers.

It is also useful to determine product and service identifier formats that may overlap or require coordination within your company to assure uniqueness. If identifiers are not unique, this may drive a requirement for maintaining a hierarchical relationship of product/service to identifier in the new billing environment for all processing.

Billing basis

Is the billing basis for a very small percentage of instances of services very much more complex than the billing basis for the majority of instances of services? If so, are these services historical artifacts or do they represent a strategic direction? If they are historical artifacts, could they continue to be supported in the old billing environment while being phased out? An important consideration when addressing this question is whether or not these services are billed with other services that will continue to be part of the company service offering. If so, you will probably need to accommodate them in the new billing system.

How much of your billing is usage based? Is most of your billing fixed rate? Do your customers expect to see call detail for usage-based billing? If the majority of your billing is fixed rate and customers do not expect a detailed bill for usage-based charges, this may impact your desirable system architecture. You may want to structure your primary billing system to handle a single billing line for each service in an account. You can then use a separate system processing usage for any usage-based services that would then supply a single line item to the main billing system for those services.

On the other hand, if your customers or regulatory agencies require detailed billing, your architecture should incorporate detailed billing as a primary requirement for the new billing environment.

Billing information sources

How many different locations will be supplying billing information and where are they? This may have significant implications for your data networking group, especially if they are not currently getting billing information from those locations. New locations may be identified for new products and services, for acquired books of business or if interfacing systems (e.g., customer order, provisioning, middleware, mediation) are also being replaced.

How many formats are being used for call detail information? Since many mediation systems support multiple formats, this may be a good time to consider minimizing the number of formats directed to your billing system.

Geographic market

Develop a consolidated list of geographic markets, identifying the number of products and services offered in each and the count of instances of products and services in each. This is useful in evaluating support for various taxing authorities, regulatory jurisdictions, currencies, languages, accounting standards and technical standards. It also may be of use to the accounting, marketing and sales people as report parameters are considered and adopted.

Customer market segment

What customer market segments does your company serve now? What market segments are you targeting in the life of the new billing system? What billing features, options and media are desirable in these market segments? What are the market segment drivers for billing? Do these market segments correspond effectively with your sales and marketing structure? What flexibility, if any, is needed in the future to move customers from one segment to another?

Price structure

How many services have a simple price structure? How many have a complex price structure? How many require different price structures or currencies for different geographic locations?

Discounts

How many different discount structures do you currently support? How many different discount structures does your company want to support? If discounts are presented to the billing system as a unique product or service, they will be considered when looking at product service codes.

Promotions

How often does your company want to do promotional pricing? How many promotional pricing options are currently supported?

Special contracts

How many special contracts are currently in force? How many special contracts does your company want to support in the future? What products and services have the most special contracts? Which customer market segments have the most special contracts? Are these contracts managed manually or via the sales or billing system? If the contract management is done online in the billing system by billing operations, how input friendly are the contracts? If the contract management is to be done online in the sales system, will there be special interface requirements to the billing system? If contracts are to be done manually when the new billing system is in place, have the manual costs been included in the total project expense evaluation?

Related products, services, and features

Do all related products, services, and features know about each other? That is, if Product K says it is related to Service L and Service M, do Service L and Service M each say that Product K is a related product? What portion of your company's offerings has related products and services? If the relationship is hierarchical, is each service appropriately identified?

Warranty and support information

How many products or services have related warranty or support information? Is the database of record for this information consistent across those products and services? How many different structures of warranty and support are there across all product lines? If the billing system is to manage these items what special capabilities must be in place?

Accounting classification

Are the accounting classification codes consistent in format across all products and services? How many dimensions of accounting classification does your company desire? Are there dependencies between accounting classifications?

Quantity

How many product and service instances will the new billing system have to support—at conversion, at one year from conversion and throughout the life of the system?

Order volume

How many orders will the new billing system have to support—at conversion, at one year from conversion and throughout the life of the system? How many of these orders will be establishing new accounts? Changing account services? Deactivating accounts?

Considerations When Acquiring or Selling Business

Acquiring or selling a business is increasingly common in the telecommunications industries. This may be as part of a merger or acquisition or resulting from divesting a particular line of business, type of customer or territory. Indeed, such transactions are often the "trigger event" for acquiring a new billing environment.

In order to adequately plan the integration of business into the buying company, the due diligence should include, at a minimum: an audited count of unique products and services; an audited count of instances of products and services; and an audited count of customer account records by jurisdiction. Ideally, the due diligence will also include gathering and providing to the buyer the information reviewed in Chapter 15 for each product and service represented in the book of business—either by the selling company or through access to the relevant personnel in the selling company—as well as pertinent financial information (e.g., accounts receivable, days outstanding of receivables). In the absence of the ideal situation—particularly difficult to achieve when the source company is bankrupt or otherwise undergoing major restructuring—detailed examination of the current billing system should be a condition of the acquisition.

Summary

In this chapter, we have surveyed the process of correlating, summarizing and analyzing product and service information across all products and services to develop business requirements for the new billing environment. In the next chapters, we will look at some external and internal drivers of business requirements.

Chapter

17

Customer Considerations

Gathering information about who the customer is and how the customer interacts with the company's billing environment is a vital step as you move toward the selection of a new billing system. In this chapter, we discuss how customer groups may be defined and we look at the various needs of those different groups. We survey billing and payment methods and how the interactions among billing and other portions of the company contribute to billing system selection. A factor of importance here is the type of reports that stakeholders desire from the billing environment—identifying these needs early in the selection process avoids ugly redesign after implementation.

The information gathered in Chapters 15 and 16 is useful in dealing with the customer-oriented issues here. For example, plans for new product and service offerings may make changes in customer segmentation desirable.

Customer Account Records

The customer account record contains the plethora of information that a billing system needs to produce a bill. In earlier chapters, we have discussed the service and product data it contains, but have spoken only in general terms about the contents that describe the customer and customer preferences.

What are these non-product/service data and how are they used? The customer account records in most billing systems contain the following items:

- **Account identifier or number**—The account number is the primary identifier of the customer account record. It may be the same as the customer's telephone number or a fictitious number, but it will be unique and used on all transactions for the customer throughout all systems.
- **Account name**—The account name identifies the person or entity responsible for paying the bill. This may be an actual individual or a corporation (legally an individual, although a fictitious one). Businesses other than cor-

porations should be billed in the owner's name or, for partnership companies, in the name of one of the partners.

- **Billing contact**—The billing contact name may be the same as the account name; this is the person who is authorized to discuss the account or to whom collection calls may be directed.
- **Billing address**—This is the address on the bill and it may be different from that of the service(s).
- **Directory listing appearance**—For companies that provide telephone directory services, directory appearance information—or advice not to publish—for white and possibly yellow page directories is maintained. Chargeable directory services (e.g., additional listings or boldface type) are handled as any other product or service and may appear in a record ancillary to the main customer account record.
- **Final billing address**—A billing system may or may not be configured to contain a unique field for the address to be used on bills rendered subsequent to discontinuance of the account. For systems without the unique field, the billing address is used even after the order to close the account.
- **Credit information**—Ideally, the information contained here will provide adequate support to collect outstanding balances and overdue balances. For businesses, it should contain information on the type of ownership (sole proprietorship, partnership, corporation), the state or country of incorporation, partner names or corporate officers, years established and information on other accounts, if any. The type of information to be collected and maintained usually is determined by marketing/sales with advice from legal and possibly regulatory.
- **Credit classification**—The credit classification indicates the company assessment of the customer's creditworthiness. It is used to support the need for advance payments/deposits and to schedule collection efforts.
- **Credit history**—Generally, a record of returned check activity, late payments and collection actions taken is retained for a number of months, as specified by company needs.
- **Deposit information**—The information contained for accounts secured by a deposit usually includes the amount, date received, date to be returned and interest percentage (if any is to be paid).
- **Bill cycle**—This identifies when the customer account bill will be prepared.
- **Balance and payment history**—Shown or accessible are items such as current balance, number of months represented by current balance, adjustments pending reflection on bill, recent payments and dates of those payments.
- **Primary bill medium**—The customer's preferred billing medium is identified here.

- **Preferred payment method**—The choices may include direct (i.e., the customer sends a payment to the company's remittance document address or walks into a company office with the payment), credit card, debit card, EFT, and online bill payer. For customers selecting credit or debit card payment, the customer record will contain or be able to access the card number expiration date and any other relevant information.
- **Market segment**—The market segment is applicable to the entire account.
- **Tax jurisdiction**—The tax jurisdiction contained at the account level is generally the one applicable to the primary billing address.
- **Tax exemption indicator**—Customers who have provided confirmation of tax-exempt status are identified, which precludes application of taxes during bill preparation.
- **Third-party notification indicator**—The company may offer to notify a person designated by the customer in the event it is unable to reach the customer for things such as possible service interruption. In addition to the indicator, the third party's name and contact information will be available.
- **Master account/slave account indicator**—In billing systems that permit individual accounts to be combined for billing or report production, the customer account records will identify associated accounts and the hierarchical relationship of each account. See Figure 7-3 for an illustration of hierarchical accounts.

Segmentation

Is segmentation necessary or may all customers be lumped together and handled with common processes and practices? Some of the factors that go into answering that question are company strategic directions, regulation, sales force configuration, revenue protection and numbers of accounts.

A new company with very few customers may not segment at all, but most companies wish to identify a customer as a member of several groups. Modern billing systems accommodate a number of indicators in customer records—along with efficient report production capabilities—to be able to identify subsets of customers for unique types of interaction.

Market segment

Market segments are based on customer characteristics such as the type of service, projected monthly account billing, service address, type of business or a combination of factors. Regulators define some segments; regulated jurisdictions still require "business" and "residence" market segments. Company-defined market segments primarily are used to match sales forces, offerings and customer interfaces to customers, but they also may exist to categorize data for the company.

A number of groups may be involved in defining market segments and the questions to be answered are varied.

- Are there regulatory requirements to satisfy?
- How is the sales force organized to sell to the customer and is that how the segments should be structured?
- Will planned services require unique segments?
- Will billing media or other billing options require unique segments?
- Does management need financial reports for subsets of customers within a regulated segment?
- Do other customer-facing groups within the company, such as collections, have needs for unique segmentation that cannot be satisfied with other indicators?

Your company may have well-established market segments that will not change in a new billing environment. If that is the case, you may wish to skip the next paragraph.

An example of how these questions may be used to determine market segments is shown in Table 17-1. Assume a company operates in two states and is providing regulated business and residence local services, wireless service and Internet service. Additionally, the company plans to introduce combined billing for local/Internet residence service sometime in the future, initially in one state; this will be the company's first instance of combined billing. Each state regulatory commission requires the segmentation of local service by business and residence. But the sales organization wants to have the local business market segmented further, with one state maintaining two separate segments for customers and the other state maintaining three. To add to the fun, the definition of the local business segments is different in each state. The decision is made that the market segments will be determined by line of business, subset by state and business segmentation. The proposed new service offering is to be assigned its own unique segment. The result is a set of 15 market segments to categorize the six types of service the company is or will be offering.

Stakeholders

- Product management
- Sales
- Regulatory
- Legal
- Billing operations
- Accounting standards

TABLE 17-1 Market Segment Worksheet

Line of Business (Segment Characteristic)	Local Residence Regulated	Local Business Regulated	Wireless Residence	Wireless Business	Internet	Planned Combined Internet/Local Residence Billing
Segment:	LRX: State X	LBXSM: State X Small and	WR: State X and State Y Medium	WBX: State X	NB: Business All	CNX: State X
		LBXL: State X Large		WBY: State Y Small	NR: Residence All	
				WBY: State Y Medium		
	LRY: State Y	LBYS: State Y Small		WBY: State Y Large		
		LBYM: State Y Medium				
		LBYL: State Y Large				

State X:
- Small/Medium Business = 1–100 lines
- Large Business = 101+ lines

State Y:
- Small Business = <$200/month
- Medium Business = $201–$1000/month
- Large Business = >$1000/month

Credit classification

Credit classification is another type of segmentation. Some companies, especially when beginning business or establishing a new line of business, opt to have one set of procedures that apply to all customers. Most companies use a range of credit classifications that determine such things as the need for payments prior to service establishment, the interval a customer is allowed for payment prior to being considered overdue and the type of collection steps to be taken if the account becomes delinquent. These are strategic decisions that should be reviewed, confirmed and documented before selecting a new billing system.

Establishing the credit classification. Credit classifications are applied to a customer account when it is established and are updated as needed during the lifetime of the account. Creditworthiness may be based on an internal company formula manually applied by sales or collectors, by accessing external sources such as credit agencies or credit card companies or by providing information to the billing software for a credit scoring.

Documenting the credit granting and update processes is an important step toward determining what, if any, interfaces are needed for the billing system. Where in your company's process is the credit classification assigned for the new account? If it is to be done in the billing system, what types of system interfaces need to be established? If the billing system is to interact with credit agencies or credit card companies, the contacts in those companies need to be identified and the interface standards employed must be documented.

Some companies will not establish accounts for customers of unproven credit without an advance payment or a deposit. Modern billing systems generally can accommodate these types of transactions. The conditions that cause your company to collect such monies and, what, if any, special requirements the company uses must be part of the process documentation. These requirements may impact customer account creation if they call for notification to sales or provisioning to hold the order until the money is received. For example, company processes may currently initiate provisioning before the advance money is received, but require that the check must be confirmed as paid by the customer's bank prior to completion of provisioning. When the deposit is not paid (e.g., check returned unpaid), the company will suspend the provisioning work until the money is received or the order is canceled. The billing environment must be able to accommodate this workflow pattern or the process must be changed prior to implementation of a new billing system.

Establishing collection parameters. Billing systems can have different collection parameters for each credit classification. One of the strategic choices to be made is how much of the collection activity the company wants to delegate to the billing system. One choice is to have all overdue accounts referred to the collection organization for action. Alternatively, all contact with the customer may be generated by the billing system in the form of billing messages, notices and letters culminating in discontinuance of the service without any customer care intervention at all. Most companies select a mixture of written and verbal contact for the majority of the customer base. Table 11-1 in Chapter 11 shows a sample of a collection approach for a single credit classification. Often companies assign several credit classifications to each market segment. For example, a local residence account may be identified as excellent, acceptable or poor; each of the classifications could use different dates and types of collection actions.

The collection parameters for each credit classification should be reviewed and analyzed.

- What types of billing system outputs are required to support the collection activities?

- If referrals are to be made to customer care for action, what form will those referrals take: online interactive notification to customer care systems, work queues in billing systems that can be accessed by customer care or printed reports?
- Will all customer care activities be entered into the customer account records and maintained there? If so, are there requirements for length of notes?
- How long will such treatment notations be retained and which company groups will have access to the notes?
- If the billing system is to generate written notification of delinquency to customers, will that take the form of billing messages and, if so, what is the estimated increase in bill lines per message?
- If notices and/or letters are used, how many types will there be and where will they be output? If they are to be printed, where and by whom? What print-specific formats are required? What are the estimated volumes and timing of print production?

The answers to these questions are needed for selecting the best billing system for your company and for ensuring good estimates for sizing of work groups and budgeting of other resources. All interfacing software and/or hardware managers need to be identified and compatibility issues need to be resolved.

Stakeholders

- Product management
- Sales
- Billing operations
- Desktop/workstation support
- Carrier relations/strategic partner relations

Differentiation Strategies and Customer Values

There is some truth in the old adage that says the customer's primary view of a company is the monthly bill. Not all that long ago the vast majority of bills were transmitted on paper, and each customer bill showed the same format, in good part because of billing system limitations. That is no longer the case. Today there are a number of ways to use the bill, associated billing reports and the resources of the billing system to differentiate your company from another—and enhance customer value in the bargain.

Bill mediums and formats

You may want to offer your customers the opportunity to choose a billing medium from a full array of possibilities (e.g., paper, CD, Internet, email, EDI, as described in Chapter 9), or from a limited number of choices. Your strategic direction for a given market segment should dictate what you offer, as the various ways of transmitting the bill will have differing degrees of appeal to different groups of customers.

While some companies are willing to provide billing information to customers in more that one way without charge, it is becoming more common to charge for everything after the first, or primary, bill. All bills or supplemental reports after the primary version may be coded as a product and charged accordingly. A notable exception is detailed usage statements—some companies provide usage detail without charge, but only at the express request of the customer. The standard bill format contains a line stating only the total usage amount and the customer is able to forego dealing with the call detail unless some need arises for the detail. As customers become more concerned with ecological issues, this type of approach is viewed to be of high value.

Customers also place high value on getting combined bills. A combined bill may show all of the products and services received from the company for one account or it may show multiple accounts, with all associated products and services for each one. Most modern billing systems are capable of creating such bills and getting this capability is often a primary motivation for a company to move from existing legacy billing systems that are generating multiple bills every month for customers with more than one account and/or type of service. The difficulty of a conversion is greater when migrating from stand-alone billing to combined billing: a strategic decision may be to implement the new billing system on a stand-alone basis and gradually implement combined billing by customer segment.

Stakeholders

- Product management
- Sales
- Regulatory
- Billing operations
- Carrier relations/strategic partner relations

Other value added items

In addition to how the bill looks and how it is delivered, there are other billing choices that may be valuable to customers. The possibilities need to be supportive of the company's strategic direction and fully evaluated to ensure that the cost/benefit assessment has been considered.

Many customers desire to select their monthly billing date rather than have it assigned by the company. This is especially true for business customers. While most modern billing systems can accommodate customer choice, there may be limiting factors that make it an unattractive option for the company. One significant issue in the past for companies producing large numbers of customer bills was the need to balance the number of accounts in the billing cycles over the course of a month. When most bills were printed, the print room capacity and the objective interval from bill preparation date to bill mail date helped the company determine the maximum number of accounts that could be assigned to any given billing cycle. Today, given the diversity of billing media, companies are finding it less unattractive to offer the customers a choice of billing date.

Unique bill formats—or reformatted bills—are important to some customers. They want to take the billing information that your company provides and have it repackaged to facilitate internal record keeping. Recouping the cost of customizing reports and maintaining those reports has often been non-cost effective for telcos in the past, but there are vendors that will contract with customers to take an electronic version of the bill (provided by you) and produce reports. This type of arrangement is often treated as a product: it is included in the product catalog, assigned a product/service code, a charge is developed and it is negotiated and ordered through sales.

Modern billing systems allow access to unbilled usage charges for the customer, usually via a query to the customer care organization. This is a very valuable service to the customer who has been a victim of fraud or who is attempting to evaluate the impact of an extraordinary one-time usage event. The billing systems usually allow an unbilled call to be adjusted without appearing on the next regular bill, which can mitigate the customer dissatisfaction with other billed but fraudulent charges.

If you are very lucky, product management may have customer satisfaction survey results by market segment that give specific information about what types of bill media have the highest value. If such surveys have not been done, a team consisting of sales, marketing, customer care, and billing operations personnel would have the best background, skills, and experience to answer questions about formats, billing media, and other value added options.

Payment Options

Payments are directed to billing systems in a number of different ways and the best way for one customer may not work for another. For the company, whichever method is used should be accurate, auditable, efficient, allow posting of the payment to the customer account, place the cash in the proper financial institution, and properly update the accounting information in the corporate books.

Most billing systems efficiently and effectively handle payment processing when information is received by posting the payments to the customer account, applying the accounting classification, and notifying corporate accounting to update the corporate books. How the information is provided to a new billing system from lockbox banks, via EFT from online bill paying services and from credit/debit card will require the same type of coordination as other third-party interfaces: system contacts must be identified and interface standards and controls must be documented.

As customers are given new options for paying their bills, the company must make sure that it updates and maintains the internal processes for investigating lost payments, dealing with misapplied payments and handling payment discrepancies in general. There should be documentation showing conflict resolution contact information for each new source of payment data, with clear escalation procedures documented.

Stakeholders

- Product management
- Sales
- Accounting standards
- Billing operations
- Carrier relations/strategic partner relations

Summary

In this chapter, we have looked at how information about the customer is gathered and used to organize a billing system in segments. These segments give the company the ability to meet customer needs and to track company objectives. We have identified where processes need to be developed and implemented to support additions or changes to billing systems.

Chapter

18

Regulatory and Tax Considerations

No plan for telecommunications services and products is complete until it has been reviewed and analyzed for tax applications and possible regulatory requirements. Regulation has been significantly reduced since the 1980s but it is far from a thing of the past. In this chapter, we look at the strategic plans developed in Chapters 15 and 16 in order to identify applicable regulatory and taxation jurisdictions. We discuss who should analyze the jurisdictional requirements and determine how to implement them.

Regulatory Authorities

There is still both federal and state telecommunications regulation in the United States. It exists in varying degrees, depending on the type of services and products provided and the jurisdiction in which they are offered.

Every state in which your company provides service—that is, every state that appears in your customer files as a service address—must be considered to be a potential regulator until its regulations have been researched and evaluated. That holds true for interstate services as well; each type of service should be researched and evaluated to confirm which regulations, if any, apply to your company. We strongly recommend that this work be documented and maintained for future use.

Countries outside the United States in which you do business may have their layers of regulations, too. Each location must be evaluated for national level requirements and other jurisdictional regulatory bodies analogous to U.S. state and local commissions. Again, documentation is recommended.

The company regulatory or legal personnel are usually given the tasks of research and evaluation.

Stakeholders

- Regulatory
- Legal

Regulatory Requirements

Regulatory requirements take a great and varied number of forms. Some bar the entry of companies into certain markets or prohibit the provision of certain types of service. Some mandate specific customer bill charges and associated verbiage. Others require the company to prepare and file periodic reports containing financial data or service quality data. Regulatory personnel must scrutinize each regulation to identify which parts of the company are affected and who must take action to bring the company into regulatory compliance.

Stakeholders

- Regulatory
- Legal
- Product management
- Billing operations
- Others as identified during regulation reviews

Taxation Authorities

In addition to federal and state tax jurisdictions, there are many local entities that tax telecommunications services and products. The same type of reviews done to identify regulators must be done to identify taxation authorities: each community, state and country in which services or products are provided needs to be researched and evaluated. Documentation of all findings is needed; information on taxable jurisdictions and the types of taxes applicable is needed by sales, customer care, billing operations and corporate accounting.

Tax accounting or legal personnel may have done this evaluation at the behest of product management when service and product planning occurred. In any event, product management usually is involved in this activity as with any other aspect of a service or product that has the possibility of creating a negative customer reaction.

Stakeholders

- Tax accounting
- Legal
- Product management
- Sales
- Customer care

- Billing operations
- Accounting systems

Taxation Structures

The numbers and types of tax applications required for a company with multistate and interstate operations may be daunting. Some companies choose to do all tax calculation and application themselves; these are usually companies with a limited number of tax jurisdictions and services. Many others make use of one of the number of quite good tax modules available on the market today. Even if you choose to take the vendor option, your system must be able to determine which jurisdiction's tax is applicable to transactions being processed in order to use the modules.

Each taxing jurisdiction should be documented, with the type of tax and rates identified, for customer contact use by sales, customer care and billing operations.

Stakeholders

- Tax accounting
- Legal
- Product management
- Sales
- Customer care
- Billing operations
- Corporate accounting
- Carrier relations/strategic partner relations

Summary

In this chapter, we have covered how and by whom regulatory and tax jurisdictions need to be identified. We have discussed what steps need to be taken to analyze specific impacts of regulatory and tax on the company processes and personnel.

Chapter

19

Cash Flow Considerations

Many of the items we have discussed elsewhere could be categorized as cash flow considerations; indeed, almost everything associated with a billing system will affect cash flow in some way. For example, the speed with which a provisioned service is identified in the billing system as eligible for billing will determine when the first bill will be rendered. Should that notification to billing be delayed, or if the billing system cannot process the notification—which means that conflict resolution between the two systems is required—the next billing cycle for the customer may be missed, thereby adding at least 30 days to the cash flow interval.

It is short sighted to see cash flow management as applicable solely in the interval from bill delivery to the customer to cash receipt. In this chapter, we discuss analyzing the cash flow effects of proposed strategic directions—as well as the effects of existing company processes. We also look at alternatives to keeping all cash flow management activities in house.

Process Impacts

The positive cash flow impact of accurate, efficient and timely processes should not be ignored, yet all too often the process definition portion of new product implementation is given low priority. The goal should be to minimize the time between a customer's receipt of service and receipt of the bill, not just the interval between billing and the deposit of cash in the company's account. Strategic plans should be reviewed for the following:

- Have objective intervals been established for the processing of the service, product or work effort?

- Is someone identified as the "owner" of each objective and does that owner have authority to identify and implement improvements consistent with company strategy?
- Are error resolution procedures identified, documented and accepted by all involved organizations? Have they been disseminated to all interfacing groups?
- Is an escalation and continency procedure identified for error resolution? Is it documented and accepted by all involved organizations? Has it been disseminated to all interfacing groups?
- Have accuracy metrics been established?
- Have procedures for error root cause analysis been identified, documented and accepted by all involved organizations?

Existing processes that will be used to support the new billing system should get the same scrutiny given to the strategic plans. To the extent possible, changes to the existing processes should be in place prior to implementation of the new billing system.

Approaches to Improvement

Let us assume that cash flow would be improved by improving the intervals between key critical events in the billing cycle. However, the nature, timing, volume and flexibility of various transactions are quite different, leading to different technical approaches and requirements.

Service turn-up to initial bill

Service turn-up is a relatively asynchronous process, resulting in single complex messages from the provisioning environment to trigger initial billing. Generally, both the provisioning management system and the billing system are directly under the control of your company. Coordinating data formats and structures to eliminate mismatches and complex translation processing can minimize the interval from turn-up to generating the initial bill. The establishment of fully matched message protocols and automated error notification to provisioning, if anything should happen to go wrong, might be appropriate and possible in this instance.

Usage to billing

The transmission of usage detail to the billing environment may be asynchronous single transactions in special cases, but is normally high-volume scheduled bulk data transfers. Increasing the frequency of these transfers may shorten the interval from usage to billing, although it may generate significant control and management overhead. In many cases the source of usage detail is not within your control, although it is a good idea to negotiate service agreements that include timeliness from the actual usage to a successful feed.

The problems in usage feeds are more likely to be entire corrupt files, rather than errors in a single data field, so controls should focus more on bulk data integrity checks. Tracking of data transfers to verify that a particular transaction or data set has been received once and only once is important in this environment.

Billing to payment

The interval from billing to booking a payment is impacted by many external factors, ranging from a customer's mood to anthrax-contaminated post offices! Some of those factors can be minimized by moving to a more controllable bill delivery environment and by making it convenient for the customer to pay the bill. Perhaps even more valuable in this environment is minimizing the opportunity for "fumble fingers" to create erroneous data entries, since each instance of manual data entry introduces increased opportunity for error.

Automating the payment process has many advantages beyond error reduction, however. An online debit processing environment will validate availability of funds before making the payment transaction, eliminating all the overhead of processing and returning "bad" checks. It also eliminates all the time involved in mailing and mail processing.

Fraud and uncollectables

The approach to fraud and uncollectables may be significantly different. First, identifying particularly fraud-prone services and creating a billing structure with a sizable, paid-in-advance recurring charge can minimize corporate risk. Take note of the evolution of pricing structures in the wireless arena! From the perspective of the billing system, this simply means that the system must handle recurring charges billed in advance—a fairly standard billing configuration.

There are, of course, lots of other ways to minimize fraud and uncollectables. Many involve prompt identification of fraud and/or uncollectable situations in the billing environment and prompt notification of the appropriate group for investigation and resolution. These interfaces are often customized to your company's specific business and operational environment.

Alternatives to Internal Cash Processing

There are alternatives available for companies wishing to outsource all or parts of cash flow management. The tradeoff for paying a third party for this service is usually the availability of cash more rapidly than would be seen with in-house processing.

The use of lockbox banks is a good example of cash flow advantages for the company, albeit not without an offsetting expense. Customers are given remittance documents that are addressed to the company, but show the bank's address. For a fee, the bank prepares information for the company showing the receipt date, the account identification and payment amount. These data are usually electronically transmitted and input directly to the billing system. Concurrently, the bank moves the cash into the company's account, giving the company same-day availability of its cash. This type of arrangement meets customer needs, too, since the payment is posted to the customer account quickly—this may be needed to avoid late payment charges or forestall scheduled collection activity.

Another option used by some companies is to sell the collectables at a discount shortly after rendering the bill. The company receives cash immediately and the purchaser does the work of collecting the billed amounts from the customer and providing information to post in the customer account.

Electronic bill paying is becoming more and more popular with customers. As with other potential contracted payment processes, the company should carefully evaluate what the costs and benefits are for each affiliation.

Summary

Cash flow is impacted by almost every step between the company's expenditure to provide a service through the receipt of cash in payment for the service. In this chapter, we have discussed how processes may be developed to improve timeliness, accuracy and efficiency—all of which will have beneficial effects on cash flow. We have also looked at possible contracted arrangements for cash flow improvement.

Chapter

20

Interfacing System and Process Considerations

Billing does not exist on its own nor is it a "black box" that has no relationship to the rest of the company's business and processes—although it is often treated that way!

In this chapter, we look in some detail at data flows into and out of billing, as well as controls on those flows. We examine issues of data integrity and synchronization between disparate systems and process functions. We also take a brief look at systems and data that are related to billing, although they do not have direct interfaces with the billing environment.

The Systems and Processes

Many systems and data flows have impact on the billing environment. Likewise, the billing environment has impact on many nonbilling processes, systems and data flows. Here we look at some of those systems that impact or are impacted by the billing environment, both directly and indirectly. This should clarify why there are so many stakeholders in anything that affects the structure, logic, format or timing of billing information.

Marketing

Marketing requires numerous data feeds from billing, including revenues by product, market segment and location. Other information includes payment performance and uncollectables. The interfaces from billing to marketing are largely informational, but may be mission critical to marketing. In addition, they may impact management and executive compensations, tariff support and other critical financial functions.

Marketing is also the primary source of information on products and services, pricing, discounts and promotions and market segmentation. This information may be manually entered into the billing environment or passed through customer order systems, provisioning systems and/or middleware.

Sales

Typically, the sales system environment includes presales functions, sales support functions and account management functions. Presales functions include sales lead management, sales inquiry and quote support. Sales support functions include order entry, service configuration, and credit management. Account management functions are reporting on account financial performance, service quality, and customer satisfaction, as well as account product and service overview and configuration access.

Presales systems. Presales systems rarely interface directly with the billing environment. However, billing basis and pricing information used for quotes in the presales environment should be consistent with the billing basis and pricing information in the customer order and billing systems. Indeed, since quotes from the presales environment often become the basis for contractual obligations, eventually that pricing information must agree with the billing environment.

The consequences of unsynchronized data can be complicated. If a presales quote is based on pricing information not currently supported in the customer order and billing environments, the simplest resolution may appear to be bringing the order and billing databases in line with the quote. However, if the quoted pricing is outdated or not yet in effect, that may not be appropriate. Depending on the circumstances of the quote, a unique "service" may have to be established or, more likely, "override pricing" will be used to overlay rates for an existing service. In either case, these options require manual intervention, usually are not cost effective and make the service and the account more difficult to administer.

Sales support systems. Sales support systems are the source for much of the information used by the billing environment. These systems initiate the customer account, classify the customer, obtain an initial customer credit rating and order individual products and services.

Whether this information is provided directly to the billing environment or goes by way of provisioning and/or middleware systems, the structure, content, validation and timing of the information can have significant impact on billing. For instance, the customer ordering system may use a single field for street address and do no parsing or validation of that street address. If the related billing system parses street address by number, optional direction indicator, street name, street type (Road, Avenue, Way, etc.), optional floor, room or apartment number and verifies that parsed address against postal guides, then address information that is completely acceptable to the customer order system may be rejected by billing. The billing system requires the parsed information to correctly establish the account for processing items such as tax and mail delivery.

One way to resolve this is to implement identical parsing and validation in the customer ordering system. This may not be practical. The customer ordering system may be an off-the-shelf vendor product that cannot be customized to meet this need—or it may be an aging legacy system slated for retirement in a short while, making it uneconomic to modify.

Obviously, you cannot just ignore the inability of the billing system to process the order record. If your company and order volume are small, the billing environment can reject the record (not process it at all), create an error log and send the order for manual resolution and entry to the billing system or the source system. Even in higher-volume environments, this is often the process while integrating new systems into the data flow. These rejected records must be tracked for status and resolution manually—often by a team comprised of both sales and billing personnel.

If there are significant volumes of orders, there is obvious value to automating both the notification of rejected orders and the problem resolution process. This can be accomplished by modifying both systems, by employing a comprehensive middleware implementation that resolves many mismatches and manages rejects and resolution, or by implementing a separate system to manage reject resolution.

Account management systems. Account management may rely heavily on information provided by the billing environment for individual customers and accounts. Information such as payment history and billing adjustments can be very useful in identifying unprofitable accounts. It can also identify dissatisfied customers who could be turned around. In most companies, the billing system is the database of record for the configuration and contents of the customer account.

The interfaces from billing to account management systems are largely informational, but may be mission critical to account management. The synchronization of account information between billing and account management in the post-service turn-up period should be fully defined for any billing implementation.

Some account management groups may be involved in billing queries and dispute resolution. While this may be a function more commonly performed by customer care, customer contacts and queries may cause account management to submit adjustments and/or notes to the customer account record in the billing system. This is especially the case for very large customers.

Sales compensation systems. In many companies, sales compensation is based on booked revenue from the sale. Typically, billing will provide a feed of revenue by service or by account for each billing cycle to the sales compensation system, which then will calculate the appropriate sales compensation payments. This interface should be subject to stringent quality controls appropriate for financial transaction data.

Provisioning

Provisioning is the process by which products and services become available to the customer. In Section I, we looked at provisioning and the provisioning systems as they are directly involved in the billing data flow. We have noted frequently that virtually all account and product/service information may flow through the provisioning systems to the billing environment. In addition, the provisioning process creates some billing information. Also, much of the information used by related provisioning functions should be consistent with data in the billing systems.

As with the sales support systems/billing systems interface, there is a need for defining error resolution processes, which may be manual or automated. Timely error resolution has a major impact on customer satisfaction, on service resource utilization, on the cost of provisioning and on cash flow, as we saw in Chapter 19.

Configuration/engineering. Service and equipment configuration and engineering utilizes information also used by billing. This information includes product or service code, service address, and service due date. Configuration and engineering will also assign the service identifier (e.g., circuit number or IP address) for many services.

Configuration and engineering determines what resources (e.g., transport facility, central office equipment, server, gateway) will be used to provide a service. This information may be used to apply or verify billing adjustments for major infrastructure failures.

Implementation management. The implementation management function tracks when things are done during the provisioning process, as well as capturing what is done and what resources, such as time and materials, are used. Implementation management is the source of the confirmed installation date, as well as of any installation-related time and materials charges.

Implementation management also has the responsibility of overseeing the timely execution of disconnect orders. Unfortunately, disconnect orders often are not given adequate attention, resulting in bloated network inventory and lots of service engineering rework when resources assigned to a new service have not been released from their old assignments. Billing will stop billing on the requested disconnect date, even if service is allowed to continue and is used after that date.

Implementation. Service implementation generally involves configuring network equipment, network access and customer premises equipment to supply service for the customer. Although service implementation systems usually do not interface directly with billing, it is critical that the network information and the billing information be synchronized, especially for usage-based services. If the network has the wrong customer telephone number, for instance, calls will not be billed to the correct account.

Maintenance

We have not focused on maintenance functions previously. Billing has limited interfaces with the maintenance environment. However, maintenance may generate some information directly to billing.

Maintenance management. Typically, if there is any direct interface between the maintenance environment and the billing environment, it will be from the maintenance management systems. These systems would have interfaces to test management and to restoration and repair.

Test management and professional services. In today's multivendor telecommunications environment, customers often request "coordinated testing"—that is, testing that involves multiple vendors testing beyond the boundaries of their own service. Most carriers charge a time and materials fee for this type of testing. While this may be initiated by a service order as a new billing element, in many companies, this request comes directly to service maintenance and charges are forwarded directly from the maintenance systems to billing. Similarly, customers may request other technical advice and assistance.

Restoration and repair. Under some circumstances, there may be a billing credit generated because of a service outage or other inappropriate denial of service. This may be generated directly from the maintenance management system to the billing system or may flow through the customer care environment. There may also be product or service repair charges if those repairs are not covered by the terms of service. These may be flat rate charges or time and materials charges.

Materials management

Materials management in this context is actually a part of the provisioning process or, less frequently, of the maintenance process. However, the functions of product ordering, warehousing, and shipping are sufficiently different from service provisioning or maintenance functions—and supported by separate systems—that they deserve separate consideration.

Order fulfillment. Order fulfillment functions receive orders for products (e.g., equipment and materials), ship those products to the customer and notify provisioning or billing directly that the products have been shipped, when they were shipped and, perhaps, the status of shipping.

Agency procurement. Some telecommunications companies act as agents of their customers to purchase products. This requires the ability to process an order for such equipment from sales, to issue an order to the appropriate vendor and to track the delivery of the order to the customer. If the customer has requested consolidated billing rather than a separate bill from the vendor, this type of procurement also must post the vendor's charges to the billing system.

Customer care

Customer care is responsible for the ongoing customer relationship—including account inquiry and billing adjustment, trouble reporting and general user assistance. Either the billing systems have a direct, on-demand (by customer care) connection to customer care systems or the customer care agents directly access the billing system for both inquiry and update.

Trouble reporting. Trouble reporting and management may be the responsibility of the customer care function for some products and services and market segments. In this role, they take trouble reports, provide updates to the customer on trouble resolution progress and verify the restoration of service or product repair. They may determine that a credit is in order for interruption of service. They may also determine that a maintenance fee is required. Credits, charges and notes to the customer account record will probably flow directly to the billing system from customer care.

Account inquiry. Customer care handles calls from customers complaining about charges or requesting clarification of billed amounts. Customer care agents need access to the customer billing record, to supporting call detail and to account history and other account information in order to handle these calls. Often these calls result in the identification of erroneous or fraudulent charges—or simply of charges that threaten customer good will. This may result in a credit memo (adjustment) to the customer account, as well as supporting notes' being entered into the account record in billing. It may also result in a referral to fraud management when fraud—whether by the subscriber or by a third party—is suspected.

Customer relationship management. Customer care manages the customer relationship. Often, the customer's perception of the company (especially when nothing much goes wrong!) is governed by billing-related interactions. The ability to view customer account history and notes may enable the customer care representative to address the individual customer's concerns in a positive manner.

Billing operations

Technically, billing operations is not an "interfacing" process; it is the principal manager and user of the billing systems environment. It is highly likely that many of you reading this book and planning a new billing system acquisition are part of billing operations. However, the process so impacts and is impacted by billing systems decisions that it seemed worthwhile to discuss some of the billing operations functions.

Billing operations is responsible for the timely and accurate processing of large amounts of information into large quantities of customer bills and the accurate and timely application of payments to those bills.

Error resolution management. Billing operations usually is responsible for identifying and managing to resolution any data fallout from any of the many interfaces between billing and the rest of the world. This may include customer account and service order information, provisioning information, usage detail information, payment information and credit information. This can range from reviewing a single customer account record for an invalid street name to investigating why a whole day of usage detail cannot be read by the billing and rating systems.

The greater the volume of billing information and detail, the more critical some automation of the problem detection and resolution process becomes. This may be the function of the billing system itself or of an "add-on" capability, but it should be part of all new billing systems requirements. Key functions are swift problem identification, notification to billing operations, ability to determine the source of the erroneous data, referral for resolution, investigation support and resolution tracking and status.

Data and processing quality assurance. Billing operations is responsible for ensuring the integrity of bill runs at both the macro level (reviewing the total billed amount for reasonable deviation from previous month's bill runs, for instance) and at the level of sampling individual bills.

The billing system should provide the capability to capture macro-level parameters from each bill run as well as supporting verification and investigation of individual bills.

Reconciliation. In Section I, we discussed the need for reconciliation processes whenever data streams representing the same reality flow separately. Billing is at the center of multiple data flows representing service to the customer and payment from the customer. Therefore, billing operations is responsible for many business-critical reconciliation functions. One of the most critical is reconciling payment information in the billing system—which may come from a third-party payment processing organization—to payments reflected in the company's cash accounts.

Another area of reconciliation that can have a significant impact on the company's capital requirements is the reconciliation of lines in service to lines billed. Obviously, any lines in service which are not being billed represent a loss to the company and may lead to unnecessary network expansion.

Identifying and applying bulk credits. Billing operations is often responsible for working with network when customers impacted by a major network or node failure are to be given credit for the time out of service. This requires good coordination between service configuration systems and the billing environment. It also may point to a need for a capability to apply bulk credits across many customers, probably by service identifier.

Contract maintenance. We have looked at the need to manage special contracts for service. Whether these are held in file drawers or in a document management environment or electronically in the billing system, billing operations is responsible for ensuring that bills are appropriately based on contract information.

"Information central." Billing operations is the resource responsible for making sure that needs for billing extracts and reports—both ad hoc and ongoing—are met with appropriate and accurate data.

Bill print operations. If your company has in-house bill print centers, this generally is the responsibility of billing operations. Not only is this a complex print shop operation, but bills must be verified for accuracy and compliance to postal, legal, and regulatory guidelines.

Accounting

Ultimately, billing is about money—getting it and getting it to the right places, both physically and financially! This means that the ties between billing and the corporate accounting systems are especially tight—or should be.

Accounts receivable. Billing either directly updates the accounts receivable ledger with payment and adjustment information or, in some cases, may actually act as the accounts receivable ledger for the corporation. This is obviously a critical linkage. Less obvious, however, is the importance of good support for coordinated management of accounting classification schemes.

It is important to investigate compatibility of any new billing environment with the corporate accounting systems. It is also critical to develop very strong requirements for support of this interface.

Accounts payable. Billing also triggers advisements to accounts payable, largely for payment to taxing authorities, but also for refunds of overpayment on final bills and other customer payments such as interest to be paid on customer deposits. Again, this interface is critical and should have well defined requirements for the new billing environment.

General ledger. The billing system may directly post the general ledger or simply impact it through accounts receivable and accounts payable. This is an example where examining the systems "at one remove" from billing may prove valuable in refining requirements.

Network

In looking at usage data sources in Chapter 15, we have reviewed many of the network interfaces to billing. It is important to understand the data generation and collection capability of the network elements, as well as the relation of the network elements to mediation and collection systems in order to develop good requirements for usage interfaces. This is also valuable information for those who have to troubleshoot those interfaces.

Third-party vendors

Many companies outsource specific pieces of the billing and payments process to specialized third-party vendors. In establishing a new billing environment, you may be working with an existing vendor or considering a new vendor for these specialized functions. This will impact your requirements for the rest of the billing environment, as well as, perhaps, drive the definition of requirements for these interfaces.

Tax. For a small company with wide geographic markets, outsourcing tax application and management may be a boon. Buying comprehensive tax databases for many jurisdictions and keeping them up to date may be prohibitive for small companies.

Of course, this places a third-party vendor in the critical path of the billing cycle. This requires very good controls on the interface and very specific performance guarantees in a comprehensive service agreement. Also, this places the burden for compliance with tax legislation and administration on the tax vendor and the contract should indicate this as a legal liability of the vendor.

Banks. The traditional lockbox bank function of payment processing is perhaps the most venerable of third-party relationships to telecommunications and other utility billing. These interfaces work best when fully integrated with the corporate billing and accounting systems environments.

Credit card companies. Credit card companies offer convenience to the customer and speed and reliability of payment to the service provider—for a fee! Obviously, if you are a large-volume company, you can negotiate more favorable rates with the credit card companies. Credit card companies may or may not be willing to directly integrate with your environment. Since they are by definition high-volume payment processors, they may provide some standard interface or which you can build, rather than customizing their interface to your systems.

Credit agencies. Credit agencies provide credit rating information on customers to the sales process. They also receive payment history information from the billing and payment process.

Collection agencies. Collection agencies generally buy collectables that the company has identified as bad debt at a deeply discounted rate, providing immediate cash flow to the company and relieving the company of the overhead of pursuing delinquent customers.

Strategic partners. It is common in the telecommunications industry for a carrier to provide billing services to another carrier. Local exchange carriers will, for a fee, include long distance charges in their bills to their customers along with their own local charges. They then generally provide advisement of payment and bulk funds transfer to the other carriers' accounts.

These services may include rating and compiling charges, but they may get fully calculated billing entries and simply provide printing and mailing along with their own printing and mailing. This type of arrangement may be attractive because it is gotten at a price less than the other carrier can do bill print and mailing on a stand-alone basis or because customers demand a combined bill.

Controls on all interfaces in these processes and strong support for reconciliation, verification and audit are critical.

The Systems Information

What questions should be answered about interfacing systems and how should that information impact billing systems requirements? Who are the stakeholders for that information?

Direct interfaces

The gathering of systems information and the management of systems considerations in requirements for a new billing environment are obviously most extensive for those systems that directly interface with the billing environment.

Questions

- What data flow from this system to the billing environment?
- What is the format of the data? Is it parsed into subfields? Do these subfields contain intelligence (such as using the postal state abbreviation to indicate a geographic market)?
- What validation is done on the data before it gets to this system? In this system? Before it is sent to the billing environment?
- When and under what conditions is data transmitted to the billing system?
- How does this system know that data transmissions are successfully received by the billing system? What does it do if that does not happen? When does it do it?
- What happens if the billing system cannot process the data this system sends? How is this system notified of the problem? How are problems resolved? How are problem resolutions coordinated with earlier transactions (e.g., version tracking)?
- Who is responsible for problem resolution?
- What are the standards for problem resolution (e.g., time to resolve, number of retries)?
- What data flows from the billing environment to this system?
- What validation is done on the data before it gets to the billing environment? In the billing environment? Before it is sent to this system?
- When and under what conditions is data transmitted from the billing system?

- How does the billing system know that data transmissions are successfully received by this system? What does it do if that does not happen? When does it do it?
- What happens if this system cannot process the data the billing system sends? How is the billing environment notified of the problem? How are problems resolved? How are problem resolutions coordinated with earlier transactions (e.g., version tracking)?
- Who is responsible for problem resolution?
- What are the standards for problem resolution (e.g., time to resolve, number of retries)?
- What is the process for managing upgrades to the interfacing system? How will interaction be managed and verified?

Stakeholders
- Billing operations
- Systems management for system in question
- Users of system in question

Indirectly related systems

This chapter opened with a description of the impact some presales systems can have on the billing system, even at several removes from a direct interface. Elsewhere, we have pointed out the value of a consistent data catalog across all the company functions and processes that deal with customer account and service information. However, data catalogs are relatively static—traditionally addressing data formatting and definitions. Of equal or greater importance is the ability to manage the synchronization in time of related data across multiple platforms and functions.

Questions
- What data originate in this system that eventually become part of a customer account or service record?
- What data originate in this system that eventually result in charges to the customer or credits to the customer account?
- What data originate in billing that are eventually used by this system?
- What controls exist on the timing of creation of billing-impacting data in this system?
- What controls exist on the quality and format of billing-impacting data in this system?

Stakeholders
- Systems management of this system

- Systems management of all systems between this system and billing
- Users of this system
- Users of intervening systems
- Billing operations

Desktop and Data Networking Platform

Most modern systems rely on desktop environments for user access and session management. While browser-based interfaces are becoming somewhat less specific to a desktop environment, most applications expect a specific environment or set of environments. If your entire company is equipped with PCs, it would not be cost effective to deploy a billing system that only runs on UNIX workstations.

Questions

- What is the desktop environment in billing operations? In customer care and account inquiry?
- What is the networking environment?
- How much capacity does your network have?
- What are your networking protocols?

Stakeholders

- Desktop support
- LAN/WAN support
- Billing operations
- Customer care
- Customer care systems
- Platform or architecture standards

Summary: Where We Have Been and Where We Are Going

In this section, we have reviewed many different kinds of business information and identified ways in which this information can and should impact the requirements for a new billing environment. This information ranges from a careful look at overall corporate strategy and product strategy through customer account and service information to definition and control of interfaces between billing and other systems.

This information will form the basis for good business requirements for your new billing environment. In Section III, we will walk through the entire process of selecting a billing solution, from identifying vendors through the RFI and RFP process through qualifying and selecting vendors and negotiating contracts that will facilitate successful projects.

Section

III

Selecting a Billing System Solution

Selecting a telecommunications billing system solution involves a great many considerations, ranging from your business strategy, product and customer mix to system capacity, features and flexibility. The most elegant billing solution in the world is wrong for your company if it doesn't match your business and technical strategy, if it costs too much or can't handle your volumes, or if you have little or no IT resources and the vendor does not provide ongoing support.

In Section III, we will examine many of the criteria that should inform any decision concerning the acquisition of a billing system—or, as is often the case in mergers and acquisitions, the choice between existing solutions. We will also examine the process of identifying and qualifying vendors and solutions, assessing proposed solutions and negotiating vendor contracts.

This section is organized to walk you through the vendor and solution selection process in the order that things would occur in the "real world." Chapter 21 provides a template for managing requirements and relating them to the decision-making process. Chapters 22 through 28 provide more detailed information on specific requirements issues. Chapter 29 provides an overview of the various types of vendors and products available for billing solutions. Chapter 30 discusses qualifying vendors. Chapters 31 through 35 provide an overview of various documents that can be part of the system acquisition process and provide models for each type of document.

In order to illustrate how these tools relate to a real-world situation, we have created the fictitious company "From the Ashes Communications, Inc." As we walk through the development and use of requirements, RFIs, RFPs and contracts, you will become much more familiar with our "new venture."

From the Ashes Communications, Inc. is a "post-meltdown venture." It is being created through the acquisition of assets and accounts of several bankrupt companies—two companies focusing primarily on broadband interstate services, a regional Internet Service Provider (ISP) and a Competitive Local Exchange Carrier (CLEC). Its target market is small to mid-size businesses. It has existing customers from the various accounts acquired, as well as billing systems from one of the broadband companies and the CLEC. The second broadband company was using the invoicing capability of its PC-based accounting system to bill its limited number of customers and the ISP was using monthly flat-rate credit card billing.

From the Ashes plans to offer integrated voice and data services, business ISP and Internet presence provider (IPP) services and Web site design and management services. It intends to be high value-added for companies who do not want to—or are too small to—maintain their own information technology organization. It is currently seeking to acquire a local area network (LAN) installation and consulting firm. As a post-meltdown venture, it places high value on financial and process controls and management tools—learning from the mistakes of its predecessors.

A team of billing and systems managers, including some personnel hired from the bankrupt companies, has been collecting information about the existing customers, services, billing systems and processes since the early stages of the due diligence for the acquisitions. Of course, they have been using the tools and questions introduced in Section II.

Chapter

21

Business Requirements and Billing Solution Selection

In order to determine which billing system is right for your company, you need to have a very clear picture of what your company requires. This means defining what the billing solution will have to do and support at a fairly detailed level. Chapters 22 through 28 will walk through the development of major requirements areas, based on data gathering as discussed in Section II. Knowing what your company requires also means understanding what is essential and fixed, what is desirable but flexible in implementation and what is optional or of little importance for the billing system to do. (There may be things that your company must do, but that may be done manually or otherwise outside the billing system proper.)

Who

In general, the development of business requirements should be done by internal company management. You may choose to utilize consultants to facilitate the process (we do like to eat and pay the mortgage!), but the key decisions should be made by your company's executives, managers and strategists. You will notice as we go through the acquisition process in our fictional company that the acquisition management team and decision makers are the VP of Marketing, the Chief Technology Officer, the Chief Information Officer and the Director of Billing Operations. They are supported by subject matter experts and line managers from their respective organizations.

What

Just as we recommended good documentation and change control for data collection in Section II, we strongly recommend good documentation and change control for requirements. Requirements documentation will eventually form part of your contractual relationship with a vendor or vendors. Poorly documented requirements are a major cause of project disasters for both vendor and customer. It is even more critical to obtain stakeholder buy-in on documented requirements than it is to do so on data gathering documents. Each stakeholder group needs to identify and assign responsibility to an individual for this project. This might be a good time to review the stakeholder list in Chapter 14, as we will refer to stakeholders for each requirements area.

When

Initial business requirements should be formulated before issuing even a very simple request for information (RFI). We have structured this section of the book to look at requirements at the point they are initially formulated. In the examples, we indicate areas of expected requirements change or refinement. We recognize that all data may not be available at the time an RFI/RFP document must be finalized, and we discuss how to work around having less than a full set of facts at that time. This highlights the importance of change control throughout the systems acquisition process and beyond.

Requirements may be modified throughout the acquisition process—either due to changing business conditions or due to decisions made in the acquisition process. However, once the contract is executed and development begins, changes to requirements can have significant impacts on both the time and cost of implementing your new billing system.

Why

Any requirement should contain descriptions of the business reason for that requirement. Then, if the business driver changes, you can review the requirement to determine if it is still relevant. If there are known areas of uncertainty in your business, it may be a good idea to index the requirements to those areas of uncertainty.

In our fictitious company, From the Ashes Communications, some requirements are based on the assumption that specific books of business and assets will be acquired by the company and will be migrated to the new billing system. At the time initial requirements are developed, two of these acquisitions have not been completed. Also, information on existing services in those two companies is incomplete and of questionable reliability. (This situation may be analogous to that of a company with a legacy billing system module that is not documented and not yet completely analyzed.) Therefore, it is important to identify which requirements are driven by the current book of business in those two companies in case the negotiations do not conclude with the planned acquisition.

How Specific Should Requirements Be?

Good requirements are specific enough to ensure that your business gets what it needs, but not so specific that you force the vendor to do unnecessary custom development. Some of the worst systems acquisition disasters the authors have seen in many years of untangling systems acquisition disasters came from a company specifying a specific data base management system (DBMS) to a vendor who had an otherwise fully-conforming solution using another DBMS. This led to massive customization at the expense of the purchasing company. Even worse, it led to unstable, underperforming applications!

If your company has a strategic driver that makes a specific DBMS a critical factor, then you may want to prioritize having a native platform (the one for which the application was developed or one on which the application is supported for general availability) that conforms to your DBMS requirements. Many vendors will agree to doing a platform migration at a cost—even if they do not have the requisite platform experience!

Some companies specify a specific platform element such as a DBMS because they have in-house expertise in that technology. Usually, this is an economic rather than a strategic consideration. Generally, it is better to find the solution that best meets your other business needs and factor in the cost of obtaining or training the appropriate support resources.

"Just enough" specificity applies not just to platform requirements, but to all other areas of requirements. For instance, a requirement for position and supervisor workflow management reflects a business need for managing workflow. This requirement could be met by text-oriented "work queues" or by a graphically oriented workflow management interface. If the requirements indicate text-oriented work queues (usually because that's what the users had in the old billing environment), you might disqualify a vendor who offered state-of-the-art graphic workflow management. On the other hand, if you allowed the romance of the "bleeding edge" to lead you to specify XML interfaces, you might reject an otherwise perfect solution using HTML for HTML-appropriate things.

Typical Tradeoffs

All acquisition decisions require evaluating tradeoffs of some desirable properties for others. Here are a few of the most common tradeoffs.

Custom features versus shared development costs

There are a number of telecommunications billing system products on the market today that range from turnkey solutions for a specific type of business to highly customizable system modules. Buying a product "off the shelf" largely moves development expense and risk to the vendor. It may provide for a continuing upgrade path and for commonly used interfaces. Building or contracting for a custom billing system allows for greater market differentiation through billing and may make integration with other legacy or custom systems easier. However, a custom solution means that you are responsible not only for the initial development costs, but also for the effort and cost of updating the system for platform upgrades, standards changes and new capabilities.

A few years ago, many companies were driven from legacy workhorse billing applications because their computing platforms—hardware, operating system, database management system—were not Y2K compatible and were no longer supported by the vendor. Other companies have been motivated to move to new billing systems for combined billing capabilities or for increased compatibility with other company systems. Now, many companies are looking at new billers in order to support Web-based billing and payment capabilities. Obviously, supporting sweeping new platform directions such as Web enablement may be prohibitively expensive for a single implementation, but easily affordable when development costs are spread among many companies, as they are for the major billing system vendors.

Leading edge versus stability

In the software industry, it is almost axiomatic that new software will not be perfect. Even if it is entirely "bug free"—a highly unlikely possibility—it will not have been tuned by experience and user feedback to its optimum utility. Mature software is generally stable, predictable and manageable. In addition, it is easier to find expertise in mature software—both for support and on the user side.

However, if your communications enterprise is targeting the "early adopters" market for communications services, you may have to choose newer platforms that fully support new technologies, capabilities and services. This usually means that more sophisticated in-house analysts and systems support staff are needed to work with vendors and company users.

Simplicity and speed versus flexibility

Large and complex data structures generally take more processing capability and more complex management structures than simple structures or tables. Rule-based service and pricing specification is generally more complicated and requires more input than simple tables and fields. However, these more complex structures may provide the flexibility required to support innovative services, technologies, and billing arrangements.

Companies that have a strategic commitment to value-added and innovative services should give significant weight to requirements for flexibility and configurability.

Open standards versus proprietary implementations

Open standards allow many vendors to create "plug-ins" and interfaces, utilizing their unique expertise. Proprietary implementations may be more consistent and stable.

Real time versus batch

While this is one of the oldest tradeoffs in computing, it has new relevance in billing for convergence services. Real-time or near real-time usage detail collection, rating and display is essential for some "convergence" services such as on-line vending payments. Real-time usage processing also controls opportunities for several types of fraud, especially for out-of-network calls (e.g., cellular roaming, credit card). Real-time transaction-based collection, rating and display, however, entail a significant amount of networking and processing overhead and require real-time error correction and 24/7 management.

Ideally, only transactions that require real-time processing will be handled one at a time, while transactions that can be processed in bulk will be handled in bulk. This, of course, requires a very flexible platform. If your system processes each transaction in real time, you will incur significant unnecessary processing overheads for transactions that could and should be processed efficiently in bulk. If your system processes transactions only in bulk, you may limit your strategic service options.

While most platforms can be made to handle one-at-a-time transactions and many-at-a-time transactions, they are often optimized for one or the other. Examining your company's strategic needs and determining the mix of one-at-a-time versus many-at-a-time can help in identifying the right solution for your company.

Custom input requirements versus retraining workforce

The input requirements to establish customer accounts, products and services for most off the shelf billing systems are similar, but not identical. The requirements may be quite different from those used in the company's existing sales and provisioning systems. Those companies that desire to implement a new billing system without any alteration to existing workforce processes and training will require customization, which may be extensive and expensive to maintain.

Frequently, one of the values of implementing a new billing system is the opportunity to review and improve existing sales and provisioning processes. The cost of this activity, along with the force training expense, may well be offset by added efficiencies and improved customer satisfaction.

Summary

In this chapter, we took a high-level look at what system requirements are and how they are used. In Chapters 22 through 28, we look at the process of going from data collection as detailed in Section II to appropriate system requirements. In Chapters 32 through 36, we look at using requirements within the acquisition process.

Chapter 22

Customer Level Requirements

Customer level requirements encompass the ways in which the customer is defined and maintained in the billing system. Those customer considerations that were discussed in Chapter 17—who the customer is and how the customer interacts with the company's billing environment—now must be stated as requirements. In this chapter, we combine the data regarding the existing billing system with the understanding of the company's strategic direction to develop requirements for a new billing system. We illustrate requirements development with examples from our fictional company, From the Ashes, Inc. (FtA).

Account Structures

Account structure, at its simplest, consists of a customer with a single service address and a single service or product. The more complex structural arrangements that combine multiple accounts, services, jurisdictions, and report structures may severely test off-the-shelf billing systems.

Account structure variations

There are a number of data items that will be common to all accounts—from simple to complex—in the billing system. These nonproduct/service data, detailed in Chapter 17, include both required items and optional information. The required items are generally available with little or no modification required in modern billing systems. Among the required items are account identifier, account name, billing contact, billing address, credit classification, bill cycle, primary bill medium, tax exemption indicator, tax jurisdiction, and preferred payment method.

Your company processes, practices or strategic direction may require you to include a provision for additional data items in every customer account record. Such data include credit card number, balance, and payment history (actually included in the customer account record versus in a separate, linked file), credit information, credit history, market segment indicator, directory listing appearance, deposit amount, final billing address, third-party notification indicator/information, and master account/slave account indicator. As with the required items mentioned above, these items are generally available in modern billing systems, particularly those directed to the local exchange and interexchange carrier markets.

Accounts may be combined on bills or perhaps just on reports. The rules for eligibility for multiple account output are defined by your company, to meet the marketing and corporate objectives. Combined billing may be limited to accounts for the same customer bill name or jurisdiction or any other identifiable business rules. The important thing here is that the rules be clear and, to the extent that they are to be administered by the billing system, be included in the requirements. For example, the billing system may be expected to administer the rule if all multiple account billing must have the same account name—ABC Corporation—but if subsidiaries of a parent corporation with different account names may be combined, some additional information must be given to the billing system to verify qualification—or verification must be done prior to submission to the billing system.

Appendix A contains a customer information worksheet form that may be used to capture the customer account record data.

From the Ashes (FtA) account structure

FtA believes that it currently has no combined account billing, nor does it have any combined services billing. This information is only preliminary, since the CLEC data are unverified. Although only paper billing and limited credit card billing are in use today, the FtA strategic plan calls for the use of all forms of billing media and preferred payment methods. FtA will provide directory listing service for white pages only and will be required by regulators to offer third party notification for some local consumer accounts.

FtA's strategy calls for combined account billing and additionally requires the ability to produce combined reports for disparate accounts, even when those accounts are not billed together. Rules have been identified at a fairly high level for account combinations and these high level definitions are adequate for preparing the RFI, but will need more work prior to the RFP phase. Figure 22-1 shows a portion of the account structure requirements for the FtA billing system.

Account structure stakeholders

- Product management
- Billing operations
- Regulatory

Customer Account Requirements
January 20, 2003

8. Market Segment
The account's market segment must be part of the customer account record. This code will be between 3 and 8 characters in length and will be submitted as part of the service order entries. The character set includes A–Z and 0–9. It should be edited against a billing system list of acceptable segment codes.

9. Tax Jurisdiction
The account's primary tax jurisdiction must be part of the customer account record. This field will be submitted as part of the service order entries, and should be edited against a billing system list of acceptable jurisdictions. This field consists of two uppercase alphas.

10. Tax Exemption Information
FtA must be able to designate an account as tax exempt; this information will be submitted as part of the service order entries. The customer account record needs to have a field for the tax exemption reason—either as a code or a text field—and show both the date certification was received and the file location within the company of the certificate.

11. Third Party Notification
FtA may be required to support third party notification, so capability must exist in customer account record to obtain and retain information about the third party. Information must include name, address, can-be-reached telephone number and the relationship of the third party to the customer/billing name on the account.

12. Account Structure
FtA requires capability to combine accounts for billing and/or for producing customer reports. The customer account structure must allow for designation of an account as a master or slave account for combined billing, and allow for FtA to identify hierarchical positioning of accounts that are to be combined. Additionally, FtA must be able to combine data for some accounts for the purpose of producing reports, even when the billing is not to be combined.

Page 7 of 7

Figure 22-1 FtA customer account requirements.

Billing Options

Billing options indicators in the customer account record, broadly, direct the billing system on how to do things, where to go to find information or where to direct outputs. We have discussed customer account information for some of these items—for example, the customer account record has a deposit field that may show that a deposit is held for a customer, but the billing option information will clarify how long it is to be retained, if interest is to be paid and under what circumstances—as well as how—the deposit is to be returned.

Billing options requirements

Billing options are items as diverse as combined billing of services/products and the inclusion of marketing/sales messages as part of the bill. The information gathered in Chapter 17 that answers "why" or "how" questions is included in this category. Among the most common are combined billing of services/products, rounding conventions, payment options, the method and extent of treatment options, the use of the billing system as a note repository and including marketing/sales messages as part of the bill.

FtA billing options

The FtA team has decided that all customers to be migrated into the new system should be established with the same type of account structure options as are contained in the old billers, but that the customer account records should be able to accommodate more sophisticated convergent billing of both products/services and accounts. This will facilitate a cleaner conversion and reduce the expected period for parallel processing of billing. Implementation of combined product/service billing and combined account billing will be scheduled for a period subsequent to conversion to the new billing system.

Billing Options Requirements

January 20, 2003

8. Rounding

FtA requires documentation on how vendor software rounding is accomplished for product/service transaction compilation, tax processing, etc. FtA is still researching the methodology in use in existing systems, and needs explicit information from vendor in order to prepare testing and conversion plans.

9. Deposit Processing

FtA must be able to receive and post a deposit to a customer account record prior to the completion of an initial service turn-up or product delivery for the customer. The customer account record must contain information showing the date that the deposit was received, the reason the deposit was taken, the amount of the deposit, the date scheduled for deposit return, whether or not interest is applicable, when interest is to be paid, and the manner in which the deposit it to be returned, if at all.

10. Convergent Billing

A customer account bill must be capable of containing charges for all services that a customer is using. These may include local, LD, broadband, wireless, ISP, services for which time and materials are charged and products for which there are one-time charges. FtA currently does not have convergent billing in any of its existing billers; it expects to convert without convergent billing but will need to implement convergent billing within the first 6 months the system is in production.

11. Bill Cycles

FtA requires multiple bill cycles, the number to be determined at a future date. The billing system must allow FtA to identify the number and dates of the cycles. FtA wants the billing system to be able to automatically assign the bill cycle to a customer account or allow the bill cycle to be input via the service order or by billing operations. The billing period of a bill cycle must be flexible, and at FtA's discretion be monthly, bimonthly, quarterly.

Page 7 of 7

Figure 22-2 FtA billing options requirements.

The team has enough information to input to the RFI, but definitely needs more specifics regarding the CLEC before it can prepare the RFP documentation. Figure 22-2 shows a portion of the billing options requirements for the FtA billing system.

Billing options stakeholders

- Product management
- Billing operations
- Regulatory
- Accounting standards
- Desktop/workstation support
- Legal

Geographic Markets

Geographic markets are usually defined as the locations in which your customers take delivery of your services, although this may actually identify the customer's billing address—as in the case of the wireless service user.

Geographic markets requirements

The number and variety of geographic markets need to be identified. Domestic United States markets and international markets should be differentiated, as they require different types of legal, regulatory and financial support.

FtA geographic markets

FtA will be doing business in three state geographic markets to begin with: Arizona, California and Nevada. At the present time, there is no plan to expand to additional states. Figure 22-3 shows the geographic market requirements for the FtA billing system.

Geographic markets stakeholders

- Product management
- Billing operations
- Regulatory
- Legal
- Accounting systems
- Accounting standards
- Tax accounting

Geographic Markets Requirements
January 20, 2003

Geographic Markets
FtA is currently doing business in Arizona, California and Nevada. It does not anticipate expansion to other states or into international jurisdictions in the foreseeable future.

Page 1 of 1

Figure 22-3 FtA geographic markets requirements.

Market Segments

Market segments are based on customer characteristics such as the type of service, projected monthly account billing, service address, type of business or a combination of factors. Regulators define some segments: regulated jurisdictions still require "business" and "residence" market segments.

Market segments requirements

The company must define completely what characteristics will comprise each market segment. The market segments will generally be aligned with the company sales force and management structure, with adherence to any regulatory requirements. An illustration of a simple company's market segment definitions is shown in Table 17-1.

FtA market segments

The FtA team has determined that it will begin operations with four market segments. The Arizona CLEC must have separate business and residence segments. California and Nevada will each have one segment. Further segmentation is planned to accommodate integrated voice and data services, business ISP and IPP services and Web site design and management services. FtA wants to be able to reassign customers to market segments in a fairly graceful manner in the future.

Figure 22-4 shows the proposed initial market segments for the FtA billing system. Based on this proposal, FtA requires that the new billing system support a three-character market segment designation specifiable by FtA personnel from a user interface. It also requires that products and services be assignable to one or more market segments. FtA requires that service order processing in the billing system verify that the ordered product or service is appropriate to the customer's market segment.

Market segments stakeholders

- Product management
- Billing operations
- Regulatory

FtA

Jurisdiction	Arizona	California	Nevada
Line of Business (Segment Characteristic)	ALR: (AZ Residence: including local, LD, ISP)		
	ALB: (AZ Business: including local, LD, ISP, broadband)	CBC: (CA Business: currently broadband with future ISP, IPP)	NBC: (NV Business: currently broadband with future ISP, IPP)
	AWS: (AZ Web site design and management: future product)	CWS: (CA Web site design and management: future product)	NWS: (NV Web site design and management: future product)

Figure 22-4 FtA proposed market segments.

Media Options

There are a number of ways to direct bill information to a customer; no longer is paper the only way to get this information into the customer's possession. Most companies are moving toward the ability to offer all modern media options, both for reasons of cost control and for customer satisfaction.

Media types and requirements

With the exception of CD and EDI, the media types available for use might be requested by all categories of customers and would probably be offered to all customer segments.

Print. The print medium is still the most widely used method of getting billed information to the customer and most modern billing systems have the ability to format billing information completely for printing or otherwise prepare it to go to a print vendor for formatting and delivery. Some fundamental decisions that need to be made are where the printing is to be done, how it will be prepared for mailing/delivery, the type—if any—of billing inserts to be expected and what volumes will be expected at what times. If a print service bureau is to be used, any unique procedures and/or output file capabilities need to be included in the billing system requirements. If printing is to be done inside the company using an existing print room, all format, bill enclosure requirements and any sorting requirements should be part of the billing system requirements.

Internet. Internet billing may be formatted by the billing system or it may be handled in much the same way that a printed bill given to a print vendor is treated. An unformatted file containing the customer billing data may be submitted to software—owned and operated by the company or by a third -party vendor—to be put into XML for "Electronic Bill Presentment and Payment" (EBPP). The decision on who is to do the XML formatting is a key one, for this is a fairly new delivery method and all billing systems may not be prepared to do this without customization. Additional decisions must be made about volumes, scheduling and whether there is a need for a hardcopy summary invoice document to be mailed separately to verify billing.

Email. Email billing may involve sending a complete detailed invoice to the customer via email or it may simply be a notice of new invoice availability on a Web site. A common "user-friendly" combination of email and Web technologies utilizes an email that automatically invokes access to the user's secure invoice Web page or provides a "hotlink" URL for the secure invoice Web page, which the user can "click" to access the bill. Pure email billing, because of the significant privacy and security issues coupled with the difficulty of formatting a readable bill for mail readers, is a fairly unattractive option.

CD media. Large business customers have received a copy of voluminous bills via magnetic tape in the past. The compact disc (CD) storage medium has replaced magnetic tape for most, if not all, of these customers. As with other media types, the company must decide how much formatting is to be done by the billing system, if that formatting is to be unique for the customer or if it is to conform to the company billing standard output and where the CDs will be created. There may be a need for the company to send a separate hardcopy summary invoice—usually with a remittance document—to the customer to serve as proof of billing.

EDI. Business customers often prefer invoices that feed directly into their accounts payable processing systems. Standards for business-to-business EDI have been around for decades and many companies have implemented EDI for receipt of invoices. The company must determine where the formatting is to be done, timing of preparation and volumes. As with CDs, there may be a need for the company to send a separate hardcopy summary invoice—usually with a remittance document—to the customer to serve as proof of billing.

Commercial credit card. An initial customer service agreement may include authorization to bill account charges to a commercial credit card. The company must establish a "merchant account" with the credit card company or with a credit card clearinghouse. The company must be prepared to provide the credit card clearinghouse with relevant information when the customer account is established. The data to be provided include verified card number, name on the credit card, card billing address and card expiration date.

FtA media options

The FtA team has confirmed that the two broadband companies and the CLEC presently provide only printed bill media. The ISP bills via commercial credit card exclusively. In the future, FtA will offer print, commercial credit card, Internet EBPP, email notification of Internet billing and CD. Figure 22-5 shows a portion of the billing media requirements for the FtA billing system.

Billing Media Requirements
January 20, 2003

2. Commercial Credit Card

FtA will offer customers the option of receiving their charges on commercial credit cards. During the bill cycle processing the billing system must be able to format and direct the total amount of the dollar amount of the bill and the description of the charge to the credit card company in compliance with the credit card company's established protocols.

FtA will offer customers Visa, MasterCard and American Express billing options. Currently, this type of billing is done solely for ISP accounts, which have only a flat rate monthly charge and no detailed charges. FtA desires to have credit card billing available to all customers, however, and requires the billing system to be able to direct a detailed bill facsimile to a *password-secured* read-only Web site when the bill total amount is directed to a credit card.

3. Internet EBPP

FtA requires that the billing system be capable of Internet EBPP, if not at conversion at least within the 18 months following contract signing. FtA envisions offering Internet EBPP to its entire customer base in the 24–30 month time frame, and expects that approximately 15,000 accounts will use the medium. Formats and specifications are to be developed in the 12–18 month time frame.

4. CD

FtA requires the billing system produce bills on CD. The CD material is to include all items present in printed bills and additional reports as indicated in the customer account record. Accounts receiving bills on CD, and no other alternative bill, must have the billing system generate a separate remittance document that will be printed—as part of the printed media output—and mailed to the customer with a return envelope.

Page 6 of 7

Figure 22-5 FtA billing media requirements.

Media options stakeholders

- Product management
- Billing operations
- Regulatory
- Carrier relations/strategic partner relations
- Server resources management
- Legal

Fields/Formats

Fields and formats information about customer account items comprises the definition of any specific required data fields and their formats. In a well-controlled environment, this will correspond to data catalog information.

Fields/formats requirements

Information about customer account fields and formats—derived in Chapter 17—is used in summary form to define billing solution requirements ensuring that a new billing environment can accommodate and manage current and planned customer record formats. This information is also used in more detail to define requirements for conversion of current accounts from an old billing system into a new billing system.

Customer Account Record
Fields and Formats Requirements

January 20, 2003

Account Name Fields

1. Company Name/Last Name
- Length = 35 characters, including spaces
- Character set = a–z, A–Z, 0–9 and the following special characters: – ‘’ . , &
- Rules = maintain capitalization characters and spacing as received from the source
- Edit rules = reject/return to source unauthorized characters and anything in excess of 35 characters
- Sources = sales/provisioning, billing
- Required

2. First Names and/or Initials
- Length = 20 characters, including spaces
- Character set = a–z, A–Z, 0–9 and the following special characters: – ‘’ .
- Rules = maintain capitalization characters and spacing as received from the source
- Edit rules = reject/return to source unauthorized characters and anything in excess of 20 characters
- Sources = sales/provisioning, billing
- Optional for business

3. Title
- Length = 10 characters, including spaces
- Character set = a–z, A–Z and following special characters: – ‘’ .
- Rules = maintain capitalization characters and spacing as received from the source
- Edit rules = reject/return to source unauthorized characters and anything in excess of 10 characters
- Optional

Page 2 of 43

Figure 22-6 Sample FtA customer account record fields and formats requirements.

FtA fields/formats

Figure 22-6 shows a section of the FtA customer account record fields and formats requirements. Note that some formats are indicated as optional. The specific formatting of data may be based upon formats in the legacy billing systems. Other formats may require some additional conversion activity and create additional integration issues, but provide acceptable functionality within a new billing platform.

Figure 22-7 shows a section of the FtA customer account conversion requirements that deals with the conversion source data. This is important information to a vendor attempting to estimate and bid upon a conversion project.

Customer Account Record Conversion
Fields and Formats Requirements
January 20, 2003

Account Name Fields
Identify any account with account name fields that do not conform to the customer account record fields and formats requirements and mark the account with a unique code identifying this condition. These accounts must not be converted; they are to be directed to a non-conforming account file, which will be directed to the conversion billing operation/customer care team for resolution, error file correction and resubmission to the system. The billing system conversion process must be capable of accepting the corrected error file for conversion processing.

Billing Contact Fields
Identify any account with billing contact fields that do not conform to the billing contact record fields and formats requirements and mark the account with a unique code identifying this condition. These accounts must not be converted; they are to be directed to a non-conforming account file, which will be directed to the conversion billing operation/customer care team for resolution, error file correction and resubmission to the system. The billing system conversion process must be capable of accepting the corrected error file for conversion processing.

Credit Information
If there is no credit information contained in the old system document—and this is the anticipated case for 85 percent of the accounts to be converted—insert the phrase "No information on file on MM/DD/YYYY; update credit information at next customer contact" in the new system customer account record credit information field. Use the conversion date to populate the MM/DD/YYYY.

Credit Classification
Set the credit classification for all broadband and ISP accounts to C, which will be a special conversion classification. For CLEC accounts, insert the credit classification shown in the old system record into the new system record.

Page 2 of 20

Figure 22-7 Sample FtA customer account record conversion requirements.

Fields/formats stakeholders

- Sales systems
- Provisioning systems
- Billing operations
- Product management

Summary

In this chapter, we have looked at how to use the customer data collected in Chapter 17 and prepare requirements for the customer account record contents and billing media. As an illustration of how this would work, we have looked at sample customer requirements for From the Ashes, our fictitious company. In Chapter 28, "Managing and Prioritizing Requirements," the requirements developed here will be summarized, along with the other billing system requirements, for a requirements checklist. The requirements checklist is a tool used in preparation of the RFI.

Chapter

23

Product/Service Level Requirements

Product and service level requirements address the characteristics of current and future product and service offerings that must be accommodated and represented in the billing environment. In Chapter 15, we reviewed collecting data about current and future services and products. In Chapter 16, we looked at the process of consolidating that information across all products and services. In this chapter, we take use consolidated information, industry knowledge and business strategy to develop business requirements for a new billing system.

Product and Service Categories

Communications billing solutions are often targeted at specific types of services with unique information formats and/or unique usage detail structures and data. Determining which of your company's services fit into understood service categories can help you and your vendors determine how many of your product and service driven requirements can be met by off-the-shelf software or data templates and which requirements may require some customized accommodations.

Industry definitions

There are a number of widely understood ways to categorize products and services. Some are related to particular industry segments while others are more grounded in technology.

Local exchange carrier (LEC). This category includes services and bundled products that are typically offered by a local exchange carrier, whether a competitive local exchange carrier (CLEC) or an incumbent local exchange carrier

(ILEC). Services in this category are traditional dial services such as single and multi-line POTS, "toll-free" services such as 876, caller-paid 900 services, Centrex services, all carrier access services, point-to-point private line of all bandwidths and all associated features and terminating equipment, both leased and purchased.

Since DSL is frequently offered by LECs, many billing solutions targeting the LECs include billing for DSL. However, since this may not be included in some older solution packages, it is wise to specify DSL services separately if required.

Interexchange carrier (IXC). This category includes services and bundled products typically offered by a long distance company (IXC). Services in this category are dial long distance, inter-LATA toll free, enhanced 800, and caller paid (900, et al.) services, distance-sensitive private line of all bandwidths, virtual private networks (VPNs), frame relay, ATM (Asynchronous Transfer Mode) and other configurable bandwidth services, along with any associated features and terminating equipment.

Wireless. This category focuses on voice-grade dial wireless services. It includes the ability to identify serving network, call duration and associated long distance charges for legacy through 2G (voice grade dial) wireless services. If your company has any 2.5G (simple broadband wireless) services, it is wise to identify these separately.

3G wireless. As described in Chapter 27, 3G wireless services have unique billing considerations. At the present time, these services should be identified separately and specifying significant detail on intended services is highly advisable.

Internet service provider (ISP)—Legacy. Legacy ISP billing is flat rate based on connect time through a dial-up port. There are billing and payment solutions targeting this market. In some instances, there may be setup charges or email charges that are flat rate or usage-based.

Internet presence provider (IPP)—Legacy. Legacy IPP billing may include server capacity, domain registration, setup fees and some usage component based on network capacity utilized by end users accessing a site or other service.

Cable—Legacy. Legacy cable services include various flat rate "packages," equipment rental and billing for on-demand services.

Cable data services ("cable modem"). Cable data services generally include a flat rate Internet access fee and equipment rental.

Time and materials. Typical telecommunications billing systems include limited support for recording technician time and may include the ability to select specific prepriced materials. Often professional services time and expense billing is more complex and requires contract-definable rate schedules and broad support for expense pass-through to the customer. When specifying time and materials support, it is wise to be clear about the expected extent of that support.

Stand-alone product. Products may be sold outright or leased. Generally, product sales support will include one-time and recurring charge support, along with service option sales and tracking and warranty tracking. However, it is good practice to specify required support.

FtA product and service categories

FtA believes that all of its current services fit into standard industry categories. Table 23-1 shows the compiled listing of current services. It appears that all services provided by the CLEC are typical LEC business services. The services offered by the broadband carriers are typical for IXCs. The ISP offers only dial-up connections to the Internet and email services.

TABLE 23-1 Current FtA Services

Service	Billing Basis	Interstate	Intrastate	Business	Residence
Broadband point-to-point	Monthly service at flat rate and in arrears	X	X	X	
One-time on-demand services	Mainly time and materials			X	
Internet connection and email	Monthly service at flat rate and in advance			X	X
POTS	Usage-based plus basic in arrears			10%	90%
Centrex	Flat rate in arrears			X	
Long distance	Usage based			X	X
Private line	Monthly service at flat rate and in arrears	X	X	X	

FtA's strategic plans, on the other hand, include the addition of legacy IPP functions, Web site development, maintenance and management and the provision of application services for a variety of convergence applications. While not planning to enter any wireless markets directly, it is currently exploring strategic relationships with 3G and commercial wireless LAN (WLAN) vendors to provide application and "content" services to the wireless infrastructure. Table 23-2 summarizes these services.

TABLE 23-2 Planned FtA Services

Service	Billing Basis	Interstate	Intrastate	Business	Residence
Internet host—virtual server	Monthly service at flat rate and in arrears			X	
Internet host—dedicated server	Monthly service at flat rate and in arrears plus usage-based (megabytes) communications charge			X	
Internet access—dedicated	Monthly service at flat rate and in arrears			X	
Applications services	Contract basis (to reseller)—usually T&M setup plus recurring plus usage			X	
Management services—remote	Contract basis—usually recurring plus demand			X	
Broadband bandwidth on demand	Usage-based (plus access recurring)			X	

Table 23-3 contains the services and products that the FtA procurement will address. FtA will sell equipment and training associated with specific service offerings, but will not sell stand-alone products.

TABLE 23-3 Summary of FtA Services and Products

Service	Billing Basis	Interstate	Intrastate	Business	Residence
LEC Type					
POTS	Usage based plus basic monthly service in arrears		X	10%	90%
Private line	Monthly service at flat rate and in arrears		X	X	
Centrex	Monthly service at flat rate and in arrears			X	
IXC Type					
Broadband point-to-point	Monthly service at flat rate and in arrears	X		X	
Long distance	Usage based	X	X	X	X
Private line	Monthly service at flat rate and in arrears	X		X	
Broadband bandwidth on demand	Usage based (plus access recurring in arrears)	X		X	

continued on next page

TABLE 23-3 Summary of FtA Services and Products (continued)

Service	Billing Basis	Interstate	Intrastate	Business	Residence
ISP Type					
Internet connection and email	Monthly service at flat rate and in advance			X	X
IPP Type					
Internet host—virtual server	Monthly service at flat rate and in arrears			X	
Internet host—dedicated server	Monthly service at flat rate and in arrears plus usage-based (megabytes) communications charge			X	
Internet access—dedicated	Monthly service at flat rate and in arrears			X	
APP Type					
Applications services	Contract basis (to reseller)—usually T&M setup plus recurring plus usage			X	
Professional Services & MSP					
Management services—remote	Contract basis—usually recurring plus demand			X	
One-time on-demand services	Mainly time and materials			X	

Product and service stakeholders

- Product management
- Sales systems
- Provisioning systems
- Carrier relations/strategic partner relations
- Regulatory
- Network equipment engineering
- Accounting systems
- Tax accounting

Fields/Formats

Fields and formats information about products is exactly that—the definition of any specific required data fields and their formats. In a well-controlled environment, this will correspond to data catalog information.

Deriving requirements

Information about products and services fields and formats—derived in Chapters 15 and 16—is used in summary form to define billing solution requirements, ensuring that a new billing environment can accommodate and manage current and planned products and services. This information is also used in more detail to define requirements for conversion of current products and services from an old billing system into the new billing system.

FtA fields and formats requirements

Figure 23-1 shows a section of the FtA product and service-related fields and formats requirements. The specific formatting of data may be based upon formats in the legacy billing systems. Other formats may require some additional conversion activity and create additional integration issues, but provide acceptable functionality within a new billing platform.

Figure 23-2 shows a section of the FtA product and service conversion requirements that deals with the conversion source data. This is important

Billing Systems Requirements
January 20, 2003

1. Product/Service Code

FtA requires that a unique identifier be associated with each product, service, feature or other billable element.

This identifier must be at least five characters long and allow numeric (0–9) characters and alphabetic characters (A–Z). No special characters or punctuation will be allowed.

Input of both lower- and uppercase alpha characters should be allowed. Lowercase input from any source should be translated to uppercase characters for database update or search.

The field should be left justified for display and storage. Input should be permissive—automatically left justifying any input.

The product/service code is the key field for the product/service entity. It is used to retrieve information about a product or service such as description, rate structure, billing basis and source, effective date, market, and accounting classification.

The product/service code table is part of the product catalog, which has a major release each quarter. Incremental updates to the product catalog are made as needed, but not more often than weekly.

FtA has identified 97 unique product/service codes for current products, services and features. This number is expected to grow to over 500 during the life of the new billing system.

2. Product/Service Name

FtA requires that a product/service name be associated with each product, service or feature. The product/service name is the service description that appears on the customer's bill. This field is not unique.

The field must be 40 characters in length. There is no fixed format for the field. The field allows input of numeric characters (0–9), upper- and lowercase alpha characters, embedded spaces and hyphens.

Page 1 of 9

Figure 23-1 Sample FtA product/service fields and formats requirements.

Product/Service Conversion Requirements
January 20, 2003

Contents:

1. Background
2. Sizing Information
3. Assessment of Data Quality
4. Current Formats and Data

1. Background

FtA wishes to convert billing currently being performed on four different platforms to the new billing system. The four old platforms have unique formats and data structures and will require individual conversion efforts.

2. Sizing Information by Source System

Source 1:
2 Service offerings
- One standard service offering by contract
- On demand technical support at hourly rate (no service record)

Service records 37

Source 2:
2 Service offerings
- One standard service offering by contract
- On demand technical support at hourly rate (no service record)

Service records ~9000

Source 3:
1 Service offering
One standard service offering, published rates
Service records ~6000

Page 1 of 15

Figure 23-2 Sample FtA product/service conversion requirements.

information to a vendor attempting to estimate and bid upon a conversion project.

Fields and formats stakeholders

- Billing operations
- Sales systems
- Provisioning systems
- Product management
- Server resources management
- Accounting systems
- Fraud systems
- Regulatory
- Tax accounting

Billing Basis

Billing basis requirements identify whether services are one time, recurring or usage based. They also identify whether charges are a flat rate, based on a standard formula per instance or related to the total bill or some portion of the bill. They indicate what calculation is required in the billing system and what the source and nature of the charge information is.

Billing basis requirements

We looked at the gathering of billing basis information in Chapter 15 and the consolidation of this information in Chapter 16. Turning the consolidated information into good requirements is a simple matter of organizing the information by:

- Frequency (transaction, recurring, one-time)
- Charge calculation including whether advance versus arrears billing
- Charge basis (time, packets, distance, occurrence, expense)
- Usage source specifications

FtA billing basis requirements

Figure 23-3 shows a portion of the billing basis requirements for the new FtA platform. These requirements reflect both current and strategic requirements.

Billing basis stakeholders

- Product management
- Billing operations
- Network equipment engineering
- Server resources management
- Regulatory
- Sales systems
- Carrier relations/strategic partner relations

Volatility

Volatility requirements relate to new services creation or existing service changes or elimination. This volatility can be in pricing, conditions, discounts, promotions, billing basis, and market segment, as well as in new or phased out services. These requirements relate to support for creating or changing products, services and "assemblies" of products and services.

Billing Basis Requirements
January 20, 2003

1. Background

FtA intends to become a leading provider of innovative communications and information services to business markets. Therefore, it requires the capability for very flexible definitions of billing basis units for both recurring and usage-based services.

2. Billing Basis Indicator

FtA wishes their new billing system to accommodate an indicator for "recurring," "on demand," or "usage based" associated with the pricing for a product/service. Recurring defines services charged at the same price every billing period. On-demand services are technical and professional support functions that are charged on a time and expenses basis. Usage-based services are established by service order, but charged based on subscriber-initiated use.

3. Units

FtA wishes to define chargeable units for any pricing element. These units may be any units of time such as hours, minutes or seconds, any unit of capacity such as bandwidth or number of packets or gigabytes of storage, any unit of distance such as miles or units of currency.

4. Relationship to Product/Service

The relationship of product/service to pricing element is one-to-many (e.g., a monthly recurring charge plus a usage charge for dynamically allocated capacity). The relationship of pricing element to billing basis is one-to-one. The relationship of billing basis to billing basis units is also one-to-many (e.g., charges based on duration and bandwidth or on volume and error performance).

5. Summary of Data Fields

Billing basis indicator (fixed options)
Type of unit (e.g., time, capacity, distance)

Page 1 of 3

Figure 23-3 FtA billing basis requirements.

Deriving requirements from gathered data

If product and service volatility is high, the frequency of service modification activity should be included in business requirements. Information needed includes any requirements for maintenance of parallel new and old rate tables and other change management information, such as any unique behavior upon activation date, archiving requirements and credit calculation behavior. The allowable impact of product and service volatility on system operations and availability should also be indicated in requirements. Volatility also drives ease of use requirements for service updates in terms of desirable maximum time for service additions, changes, and retirements.

FtA requirement for product and service volatility

FtA intends to move into leading edge markets for "content" and "applications" services, including some transaction payment services. This will generate some product volatility, as well as the need for robust rate specification and change management and archiving functions. Figure 23-4 shows a section of the FtA requirements for managing service volatility.

Service Volatility Requirements
January 20, 2003

1. Background

FtA will be pursuing markets with innovative service offerings and technology. It is anticipated that new services will be added at the rate of one per month, while updates to existing services will occur at the rate of one per week. Also, complete product catalogs will be released once each quarter.

2. Bulk Update

FtA requires that its new billing system accept bulk replacement of product catalog information quarterly. This bulk update should provide adequate validation functions to ensure that errored records are not loaded and a detailed audit trail of input error information is created and accessible.

The product catalog information does not become active when it is loaded. The product catalog contains unique activation dates for each service and related rate table.

Loading the product catalog should not impact system availability or accessibility.

3. Manual Update

An experienced user should be able to configure a new product or service in two hours or less and to update a product or service in 30 minutes or less.

Service creation or update should not impact system availability or accessibility.

4. Data Catalog Structure and Format

The product management system that will generate product catalogs is in the acquisition phase. The definition of data output from the requirements for that system is attached. However, the contract and development negotiation may result in changes to the specific structure and format of this information.

Page 1 of 1

Figure 23-4 Sample FtA service volatility requirements.

Product and service volatility stakeholders

- Product management
- Billing operations
- Sales systems
- Provisioning systems
- Middleware systems
- Server resources management
- Regulatory

Summary

In this chapter, we have surveyed the process of moving from data collection to business requirements for product and service information. We have also looked at sample requirements for our fictitious company's billing requirements. These requirements are tied to a Requirements Checklist that we will review in Chapter 28, "Managing and Prioritizing Requirements."

Chapter

24

Operational and Systems Level Requirements

In Chapters 22 and 23, we used information about customers, services and products to prepare requirements that, when implemented by a billing environment, will produce the ends desired by your company. The operational requirements stated in this chapter cover the gamut of the company's billing business rules and billing operations processes. The systems level requirements include the types of databases required, and information on the timing and synchronization of various portions of the billing environment and its external interfaces. In this chapter, we will look at requirements for how and where those requirements will be met.

Functionality

Functionality requirements identify what the system is expected to do. While customer level and product/service level requirements primarily address what information the billing environment must handle and the relationship of that information, functionality requirements identify what actions and processes must be part of the billing environment.

Customer ordering and account creation

Some billing systems are capable of supporting both presales and sales order taking, but most companies process customer orders in a sales system, which then communicates with provisioning and billing. The majority of companies find that this modularization provides the greatest flexibility and support for the very different group needs of sales, provisioning and billing.

The orders that are received may be designed to contain all data required for the creation of the customer account record in the billing system. Alternatively, they may contain only some of the data required. Data not on the order, by design, must be either developed within the billing system or manually entered by the billing operations organization. If the data are developed within the billing system, they are generally derived from other information in the order in conjunction with data from tables in the billing system. Incomplete orders are identified and they generate error notifications.

Error conditions may or may not preclude billing system record creation, depending on the error condition and the company specifications. The company may determine that all errors, even potentially non-fatal ones, will cause the entire order to be rejected, thus necessitating the re-entry of a corrected order into the sales/provisioning system. Alternatively, partial posting may be permitted, with the error condition referred to an off-line group (it may be billing operations or sales) for correction via a supplemental order or by direct input to the billing system.

FtA customer ordering and account creation. The FtA team has determined that all orders will be taken by sales, input to the sales system, sent to the provisioning system and billing will receive orders from provisioning. The team expects that the sales processes and systems will be replaced too, so there is no requirement for the new billing system to conform to an existing input data scheme. There is the expectation that some new services and products may have to be brought to market before all data can be brought to the billing system in the most efficient manner, so the new billing system must be able to gracefully accept manually input data of an as yet undetermined nature.

FtA desires to control tightly any service order error conditions from sales to provisioning to billing. Therefore, all error conditions that the billing system identifies will cause the order to totally reject (no partial posting of information from the order will be done in the billing system). The billing system will retain "knowledge" that a transaction was received and the reason that the transaction was rejected without billing system action. The rejection notice will be sent to provisioning, which may return the transaction to sales for re-entry or make the correction and resubmit it to the billing system. All manually input service/product data will be allowed only from users with special security classifications.

Customer ordering and account creation stakeholders

- Sales systems
- Provisioning systems
- Product management
- Billing operations

Customer account inquiry

The individual customer account information in the billing system is frequently needed by groups within the company, either to respond to customer queries or for administrative purposes. Who is allowed access to the data and what methods of access are utilized depends on company processes and procedures. Many companies, bearing in mind that the information is sensitive, limit accessibility to all or part of the information.

The primary company functions needing access to an individual customer account record are customer care, billing operations, and collections. Customers may have access via the Web to a segment of the data, as defined by the company. These groups may access the billing system with capability to see all data or they may, based on billing system permissions keyed to user identification, see only portions of the account contents. Other groups such as sales, with lesser needs for "real-time" access, may have access via reports or by making requests through customer care or billing operations. Most billing systems provide the capability of tailoring information availability to user identification.

FtA customer account inquiry. FtA will give sales, customer care—which will be responsible for collections, too—and billing operations personnel access to customer account records. For the time being, all data fields in customer account records will be available, but this approach is under review, so the selected billing system must be capable of allowing selected access for inquiry. The company does plan to offer Web-based customer account access in the future and the billing system must be able to provide for FtA to define the allowed fields for access.

Customer account inquiry stakeholders

- Product management
- Billing operations
- Sales
- Customer care
- Collections

Bill compilation

Bill compilation is that set of steps that gather all charges and calculate a total bill amount. This occurs at a set point in the regular billing cycle or when the system recognizes the need for a final bill or other special bill. Bill compilation is often the reason that companies contemplate purchasing a new billing system: combined billing has become a necessity and old billing systems handled only one type of service per bill.

The first of the steps is product/service charge calculation, which seems to be a no-brainer—a billing system must be able to calculate a bill—but the devil is in the details. The company processes that deal with sales and provisioning determine how much will be received from the sales, provisioning or message processing systems already rated or calculated. Will all discount promotions be given to the billing system as unique products/services that use different rate tables—in effect, that are prediscounted? Will the provisioning/maintenance organization notify the billing system of customers eligible for maintenance time and charges items by sending information on the items of service and the number of hours worked—or will the actual charge amount be sent? Will call detail records (CDRs) be rated or, if already rated at receipt, be rerated at the time that the bill is to be prepared? If there is to be CDR re-rating done, will all be rerated or only those that meet some type of date or other criteria?

The billing basis of the products and services the company provides and plans to provide must be evaluated along with the sales and provisioning processes to determine how much or how little the billing system is to calculate. There must always be enough information for the billing system to produce a completely stated bill; if all time and charges will be passed to the billing system as "lump sums," there may still be a need for the billing system to know that the charge represents a specific activity for which 1.5 hours were spent at a rate of $100 per hour. Most modern billing systems are capable of making all of the detailed calculations that complex billing basis formulas require, but it is wise to specify information if your company has unique or very new types of required calculations. Do not assume a billing system that gracefully calculates charges from time-based usage will be able to do similar calculations for packet-based or other basis usage.

The second major step is the application of taxes. Tax calculations may be done within the billing system, but most frequently, the billing system employs software modules that are able to deal with the diverse and sometimes complex differences from tax jurisdiction to tax jurisdiction. The billing system must be able to identify the appropriate tax jurisdiction(s), interact with the tax module and obtain the tax information to place into the bill.

All account level calculations may now be completed. Accounts with more than one product or service now must have those calculated amounts combined. Account level calculations may include discounts that are applicable only when multiple products/services are found in a customer account, tax rounding that is required when product/services are brought together, application of regulatory surcharges, late payment charges and calculation of balance due.

The company may do business in jurisdictions that use different currencies. Currency conversion needs should be clearly stated when preparing an RFI or RFP.

The billing system must allow for the compiled bill data to be reviewed for accuracy by billing operations prior to rendering and distributing the bill.

FtA bill compilation. FtA has not completely reviewed all the bill calculation requirements for the CLEC, but does not see this as a problem. It knows that it needs convergent billing and that the system must be able to handle a very flexible set of parameters for billing basis. It expects that there will often be times that speed in getting to market with a product will be essential. Therefore it wants user-configured table driven calculation. There will be three state tax jurisdictions, as well as Federal, to begin with, so FtA knows it needs an external tax module. All tax jurisdictions initially will be within the United States. Any tax calculation module compatible with the billing system selected is acceptable.

FtA will do bill data verification on a sample number of accounts along with a small number of sensitive accounts, prior to data release to the bill rendering process. The billing system must allow some sampling selection of accounts at the time of bill compilation. This sampling may be either a set routine built into the billing system software or allow for FtA to insert a sampling algorithm. The sampled bills will be verified against information in the customer account record—available through the standard account inquiry interface.

FtA does not foresee expanding operations to any international jurisdictions nor does it expect to need multilingual billing within the strategic planning period.

Bill compilation stakeholders

- Product management
- Billing operations
- Sales systems
- Provisioning systems
- Tax
- Middleware systems

Bill rendering and distribution

Bill rendering, or formatting, may be the same for all bills prepared or may differ, based on the manner in which the bills are to be distributed to the customers or the type of account. Regardless of the format, some data are common to all accounts. Those data include all pieces of information needed on the customer bill: billing name, address, product/service descriptions, balance forward, product/service charges, late payment/returned check charges, regulatory surcharges, taxes, payments received since the last rendered bill and any adjustments made since the last rendered bill to previously billed charges.

The billing system may format further by preparing output for the distribution medium to use without further alteration—for example, with all logos, directions on font size and spacing directions for print. Alternatively the billing system may output the information in the format(s) required by the distribution medium used to direct the bill to the customer. This choice is dictated by the distribution channels that the company determines it will use. Will printing of bills be done internally or by a printing vendor? If done by an outside vendor, what formats need to be used, and will those formats require some customizing of the billing system? The same types of questions must be asked and answered for each of the methods of distribution that the company plans to use.

Any company requirement for multilingual bills must be identified. For formatting done within the billing system, any special characters or fonts need to be identified or requested in the RFI and RFP.

FtA bill rendering and distribution. The FtA team has verified that both the broadband companies and the CLEC bill via printed bills exclusively and the ISP bills solely via credit cards. The CLEC currently uses a printing vendor and FtA expects to continue to contract out all printing of bills, but has not determined that it will continue to use the present vendor. The team, recognizing that speed to market is an important strategic need, has not determined how much internal control of formatting is needed, so it wants a billing system that is capable of full formatting or fielded data output as necessary.

In the future, FtA plans to use other media along with print. The new billing system must be able to support credit card, EDI, interactive EBPP, and CD billing.

Bill rendering and distribution stakeholders

- Product management
- Billing operations
- Middleware systems
- Carrier relations/strategic partner relations

Account maintenance

Account maintenance covers the regular day-to-day work that a billing system must do to ensure that accounts are kept up to date and complete. This work includes posting all payments, adjustments (both customer initiated and company initiated), order activity and any changes to the customer account record directly input to the billing system—in other words, not made via the service order. Such direct inputs include notes regarding customer contacts and corrections required because of incompatibilities between the billing system and sales/provisioning systems. Most modern billing systems are capable of all these functions.

Business requirements for account maintenance should identify the source of these transactions. Is the information entered directly into the billing system or does it come from another system? If it comes from another system (e.g., customer care system, lockbox bank, sales system), does it do so as a real-time transaction or as a batch feed? What data will be included in what type of transaction? How quickly must the information be available for query or modification?

The sales, provisioning and customer care processes in use by the company will impact the types of account maintenance functions needed. Will the sales and provisioning systems be gathering all data in the format needed for the billing system or will billing operations need to take actions to complete the customer account record? For example, if the sales/provisioning systems do not parse addresses, but the billing system requires parsed addresses, manual intervention is needed to correctly establish the record in the billing system. Will all changes to the customer's name and address—not associated with service or product orders—come into the billing system on service orders or can such changes be made directly into the billing system? The processes must be understood, the activities of the groups interfacing with the billing system must be analyzed and all expected types of customer account interactions must be defined.

Often companies use the billing system as the repository of record to document customer contacts for both queries and complaints. The amount of account data the company plans to hold in the customer account record must be determined. Will all customer contacts subsequent to the establishment of the account be documented and the record maintained in the customer account record in the billing system? How long will the information be retained—months, years? How quickly must retained data be available? This drives requirements for on-line or off-line archiving.

FtA account maintenance. The FtA team expects that the sales and provisioning systems will be compatible with the new billing system and that there will be data edits and/or middleware in place that will avoid the need for extensive manual reentry of data to establish accounts. FtA recognized the need for billing operations to be able to use extraordinary overrides to input information occasionally, so it wants the capability of individuals with special security classification to establish some—or all—order data directly into the billing system. FtA realizes that this is a very risky process and it is likely that it will not be in the final process design.

Further, it is anticipated that most changes to the customer account, including changes solely to customer account record data, will be input on service orders. Customer contacts will be documented in the customer account record and FtA wants to be able to see three to six months of such history on-line.

The billing system will be the database of record for the customer account record data, as well as for in-service dates of products and services. The billing system will need to reconcile product and service information with the provisioning system—if separate databases are in use in the final architecture—periodically. The billing system will also administer contract data, so it must be able to recognize end-of-contract dates and provide notification to sales and/or billing operations when actions must be taken.

Account data are to be retained online for at least 6 months—and possibly up to 12 months initially. The billing system must include offline archiving functionality, to be done either on a scheduled basis or when initiated by system administration.

Account maintenance stakeholders

- Product management
- Billing operations
- Middleware systems
- Sales systems
- Provisioning systems
- Customer care

Collections

An account that becomes overdue for payment is eligible for collection activity. The billing system may be required to do no more than identify that an overdue account condition exists, it may be required to automate all collection steps or it may direct all overdue accounts to a separate collections module for processing.

Most billing systems are capable of using the credit classifications to set collection parameters for generating notices to customers or referrals to customer care personnel. The parameters generally are identified in user-populated tables, with the capability of unique parameters for each credit classification.

The functionality questions that must be decided here center around how much collection activity is to be done in the billing system and what that activity is. Will the billing system need to provide multiple credit classifications and associated unique parameters for each? Will the billing system generate written notification of delinquency as billing messages or as separate notices or letters? If notices and/or letters are used, how many types will there be and where will they be output? Will all collection activity be done by notice or will referrals to company personnel be needed?

FtA collections. The FtA team has determined that multiple credit classifications with user-driven tables for unique parameters will be needed in the new billing system. For example, the days outstanding of the customer account balance will trigger the action to be taken as shown in Table 11-1. Although all collection processes have not been finalized, the team wants to have the new billing system provide collection functionality. It must be able to provide a full array of functions: delinquency notices on bills, delinquency notices as separate outputs, referrals to collections personnel via workstations and communications to provisioning for service interruption or cancellation.

The team has determined that it wants to be able to specify at least 15 collection events for each credit classification, with flexibility in defining the trigger date or event, as well as the nature of the activity triggered.

Collections stakeholders

- Product management
- Billing operations
- Customer care
- Provisioning systems

Reports

The billing system must produce information on the monies that are billed, collected and passed to the accounting system or general ledger system. All billing systems are able to do this; the amount of data specificity and the compatibility with the financial/general ledger system used are the critical points. Most billing systems provide finite tracking of products/services, accounts, payments, fees and adjustments.

Various groups within a company require data from the billing system. These data range from such items as which products/services are in place in which jurisdictions, to customers assigned to specific credit classifications, to numbers of days between billing an account and receiving the payment. In other words, every aspect of customer data in the billing system may need to be identified and categorized in a number of different ways to provide management information. Billing systems have standard reports—for example, many billing systems offer standard accounts receivable aging reports—and also provide for user configuration of ad hoc or standing reports.

FtA reports. FtA has not selected the general ledger system, so it will not be able to specify that the new billing system must support a specific vendor's product. This means that the team wants the new billing system to provide the greatest flexibility for user driven report configuration. FtA expects that product management, the financial department, credit/collections and billing operations personnel may all need to define and request reports, so user-friendly request and configuration procedures are important. Accounts receivable aging reports are deemed to be an important tool; they must be available as part of the standard report capability.

Reports stakeholders

- Product management
- Sales
- Provisioning
- Billing operations
- Collections
- Accounting standards

Ease of Use

The methods used to interact with the billing system for things such as establishing accounts, entering and retrieving information and changing processing parameters must be compatible with the capabilities of the company personnel—as well as fit within the processes that the company uses. Customers, too, may be using the billing system via the Web; their ease of use is vital. This includes average time to accomplish certain tasks, the need for supporting reference documents, the length of training—or step-by-step guidance needed—to achieve mastery of each interface, as well as specific user support features.

Ordering

Sales/provisioning people may interact with the billing system indirectly or directly to enter a service order. Ease of use considerations for the sales organization include the amount of data that must be provided on service orders, who provides the data, edits to ensure completeness and the methods by which data are obtained from the billing system. When all service order activity is directed to the billing system via the provisioning system this is an indirect connection—the input from the sales organization must drive both the provisioning and the billing systems. Provisioning has the same ease of use needs as sales: redundant data should not be required and edit checks should be employed.

Orders are prepared for input in a number of ways. They may be prepared by sales on paper and given to a separate group for formatting and entry; they may be directly input to the billing system by sales/provisioning; customers may input them via the Web; or input may be some combination of these methods. The billing system must be able to work with the sales/provisioning system to populate the billing records with the least amount of input from the sales force—or customer. For example, if the billing system requires that there be an entry in both the service address field and the mailing address field, even if they are the same, the sales people may be able to enter the information in one field and the system(s) will populate both fields as required.

Most billing systems today provide Web-enabled and/or graphic user interface (GUI) direct access to information. These types of interface allow user-friendly input and navigation, leading the user to appropriate choices and identifying essential data for input.

Training requirements for users are another ease of use consideration. Is material available, complete, well written and efficient? Is training configured modularly? What amount of training time per participant is required for basic skills? Does separate training exist for trainers or will all training need to be done by the billing system vendor? Is customer direction for Web use clearly stated and user friendly?

FtA ordering. The FtA team has determined that all service order activity will come into the billing system via the provisioning system and entries required by the sales and provisioning forces are to be limited as much as possible. All calculations should be done in the billing system, so that the provisioning organization will enter only components of time and charge information.

All direct billing system interactions must utilize Web-enabled and/or GUI screens, with edits to ensure that required data fields are filled and meet formatting and content criteria before the billing system will accept transactions. Training for company personnel inputting directly to the billing system for orders should be modular, in order to gradually bring personnel from simple to complex levels of the business. Directions for Web usage by customers/non-company users must be well stated, step-by-step and clear. FtA has not determined what type of training department it will maintain; therefore, the billing system vendor should be able to provide workforce training for all personnel who do direct input.

Ordering stakeholders

- Product management
- Sales systems
- Provisioning systems
- Billing operations
- Corporate accounting

Customer care and collections

Customer care is assigned the responsibility of handling customer interactions other than sales. Most of the customer-initiated interactions require access to the billing system customer account record, as do virtually all of the company initiated interactions, including collections. Customer care—which is a function that may be housed in sales, billing operations or its own separate organization—includes customer inquiry, billing investigations and customer-initiated adjustments. Ease of use considerations for customer care include how the customer account record may be accessed, the volume of data required to be entered in order to annotate or adjust an account and how the customer account records may be navigated.

Considerations for collections are the same as those of customer care, with additional needs for access to information on overdue account customers. This information, usually based on the parameters provided by the company, may be provided at a user position or on hardcopy reports—or both. Billing systems usually accept information on collection actions taken with a minimum amount of entry effort and can generate preprogrammed or on-demand form letters, containing text supplied by the company, to customers.

Customer account access may be needed in more than one way: for example, a customer calling with an inquiry may not know the account number, requiring account access by account name or by billing telephone number. Most billing systems offer flexibility of access, with a number of them allowing the company to determine how and in what ways users will be able to locate accounts, including account number, name, address, billing cycle, tax jurisdiction, credit classification, product/service billed and almost any other data item associated with the customer account record. These parameters should be specified in your company's business requirements for a billing system.

Navigation to and through the customer account record is another ease of use aspect of billing systems. Web-enabled access and GUI screens may be pretty, but still require users a number of screens or awkward steps to find even simple data. What is a billing system's maximum number of screens to access all data in the record? Must those screens be accessed in a linear fashion? What is the maximum number of screens needed to obtain customer account and product/service information?

Training requirements for users is another ease of use consideration. Is material available, complete, well written and efficient? Is training configured modularly? What amount of training time per participant is required for basic skills? Does separate training exist for trainers or will all training need to be done by the billing system vendor?

Does your company employ or plan to employ the visually handicapped? If so, your billing system will need both modified displays for those with some visual impairment and/or an audio interface for those who are totally blind. Similarly, if your company employs hearing-impaired individuals, the system should have user-settable alternatives to any auditory cues and instructions.

FtA customer care and collections. The FtA team wants the greatest possible flexibility to access customer account records. The billing system must allow FtA to determine the keys to customer account identification and the flexibility to change those keys as FtA processes evolve and change. Navigation through the customer account record and associated product/service information must be graceful; FtA wants no more than five screens to be needed for any given inquiry or operation.

The preference is for Web-enabled access, but the FtA will accept a billing system that uses GUI screens if fully tested Web-enabled access is available within 12 to 18 months of FtA's billing system implementation.

All training should be modular and because FtA has not determined what type of training department it will maintain, the billing system vendor should be able to provide training for the customer care and collections work force.

Customer care and collections stakeholders

- Product management
- Sales systems
- Provisioning systems
- Billing operations
- Customer care

Billing operations

The billing operations organization within a company facilitates the smooth running of the billing system and must be able to interact with the system readily, accurately and easily. Modern billing systems provide user-friendly access via Web-enabled screens and/or GUI screens that enable billing operations users to accomplish its tasks. Billing operations performs such things as bill cycle management, error correction management, contract management and payment management. The billing system must provide billing operations personnel reports on all aspects of the account creation and billing via reports and/or workstation interfaces to the system. Most billing systems allow the company to configure these reports to meet unique needs.

Training requirements for users are another ease of use consideration. Is material available, complete, well written and efficient? Is training configured modularly? What amount of training time per participant is required for basic skills? Does separate training exist for trainers or will all training need to be done by the billing system vendor?

FtA billing operations. The FtA team requires that the bill verification responsibilities of billing operations be fully supported by the billing system.

Billing operations in FtA will have responsibility for all CDR error correction, as well as the monitoring of all customer account service order activity into the billing system. The billing system must allow the company to define how error conditions will be displayed and all reports must be available on Web-enabled access or GUI screens, as well as on hard copy, if requested. Of course, all ease of use items needed for customer care and collections must be available to billing operations.

All training should be modular and because FtA has not determined what type of training department it will maintain, the billing system vendor should be able to provide training for the customer care and collections workforce.

Billing operations stakeholders

- Product management
- Billing operations
- Corporate accounting

Product management functions

Product management may provide information directly to the billing system or all data may come through middleware. The information on products, services, billing cycles, and credit practices—including collection step parameters—are all the province of product management. The billing systems that are most easily used allow coordinated use of product catalogs and screen entry of table data for billing cycles, credit classifications, and credit parameters.

FtA product management functions. The FtA team expects all product management data items to be entered into the billing system via middleware or by billing operations. The data must be table driven and readily reconfigured. The billing system must provide for direct user input.

Product management functions stakeholders

- Product management
- Billing operations
- Accounting standards

Ease of Management

Ease of management includes such features as management reports and on-line status, work queue assignment and management and work efficiency and effectiveness tracking. There is significant overlap in the information infrastructure for ease of management and auditability. However, ease of management requirements deals with the availability and accessibility of that information in ways that are appropriate to work management and supervisory concerns.

Most billing systems allow for data on the system to be available to supervisory personnel and for system administration personnel in real time and for some reasonable period of time. Supervisory personnel need information over significant periods of time to allow for trending of performance and quality information. Supervisors also need the ability to define and alter the definition of management data. System administration requirements are more short term and the need is more immediate. All system actions, files available for use, size of files, times of activity, location in the system of data and other audit trail data such as record counts must be available as they occur and the information must be maintained for a period of time—usually days—for error resolution or system recovery.

FtA ease of management

The FtA team has not finalized all the processes that it will use for system administration or user supervision, but it is clear that supervisory oversight reports are required on all aspects of billing operations activity. The reports should be both on demand and produced for archives/later retrieval.

System administration needs reports on all system activity; these reports must be available at work positions as the activity occurs, as well as for at least seven days. The reports, at a minimum, must identify the source of all activities: for manually initiated activities, the user/employee must be known and for automated input the source is to be identified. The disposition of actions must be retained: for example, system administration must be able to see that an action was completed without error or, if in error, be given the error condition encountered. Additionally, system errors must be displayed on workstations for system administration as soon as encountered to facilitate resolution. Actions completed without error should show statistics such as record counts, money totals or other accuracy measures.

System administrators must be able to access information in various ways. These include, at a minimum, by completed activities, by errors encountered and by time of day run.

Ease of management stakeholders

- Sales systems
- Provisioning systems
- Billing operations
- Middleware systems
- Accounting standards

Exception Handling

Exception handling is the ability of the system to identify errors and problems and to support the resolution of those errors and problems. The billing system will identify the items that are required for record creation and maintenance, as well as allow the company to identify other required data without which processing cannot continue. Billing systems may allow partial posting of the information on an input transaction or reject everything in the transaction—e.g., service order—if any one entry fails to pass system edits. Most billing systems are able to handle errors both ways and provide notification back to the originator by report or in real time at a workstation.

FtA exception handling

The FtA team has determined that all data on service orders must pass edits or nothing is to be accepted; no partial posting of errors is to be done. Because all processes have not been fully developed, the billing system should be able to partial post orders in the event that FtA alters its position in the future. FtA service order error correction will be done via sales/provisioning. Errors on directly entered data—such as adjustments—will be identified at the time of input and will prompt the user for correction in real time.

Exception handling stakeholders

- Sales systems
- Provisioning systems
- Billing operations
- Middleware systems
- Accounting standards

Auditability

Auditability requirements relate to capabilities that facilitate tracing transactions through the system, including any transactions against customer account records, pricing structures or other information related to financial calculations, reporting and postings. Auditability requirements may include automated support for database checkpoints and for data and transaction history retention. Auditability requirements may also address the information infrastructure required for ease of management.

FtA auditability

The FtA team has not identified all auditability processes, but knows that it must have every transaction in the system identified with the date of the transaction and the user that initiated it. The bill cycle process must include the ability for billing operations to identify a sample or accounts and specific accounts prior to releasing the billing cycle accounts to bill rendering. Additionally, the billing data used for each billing cycle must be retained as it looked at the beginning of the bill preparation until the bills have been certified correct and released to bill rendering.

Auditability stakeholders

- Sales systems
- Provisioning systems
- Billing operations
- Middleware systems
- Financial operations

Sizing/Volumes

The current number of accounts, the average products/services per account and the anticipated volumes of usage for those accounts need to be known to determine the size/capability of the system that will be needed at the time of implementation. The expectation of volume growth—or loss, as the case may be—within the next three to five years should be the key to the system that is required.

FtA sizing/volumes

FtA has been able to access the billing systems used by both broadband companies and to perform a count of customer accounts and services. The final contract for the purchase of assets and customers from the ISP includes audited counts of customers and services. However, sizing information from the CLEC is an estimate provided as part of the purchase negotiation, not audited information from the current billing system and process.

FtA has decided to use the unaudited information as input to the current volume requirements and estimate that the ordered system should be capable of handling twice current volumes with upgrade capability for 10 times the current volumes. The broadband companies currently comprise five large businesses and approximately 1,800 medium sized businesses. The ISP has somewhat over 5,000 small business and consumer customers. The CLEC has the majority of FtA's customers with its approximately 20,000 consumer and small business accounts. FtA is asking for an initial capability of 50,000 accounts, with the majority of them using CDR type services. The system must have upgrade capability to handle 500,000 accounts.

Sizing/volumes stakeholders

- Sales systems
- Provisioning systems
- Billing operations
- Middleware systems
- Accounting standards

Billing Cycles

All billing systems support multiple billing cycles; the questions to ask are how many billing cycles are supported and what flexibility is offered additionally. Is it possible to include the ability to specify bill dates for the entire system, for a class of customers and for an individual customer? Is it possible to use different periods of time for different bill cycles—for example having some that are billed monthly, some that are billed quarterly and some that are dependent on activity?

FtA billing cycles

The FtA team has not determined the ultimate number of bill cycles that will be used, so it requires that the number of cycles be determinable by the company. The billing system must allow automatic assignment of the billing cycle to the customer account or accept an input billing cycle assignment. The periodicity of the cycle must be flexible as well.

Billing cycles stakeholders

- Sales systems
- Provisioning systems
- Billing operations
- Middleware systems

Platform

Platform requirements address the hardware, software and network environment of the billing system. They may include the platform architecture of the billing system itself, the corporate network and desktop environment, as well as specifications for customer access devices and software. These requirements may specify PCs, PDAs, workstations, DBMSs, network hardware, and operating systems, tools, and utilities such as reports writers and graphical support and any other third-party resource required by the billing system or by interfacing functions and systems.

FtA platform

FtA plans to use Windows XP as its default desktop platform and the current version of Internet Explorer as its standard browser. While some of its technical staff use LINUX-based desktops and other open source tools, FtA would like all functional interfaces with the billing system to be supported in the XP environment. FtA would also prefer that some critical billing operations interfaces be PDA enabled, but it does not have a preferred PDA platform.

Its current network is Windows NT based, but it is evaluating other network platforms. The billing systems project is one of several that may influence the decision about network platform.

FtA has a site license for the Oracle DBMS that can be extended to new installations. It also has a site license for Crystal Reports Writer and Microsoft Office XP (developer edition).

FtA currently has 23 Sun Enterprise Server 6500 servers. Most of these are running Solaris 2.6, although two are running a trial of LINUX. Most of these systems are supporting the ISP business. FtA does not have specific platform standards, although it will select a server platform consistent with existing systems if a vendor offers the same product on multiple platforms.

Areas in which FtA has site licenses may impact the structure of a project that uses those resources. Platform compatibility with FtA's Sun Servers will add a slightly favorable rating to a specific solution assessment.

The only firm requirement for platform is using Windows XP as the primary desktop environment.

Platform stakeholders

- Sales systems
- Provisioning systems
- Billing operations
- Middleware systems
- Accounting standards

Summary

In this chapter, we have reviewed requirements for the operations and systems environment for a billing implementation. These requirements will be used to develop various procurement documents—RFI, RFP—and evaluate vendor responses that we look at in Chapters 31 through 34. They are also refined for use in the statement of work that is incorporated into the vendor contract.

Chapter

25

Data Interfaces to Billing Systems

Figure 25-1 shows the numerous interfaces to the billing environment in a typical billing system architecture. Clearly, billing systems do not exist in hermetically sealed boxes. Since telecommunications billing systems typically handle very large amounts of information across most of their systems interfaces, the accuracy, efficiency and recoverability of those interfaces are critical to the performance of the billing environment and, often, to the financial viability of the enterprise.

Types of Interfaces

The first step in developing requirements for systems interfaces is to identify what those interfaces are. First, identify the function of the interfacing system or systems. Then determine the nature and architecture of the interface or potential interface. Analyze whether that interface will adequately meet current and future billing needs of your company and specify an interface that will meet those needs.

Sometimes the interfacing system will determine the architecture of the interface, especially if the interfacing system is not under your company's control. However, if the interfacing system is internal to your company, you may be able to design an interface that is appropriate to service differentiation, as well as being efficient, effective and manageable for the volume and type of information flow.

Typical points of integration

Many telecommunications billing systems interface with the following types of systems and processes:

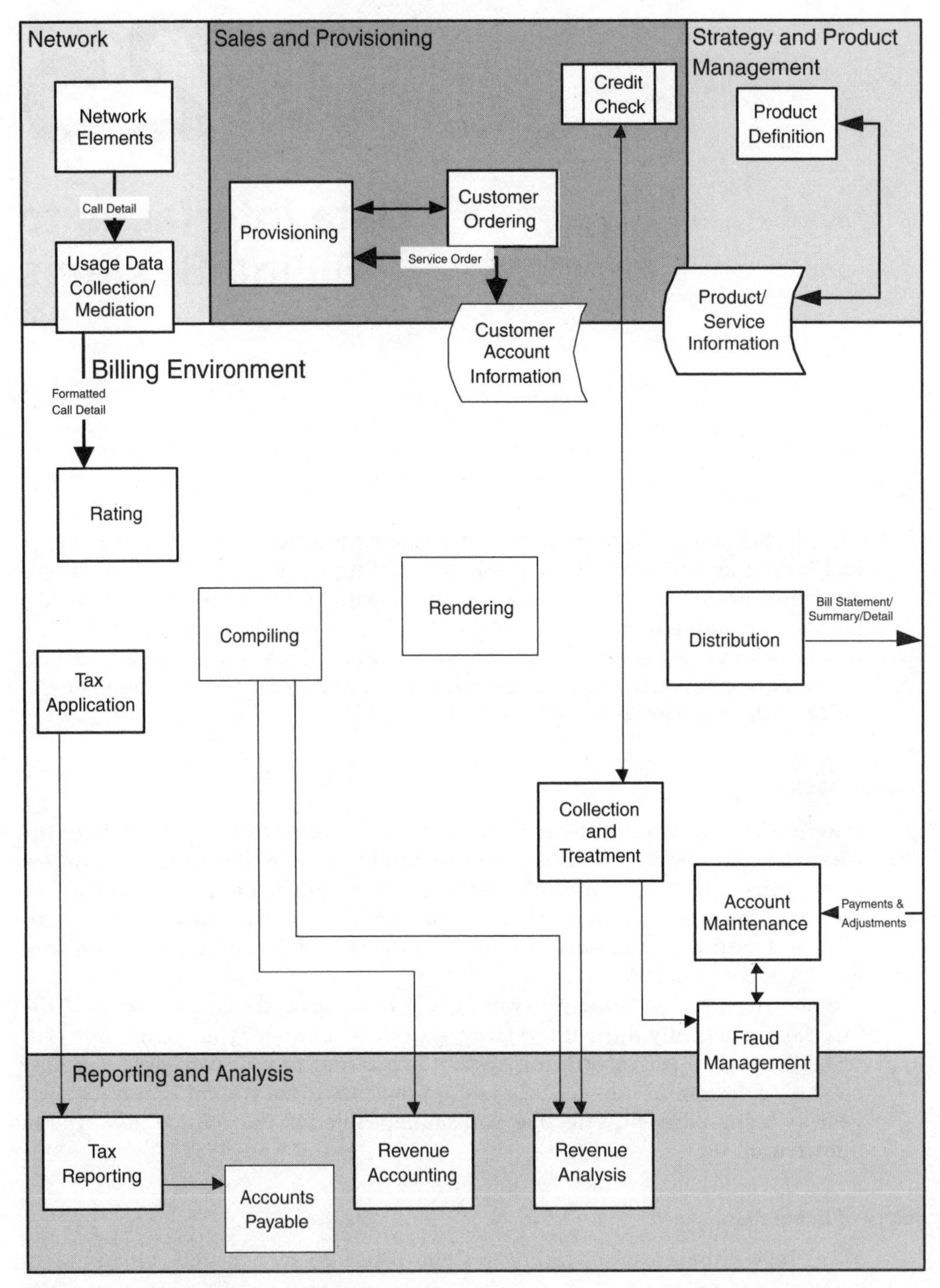

Figure 25-1 Billing environment interfaces.

- Sales support systems
- Provisioning systems
- Usage data collection systems
- Bill print systems
- Electronic bill delivery systems
- Bill/account inquiry systems
- Banks/payment reporting agencies
- Tax calculation systems
- Tax reporting systems
- Accounts receivable systems
- Collections and treatment systems
- Fraud management systems

Strategies for provisioning systems integration

In many companies' systems enterprise architecture, one of the critical interfaces to the billing system is the provisioning system. This interface can carry all account and service order information or just provisioning-related information such as in-service dates. It is perhaps the most widely varied of all interfaces in implementations. Therefore, it is a good example for various integration strategies.

"Brute force." If the order volume is small and the service complexity is minimal, it is possible to simply rekey the relevant data from the ordering to the billing system. A slightly more sophisticated version of this scenario uses workstation-based software to capture data fields from one system and enter them into another system. The contemporary version of this would be a Web form that directly updates a billing database. While this is not a good scenario for the long term, it may be appropriate for a low-volume startup selling simple long-haul broadband service, for instance.

Standard APIs. An application programming interface (API) is the specific method prescribed by an application program by which a programmer writing an application program or interface routine can make requests of another application. Most off-the-shelf billing systems come with a set of APIs for standard transactions. Data from ordering systems can be formatted and fed to one or more of those APIs. Unless the ordering system and the billing system are maintained as "modules" of the same product family and your service mix is exactly as envisioned by the developers of the systems, there is likely to be some programming required to make the ordering system APIs "fit" the billing system APIs, despite some relevant industry standards (see section on Open

Billing Forum standards overview in Chapter 26). It may also require a significant amount of system management work and reports development to ensure the reliability of system-to-system transactions.

Middleware. There is a class of software known as *middleware* that is designed to manage connections between disparate systems environments. Generally, middleware focuses on managing the reliability and synchronization of information between the sales and/or provisioning and billing systems. Often the middleware itself requires significant customization for each type of transaction, although there are middleware-oriented vendors developing "precustomized" solutions for the most common provisioning and billing systems.

Shared database. It is possible to have a sales system, a provisioning system and a billing system actually share a single database. Some off-the-shelf "solution suites" are configured in this way. Also, many custom systems for smaller enterprises utilize a single database—especially those that originated in the "common database" era of the late 1980s.

Obviously, a shared database eliminates synchronization issues, but raises many other issues of processing efficiency and utility.

Fully custom interfaces. Fully customized interfaces allow a great deal of freedom in translating ordering/provisioning information into billing information. If the data in the ordering systems are structured very differently than the data in the billing systems and if reference to various translation tables or code indices are required to translate the data, then a fully customized interface may be required. Another reason that a customized interface will be necessary is if the ordering/provisioning system(s) are custom systems with no standard interfaces.

Custom interfaces are costly to maintain, error prone and often result in less-than-perfect data synchronization. A custom interface requires a long-term commitment to the maintenance of that interface, either from the original development organization in the form of contract commitments or from the telecom company itself for permanent dedicated maintenance resources.

Bulk data transfers. Among the oldest types of system interfaces, bulk data transfers entail the originating system (provisioning, in this case) queuing transactions until a specified time or volume is reached. The transaction data are then bundled into a single formatted file and sent as a single file transfer to the receiving system (billing, in this case). This is a very efficient type of interface for very large volumes of data. If the data do not require near real-time processing and volumes are very large, bulk data transfers minimize network and operational overhead, provided the interface is equipped with excellent fault identification and recovery features.

Bulk data transfers are also very useful when converting data from another system. In that case, the records from the source system are extracted, reformatted as required and then forwarded to the bulk data input processing function.

Stakeholders

- Sales systems
- Provisioning systems
- Network equipment engineering (usage data collection systems)
- Bill print systems/vendors
- Bill/account inquiry systems
- Banks/payment reporting agencies
- Tax reporting systems
- Accounting systems
- Collections and treatment systems
- Billing operations
- Carrier relations/strategic partner relations
- Server resources management

Capacity

The required capacity of an interface may have significant impact on the appropriate architecture and implementation of that interface.

Deriving capacity requirements from data collection

Most interface volume is either fixed (driven by calendar events such as the end of the fiscal month and year, for instance) or related to service provisioning activity, to number of accounts or to usage transactions. Activity related information was collected during data collection in association with customer accounts and products and services.

Examining a service provisioning record interface to billing and likely feedback loops of information for sizing and then calculating the total volume of provisioning records will produce a usable estimate of capacity for the service provisioning interface.

Examining a typical call detail record for each service and then multiplying by the projected call volume for that service, when combined with information about all services getting call detail information from that source, will produce usable estimates of capacity for a usage data interface.

FtA capacity requirements

Figure 25-2 shows a portion of FtA's capacity requirements. Note that the requirement includes quantitative standards for interface performance, as well as general sizing information.

Capacity Requirements
January 20, 2003

1. Products and Services
FtA anticipates 500 unique service offerings within the first two years of the new billing system's operation.

2. Product/Service Records
FtA requires an initial system capacity of 500,000 service records ("lines," etc.) with the capability to grow to 1,000,000.

3. Accounts
FtA requires an initial system capacity of 100,000 accounts with the capability to grow to 300,000.

4. Orders
FtA requires an initial system capacity to handle 1,000 service orders per month with the capability to grow to 5,000 orders per month.

5. Usage Records
FtA requires an initial system capacity to handle 500,000 usage records (CDRs) per day with the capability of growing to 50,000,000 over five years.

6. Customer Care Queries
FtA requires the initial capacity to handle 10,000 queries per month with the capability of growing to 35,000.

Page 1 of 1

Figure 25-2 Sample FtA capacity requirements.

Capacity stakeholders

- Billing operations
- Product management
- Sales systems
- Provisioning systems
- Network equipment engineering (usage data collection systems)
- Bill print systems/vendors
- Bill/account inquiry systems
- Banks/payment reporting agencies
- Tax reporting systems
- Accounting systems
- Collections and treatment systems
- Carrier relations/strategic partner relations
- Server resources management

Specifications/Tools

There is a wide range of support capabilities for interface creation among billing platforms. Interface specifications and tools requirements identify the interface support capabilities your business needs.

Developing requirements for interface specifications and tools

Having identified specific interfaces, as well as the timing and capacity required for each, we can develop requirements for system interface specification and tools.

Basic interface. At a minimum, every potential interface should have an openly specified interface for record-oriented additions and deletions. Obviously, an openly specified interface for record-oriented changes is also highly desirable, but may be accomplished in many cases by a coordinated add/delete activity.

The open specifications should include information on data formats, system access, callable routines and any conflicts with other system activities. If these specified interfaces are not fully developed APIs including all expected edits and validations, the vendor should publish detailed specifications for those expected edits and validations.

Note that this basic interface information can be used to construct both a record-oriented interface and a bulk load capability if the vendor does not provide this.

Manual interfaces. Some billing platforms contain support for direct user interfaces for most or all external data input and retrieval. Other platforms provide billing functionality intended to be integrated with external customer care and billing operations management applications. Even when the intent is for integration with an external platform, direct user interfaces may be required for out-of-process error correction and management.

All billing applications should provide direct user interfaces for the following functions:

- Establish customer account
- Update customer account
- Archive customer account
- Delete customer account
- Establish product or service definition
- Update product or service definition
- Archive product or service definition
- Delete product or service definition
- Establish service instance record (service order)

- Update service instance record (service order)
- Archive service instance record
- Delete service instance record
- Apply payment
- Apply credit
- Back out payment
- Back out credit
- Input usage record
- Delete usage record

Application programming interface (API). Transaction oriented APIs should be available for all external transaction interfaces to the platform. These APIs should include provision for all edits and data filters that would be used on user interfaces for manual data entry or other fully vendor implemented interfaces. Documentation should include specifications for these edits and data filters, unless they are user specified.

APIs are valuable in interfacing legacy systems or custom development to a new billing system. They may also be used for low-volume conversion efforts. If the interface is mission critical, you may want to invest in middleware to manage the interface. Of course, the middleware will probably call the API, unless the middleware vendor has a special relationship with the billing software vendor for unique interfaces.

Some of the APIs required for most telecommunications providers include:

- Establish customer account
- Update customer account
- Archive customer account
- Delete customer account
- Establish product or service definition
- Update product or service definition
- Archive product or service definition
- Delete product or service definition
- Establish service instance record (service order)
- Update service instance record (service order)
- Archive service instance record
- Delete service instance record
- Apply payment
- Apply credit

- Back out payment
- Back out credit

Other APIs your company may need include:

- Update accounts receivable journal
- Request payment/update accounts payable
- Input usage detail
- Delete usage detail
- Identify potential fraud
- Identify account for treatment/collection
- Request tax rate
- Customer account inquiry
- Product or service definition inquiry
- Bill print output
- Electronic bill output
- Bill detail output

Standard interfaces. For some types of interfaces, there are rich standards sets that many industry systems incorporate. If your company plans only compliant interfaces, then specifying compliance requirements replaces the need for other types of programming interfaces.

The following are typical standards compliant interface points:

- Usage data input
- Electronic bill presentment
- Electronic bill payment

Bulk data loads. As mentioned earlier, bulk data loads are very valuable when data volumes are large and real-time response is not required. Similar to a good transaction API, a good bulk data interface should include provision for all edits and data filters that would be used on user interfaces for manual data entry or other fully vendor implemented interfaces. Documentation should include specifications for these edits and data filters, unless they are user specified.

Ideally, a bulk data load will identify faulty records, back out any updates based on those records and report them in enough detail for error correction. The alternative is to reject the entire bulk feed and back it out—providing enough detail for error correction, of course.

Some bulk data interfaces your company may want to specify include:

- Establish customer account
- Establish product or service definition
- Establish service instance record (service order)
- Apply payment
- Apply credit
- Input usage record

Middleware. In some ways, middleware is a sophisticated enhancement of APIs, providing transaction management, timing and regeneration support for an interface. It also serves to isolate applications from changes on the other side of the middleware interface—making adapting the problem of the middleware team. Middleware may be an appropriate approach for integration of the following types of interfaces:

- Establish customer account
- Update customer account
- Archive customer account
- Delete customer account
- Establish product or service definition
- Update product or service definition
- Archive product or service definition
- Delete product or service definition
- Establish service instance record (service order)
- Update service instance record (service order)
- Archive service instance record
- Delete service instance record
- Apply payment
- Apply credit
- Back out payment
- Back out credit
- Update accounts receivable journal
- Request payment/update accounts payable
- Input usage detail
- Delete usage detail
- Request tax rate

Other interface tools. Some vendors provide visual and other tools for specifying the details of common interfaces. This is particularly appropriate if the application is fully CORBA (see Chapter 26) compliant.

FtA specifications and tools requirements

Figure 25-3 shows a portion of the FtA interface specifications and tools requirements. Note that FtA specifies a requirement for bulk data load of customer accounts and associated product/service instance records. FtA plans further growth by acquisition and will have frequent need to perform conversion data loads as this growth occurs.

Specifications/tools stakeholders

- CIO
- Billing operations

Interface and Tools Requirements
January 20, 2003

3. Manual Interfaces

FtA requires that the following direct manual interfaces be available in the billing system. The default state of these capabilities will be "disabled." Each of these capabilities will have to be explicitly made available to a user by the billing system security manager to an individual user or to a defined group of users.

All billing applications should provide direct user interfaces for the following functions:

- Establish customer account (all fields except account identifier optional)
- Update customer account (all fields addressable; all fields except account identifier optional)
- Archive customer account
- Delete customer account
- Establish product or service definition
- Update product or service definition
- Archive product or service definition
- Delete product or service definition
- Establish service instance record (service order) (all fields except account ID and service ID optional)
- Update service instance record (service order) (all fields addressable; all fields except account ID and service ID optional)
- Archive service instance record
- Delete service instance record
- Apply payment
- Apply credit
- Back out payment
- Back out credit
- Input usage record

Page 3 of 21

Figure 25-3 Sample FtA interface and tools requirements.

- Product management
- Server resources management

Quality Tools

Some quality tools have been implied in the recommended requirements for various types of interfaces. Here we summarize requirements for quality tools for interfaces.

Developing requirements for interface quality tools

The fundamental concept for interface quality tools is that all interface transactions should be trackable and auditable, problems and exception conditions should be identified at the interface, any impact of the exception should be backed out, notification should be sent to whoever or whatever is responsible for problem resolution (person or system) and error correction should be facilitated. As always, requirements for interface quality tools are based on business needs and strategy.

Interface tracking and reporting. Every interface transaction should update synchronization and tracking utilities that are accessible for audit. Common methods of synchronization and update include unique message identifiers, message time stamps, message size verification tools such as checksums and record counts on both sides of the interface, message disposition recording and independent periodic reconciliation of both sides of the interface. Of course, all interface tracking data should be collected and stored on both sides of the interface for auditing and reporting purposes.

Identifying problems and exceptions. Requirements for identification of problems and exceptions can be quite general or very specific, both as to what constitutes a problem or exception and when that problem or exception should be identified. Some common interface problems and exceptions include:

- **Unexpected message length.** Either there is too much or too little data in the message, based either on an expected fixed message length or on message header or trailer information containing the expected message size.
- **Unrecognized message type.** Somehow the message got mangled or something unexpected got into the data stream and the system cannot figure out what to do with it.
- **Expected message did not arrive.** Relevant when an interface is scheduled, as with hourly or daily bulk feeds or when the far side of the interface is polled.
- **Data edits failed.** The message arrived and was recognized, but some of the data did not pass edits and data filters.

Note: If the failed edit was on optional fields, the system may process the rest of the information or it may reject the entire record. Failed edits on required fields should always precipitate the rejection of the entire record.

- **Downstream processing failed.** The message passed front-end format edits and filters, but encountered an error condition beyond the interface (e.g., an update transaction finds no appropriate record to update in the database).

Backing out exceptions. Depending on when in the processing stream an interface exception is detected, there may be a lot or a little bit to back out. Everything impacted by the assumption of a successful transaction will have to be backed out. Every interface transaction should update interface tracking information. Therefore, backing out a transaction that failed in downstream processing may require backing out database update transactions, decrementing the counter of successful records interfaced, decrementing the counter of successful records received and notifying the sending system to decrement the count of successfully sent records.

Notification. Notification is the most critical portion of interface quality tools. If no one knows there is a problem, it probably will not be corrected. Notification without supporting information is more frustrating than useful—requiring huge quantities of manual investigation time. All interfaces should be supported by exception notification to a designated exception management capability. This could be as simple as a printout or a queue of exception messages or as complex as notifying both an administrative position and the originating system of failure.

Exception notification messages should identify the type of interface, provide an identifier for both sides of the interface, identify the specific interface message or file and provide the timestamp of the interface. This could include start and end time on both sides of the interface for complex transactions or bulk feeds. Exception notification should identify the specific nature of the exception and the specific cause for failure. For instance, good notification would be “Zip code missing in billing address on record 02Q231” rather than “Data edits failed.”

Error correction. Ideally, error correction would take place in the source system and the transaction would be re-initiated. In that instance, tracking of resolution should be in the source system. This is not always possible for either technical or business reasons (e.g., when the source system is not under the control of your company).

The next best option is to make the data record available for correction at a user interface, along with the error notification. Then a local billing operations resource can research, resolve and correct the record and resubmit it from the error management position. The error management position would update the error log that the problem has been resolved and by whom.

Business needs and strategy. How quickly do you need to resolve error conditions? What is the business impact of unresolved error conditions? It is important to determine what is important to the business.

If your company has guaranteed customers 24-hour activation on usage-based enhanced services and your company is doing real-time automated provisioning, then establishing the customer record for billing is quite time sensitive. You might want to write requirements that indicate reporting of errors in such activation records must take place less than 15 minutes after transmission of the records and that error management personnel can resolve 90 percent of all exceptions in less than four minutes after errors are reported; 95 percent in less than 10 minutes and 100 percent in under an hour.

If your company is tightly managing cash flow, making sure all usage record exceptions are resolved within an hour may be an important requirement. That way all or almost all usage will be reflected in the current billing cycle and not be delayed for billing until the next cycle.

FtA interface quality tools requirements

FtA considers interface quality tools to be an important part of an overall policy of rigorous controls. Figure 25-4 shows a portion of the FtA quality tools requirements.

Interface quality tools stakeholders

- Billing operations
- Product management
- Sales systems
- Provisioning systems
- Network equipment engineering (usage data collection systems)
- Bill print systems/vendors
- Bill/account inquiry systems
- Banks/payment reporting agencies
- Tax reporting systems
- Accounting systems
- Collections and treatment systems
- Server resources management

Quality Tools Requirements
January 20, 2003

3. Interface Controls

All transactions at an interface will be logged immediately at the interface prior to any other processing; then a positive acknowledgement with identifying and completeness verification will be sent to the originating system. This information will be available at the billing operations console.

Transaction acceptance checks will then be performed to verify readability, recognizable formats, valid identifications and any other appropriate, noncontent-oriented verification. The results of these checks will be logged and forwarded to the sending system immediately. This information will be available at the billing operations console.

Content-oriented validations will then be performed at the individual record level. If more records are rejected than a threshold set at the billing operations console (e.g., greater than 30 percent of the total record count), the entire input will be rejected as suspect and the sending system will be notified. This information will be available at the billing operations console.

Otherwise, the system will process all non-errored records and log the records processed. Ideally, the log file should allow the immediate identification of duplicate incoming records, so they can be removed from processing. This information will be available at the billing operations console.

4. Billing Operations Process Management

FtA requires a billing operations console function for controlling and tracking overall processing status and individual transactions.

The billing operations console is a restricted function. Permission to access this function must be explicitly enabled by the security manager for the system.

The billing operations console should provide an overall view of all process queues—both internal and input and output—and error queues. Ideally, this overall information should be available graphically and in a spreadsheet-like format as a user-selectable option.

Page 2 of 9

Figure 25-4 FtA quality tools requirements sample.

Trends in Interfaces and Integration

As in all technology arenas, there are some significant trends in interface construction, management and integration. Three of the most important are Web-based integration of user and customer functions and, more generally, Web-enabled system interfaces; packaged and configurable middleware; and real time, volume rating, and usage reporting.

Web-based integrated customer interface

The trend to Web-oriented user interfaces to systems is almost universal. The majority of user interfaces written in the past few years are implemented for access via a browser. Most of these implementations are based in HTML for compatibility with the widest range of current browsers. The ease of linking these interfaces and moving data application to another has made integration of comprehensive customer care platforms at the Web interface fairly straightforward and independent of the rest of the system functionality.

Initially, these interfaces were specific functional interfaces for in-house local users of the system. However, as e-commerce becomes mainstream, both customers and enterprises have recognized the benefit of "eliminating the middleman" and allowing the customer to perform order entry and account inquiry without ever speaking to customer care personnel and without any need for customer care involvement unless an exception or error condition is encountered.

Billing platforms include specific Web pages or XML compliant interfaces with or without tool kits.

Packaged "middleware"

Middleware offers real advantages for moderate volume, mission-critical interfaces such as the one between the service order environment and the billing system, where data loss or delay has significant business impact. Some vendors have relationships with middleware vendors for partially configured interfaces.

As valuable as middleware is in providing robust transaction management capabilities, it is also yet another system that requires administration, support and maintenance. It should not necessarily be the choice for infrequent or optional interfaces.

Real-time, high-volume usage

Transaction services over "always-on" interfaces are likely to generate an orders of magnitude increase in the volume of transaction detail records. Many of the currently proposed transaction services include authorization and payment in real time in ways that may make the telecommunications or application provider become a billing services provider—or may make billing service providers from the financial services arena into application providers and, potentially, into communications providers—at least at the WLAN level. This means that usage reporting, rating and processing must become more efficient, responsive and reliable than in the past.

Summary

In this chapter, we have looked at interfaces to a potential billing application, the characteristics and appropriate application of various interface technologies, as well as the derivation of interface and interface management requirements. These requirements support the requirements checklist we review in Chapter 28 and drive the language and conditions of procurement documents, decisions and contracts as we illustrate in Chapters 31 through 36.

Chapter

26

Billing-Related Standards

Billing platforms must "talk" to other systems and technologies, oftentimes controlled by other companies or individuals. Standards play a significant role in making that feasible and manageable. Telecommunications and convergence billing systems are impacted by many standards and standards organizations and activities.

Standards Organizations

There are a number of organizations developing standards that may impact a billing environment. These range widely from coalitions of vendors through consortiums of communications enterprises to governmental and quasi-governmental organizations. Standards may be driven by geography, by services technology, by systems technology, by service category or simply by industry economics or politics. In some areas, standards are driven by regulatory requirements or guidelines.

Telecommunications billing systems can be affected by telecommunications industry standards, by "information industry" standards and by financial and accounting standards, at a minimum. As technology and business conditions change, so do applicable standards. Before finalizing any standards-oriented requirements, it is wise to confirm the current status of relevant standards. To assist you in doing so, the Web site of each mentioned standards organization is included along with a description of the organization's mission and activities.

Telecommunications standards organizations

There are numerous telecommunications standards organizations, as well as many committees and forums within the major standards organizations. Which standards apply may depend on geography and on type of service or underlying technology.

Alliance for Telecommunications Industry Solutions (ATIS). ATIS (http://www.atis.org/) is a North American standards body that is leading the development of telecommunications standards, operating procedures and guidelines through its sponsored committees and forums. ATIS member companies are North American and World Zone 1 Caribbean providers of telecommunications services and include telecommunications service providers, competitive local carriers, cellular carriers, inter-exchange companies, local exchange companies, manufacturers, software developers, resellers, enhanced service providers, and providers of operations support.

ATIS sponsors a number of subgroups that are very important to billing for North American carriers. These groups include the Order and Billing Forum (OBF), the International Forum for ANSI-41 Standards Technology (IFAST) and the T1 working group.

Ordering and Billing Forum (OBF). The OBF is arguably the most important standards body for billing in North America. The OBF addresses and establishes standards that affect ordering, billing, provisioning and exchange of information about access services, other connectivity and related matters.

The OBF has six standing committees. The Billing Committee addresses access billing related issues and maintains the Multiple Exchange Carrier Access Billing (MECAB) document, Small Exchange Carrier Access Billing (SECAB) document, and the CABS Auxiliary Report Specifications (CARS) document. The Message Processing Committee addresses nonaccess issues relative to message processing and maintains the Exchange Message Interface (EMI) document. The Subscription Committee provides a forum for customers and providers to develop common definitions and recommendations for resolution of national subscription issues and maintains the Equal Access Subscription, Customer Account Record Exchange (CARE) document. Other committees focus on ordering standards and numbering plan management.

The Wireless Workshop is an ATIS sponsored workshop, which provides an open forum for the identification, presentation, discussion, and resolution of issues impacting wireless service providers. These issues include, but are not limited to, the functional wireless areas of provisioning, message processing, exchange of roamer usage and billing. In addition, the Wireless Workshop maintains the Wireless Inter-Carrier Communications Interface Specification for Local Number Portability.

International Forum for ANSI-41 Standards Technology (IFAST). IFAST concerns itself with the ANSI-41 family of standards for international wireless roaming.

T1. T1 (http://www.t1.org/) focuses on those functions and characteristics associated with the interconnection and interoperability of telecommunications networks at interfaces with end-user systems, carriers and information and enhanced service providers. While these deal primarily with network technology, they can have impact on information available for billing.

Telcordia. Telcordia (http://www.telcordia.com/), formerly Bellcore, provides administration of standards and numbering plans at the direction of ATIS and other organizations. It also administers common language standards for designating locations, circuits, facilities and equipment.

International Telecommunications Union (ITU). The ITU (http://www.itu.int/) is a United Nations-sponsored organization founded on the principle of cooperation between governments and the private sector. With a membership encompassing telecommunications policy makers and regulators, network operators, equipment manufacturers, hardware and software developers, regional standards-making organizations, and financing institutions, ITU's activities, policies, and strategic direction are determined and shaped by the industry it serves. In the telecommunications standardization division (ITU-T), experts prepare the technical specifications for telecommunications systems, networks and services, including their operation, performance and maintenance. Their work also covers the tariff principles and accounting methods used to provide international service.

Telecommunications Standards Advisory Council of Canada (TSACC). TSACC (http://www.tsacc.ic.gc.ca/) is an industry and public sector organization whose purpose is to provide strategic advice on telecommunications and information technology standards development, both nationally and internationally. TSACC represents Canada as a member of the Global Standards Collaboration forum.

European Telecommunications Standards Institute (ETSI). Telecommunications billing standards in Europe are established by ETSI (http://www.etsi.org/). ETSI pursues the objective of developing globally applicable deliverables meeting the needs of the telecommunications/electronic communication community while still fulfilling its duty to support the European Union (EU) and the European Fair Trade Association (EFTA) regulation and initiatives. The objective of the Institute is to produce and maintain widely implemented technical standards and other deliverables as required by its members.

Global System for Mobile Communications (GSM) Association. GSM (http://www.gsmworld.com/) is the most widely implemented and deployed set of wireless telecommunications standards worldwide. The GSM Association is responsible for the deployment and evolution of standards for the GSM family of technologies, including the GSM standards for voice grade digital dial services, general packet radio service (GPRS), enhanced data rates for GSM evolution (EDGE)—sometimes called "2.5G" for high-bandwidth dial service and third-generation (3G) "always-on" data transaction services for digital wireless communications. GSMA issues standards for all aspects of these systems, including the content and format of billing and accounting information exchange and messaging.

IPDR.org. IPDR.org (http://www.ipdr.org/) has as its mission to define the essential elements of data exchange between network elements, operation support systems and business support systems. It will provide the foundation for open, carrier-grade IP-based support systems that enable next generation service providers to operate efficiently and cost-effectively.

Computing and information industry standards organizations

Since computing and information industry standards requirements are incorporated in interface, architecture and platform standards requirements, they are not included separately under "Standards Requirements." However, companies that are writing system requirements for vendor systems—or internally developed systems, for that matter—should be aware of the following standards organizations and be tracking the status and acceptance of relevant standards.

International Organization for Standardization (ISO). The ISO (http://www.iso.ch/) is a worldwide federation of national standards bodies. The mission of ISO is to promote the development of standardization and related activities in the world with a view to facilitating the international exchange of goods and services and to developing cooperation in the spheres of intellectual, scientific, technological and economic activity. ISO's work results in international agreements that are published as international standards.

The ISO 9000 series of standards provide a framework for quality management. ISO 9001 certification of a vendor's processes indicates a commitment to quality management, as well as some stability in their development and engineering processes.

Object Management Group (OMG). The OMG (http://www.omg.org/) is best known for the CORBA middleware platform, which includes the interface definition language OMG IDL and protocol IIOP. OMG is an open membership, not-for-profit consortium that produces and maintains computer industry specifications for interoperable enterprise applications. OMG's flagship specification is the multiplatform model driven architecture (MDA), recently underway but already well known in the industry. It is based on the modeling specifications the MOF, the UML, XMI, and CWM. The object management architecture (OMA) defines standard services that will carry over into MDA work. OMG Task Forces standardize domain facilities in industries such as healthcare, manufacturing, telecommunications, and others.

The communications industry has been very active in OMG and there are several telecommunications oriented subgroups including a telecommunications domain task force. However, work is just starting on many of the domain-specific standards. OMG also has a very active Internet special interest group.

American National Standards Institute (ANSI). ANSI (http://www.ansi.org/) is a private, nonprofit organization that administers and coordinates the U.S. voluntary standardization and conformity assessment system. The Institute promotes and facilitates voluntary consensus standards and conformity assessment systems, safeguarding their integrity. ANSI represents the United States in many international standards organizations.

Call Detail and Accounting Standards

For conventional (public switched telephone network or PSTN) telecommunications applications in North America, the most relevant standard may be Telcordia's "billing automated message accounting format" (BAF) standard GR-1100-CORE. This defines the formats for usage records used by most exchange switches in the North American market, as well as the exchange formats used by many telcos. If your business is PSTN switch based or if you are leasing switching capacity from conventional telcos in North America, the billing environment will have to handle BAF records.

For wireless applications, billing information is more complex. In much of the world, mobile network elements conform to the transferred account procedure (TAP) standards for roaming billing information (currently TAP3.4) set by the Global System for Mobile Communications (GSM) Association. A good background article on the TAP standards is available at http://www.gsmworld.com/using/billing/potential.shtml.

ETSI is the sponsor of the telecommunications and Internet protocol harmonization over networks (TIPHON), which is working to bring together ETSI and ITU telephony standards such as H.323—including billing and accounting messaging standards—with IP-based technology and standards. H.323 was developed for packet-switched networks and has been widely adopted for VoIP (Voice over Internet Protocol) applications.

There are several competing standards for newer transaction-based IP and VoIP billing. United States carriers are supporting the network data management-usage specification (NDM-U) developed by the Internet Protocol Detail Record Organization (IPDR.org). IPDR is working with GSM and the ITU-T to gain wider acceptance of this protocol.

Regulatory Requirements

In Section I, we looked at how to determine regulatory jurisdiction. Most U.S. telecommunications companies are subject to oversight by both the Federal Communications Commission and their state regulatory agency—usually an agency with regulatory authority over all public utilities within the state. Each country has its own regulatory structure and, perhaps, regulatory international treaty organizations.

Your regulatory organization may wish to identify specific applicable regulations that the billing platform must support. However, it is common to simply require that certain system functionality be compliant with the regulatory requirements of the jurisdictions appropriate to your business.

Some areas for which your company should specify compliance include bill formatting and content, charge calculation (e.g., what happens to partial cents?), credit posting responsiveness and minimum payment interval.

Billing Output and Payment Processing Standards

There are competing electronic bill presentment and payment (EBPP) standards. Open financial exchange (OFX) is the financial transaction standard originated by CheckFree Corp., Microsoft Corp., and Intuit Inc. and used currently by the financial industry to conduct electronic transactions. Interactive financial exchange (IFX) is the newest extensible markup language-based (XML) incarnation for EBPP and is being championed by the IFX Forum, which is led by major EBPP software makers.

EBPP standards organizations

The World Wide Web Consortium or W3C (http://www.w3.org/) develops interoperable technologies (specifications, guidelines, software and tools) for the Web. XML is a project of W3C and the development of the specification is being supervised by their XML Working Group. A special interest group of contributors and experts from various fields contribute comments and reviews. XML is a public format. The v1.0 specification was accepted by the W3C as a recommendation on February 10, 1998. IFX development and management is part of this process.

OFX is essentially a proprietary specification that has been made available to encourage other vendors to use it. However, there is no public contribution or comment process as there is in true standards activities. CheckFree, Intuit, and Microsoft created open financial exchange in early 1997 and continue to maintain the "standard" and make it generally available. OFX is a specification for the electronic exchange of financial data between financial institutions, business and consumers via the Internet. It supports a wide range of financial activities including consumer and small business banking, consumer and small business bill payment, bill presentment, and investments tracking, including stocks, bonds, mutual funds and 401(k) account details. Since 2000, with the 2.0 specification, OFX has become XML 1.0 compliant.

FtA electronic billing requirements

Since FtA intends to maintain control of bill detail delivery and it wishes to be IFX compliant, its billing output standards requirement reads as follows: "FtA requires compliance with the current approved version of interactive financial exchange (IFX) for bill detail delivery. In addition, for bill detail delivery, FtA requires a commitment to deliver and install appropriate upgrades to IFX version 1.2 compliant EBPP support within 6 months of version 1.2 adoption."

For electronic payment, FtA wishes to have flexibility in contracting with payment processing vendors and in receiving and processing electronic payments directly. Therefore, their requirements for electronic payment standards read as follows: "FtA requires compliance with the most current version of OFX (open financial exchange) and with the most current version of IFX (interactive financial exchange) for electronic bill payment. In addition, FtA requires a commitment to deliver and install electronic bill payment support compliant with the next approved version of each standard."

Summary

In this chapter, we reviewed many of the standards organizations and specific standards that can impact directly on billing environments. More and more billing products are supporting industry, national and international standards. Insisting on standards compliance can ease integration overhead and position your company for greater flexibility. Standards should be invoked when they are appropriate to your business, not just because they are available. The authors have seen systems flounder because they specified "bleeding edge" standards that were not necessary to meet the business needs.

Chapter

27

Unique Billing Considerations

If your company provides or plans to provide and charge for services other than conventional wireline telecommunications services, you will need a billing platform with some flexibility in defining both transaction data collection ("call detail") and transaction charges. This chapter reviews some of the types of services being offered or proposed today. However, if your company intends to be a "leading-edge" services provider, this list may be out of date by the time this book rolls off the presses. If you do not want your billing platform to be obsolete for your purposes just as quickly, focus on flexibility and configurability of system functions and interfaces. It is also wise to examine closely the vendor's history of commitment to supporting new technologies, standards and interfaces.

Wireless

The growth of wireless communications is an obvious fact of life at the beginning of the twenty-first century. Most of this is currently conventional voice communications, but new "third-generation" or 3G technology and applications are now being deployed. At the same time, commercial "wireless LANs" are being deployed to create local wireless "hot spots" primarily for Internet access.

Taxation for wireless services is currently based on the service address of the wireless customer as the origination location, regardless of the actual location of call origination. This leads to interesting situations such as a wireless subscriber from New York calling home from California and the call being treated as an intrastate (local) call for taxation purposes. However, the FCC has ordered that the physical location of a subscriber be identifiable by the end of 2004 for emergency services (911) purposes and newer wireless services actually involve locating the subscriber. This may result in a scramble to rewrite cellular tax jurisdiction directives.

2G and earlier

Conventional voice-grade wireless service billing has been known to cause headaches for conventional billing systems. Early wireless pricing plans included a three-charge component: *airtime*—the length of time a subscriber was connected to an incoming or outgoing call; *roaming*—usually a flat fee for connecting from a wireless network other than the subscriber's home network; and "long distance"—the regular LD charges from the subscriber's chosen long distance carrier. Some wireless providers also charge a *connect fee*—a flat charge per call. Some small wireless companies continue to use this type of structure, although it is vanishing quickly. Older wireless rating and billing platforms, however, maintain support for this type of structure. If you must support this type of rating and billing, one strategy is to use a specialized platform to rate the call and pass the rated record to the billing system for inclusion on the customer's bill.

With the advent of nationwide and even international wireless carriers, many networks have chosen to implement a "total minutes" rate structure for all calls originated by end users. Terminating "airtime" may be covered by the basic service rate—or so the CFO hopes—or, as in a number of plans, counted against the total minute allowance.

Roaming customers continue to generate call detail through another carrier or via a roaming clearinghouse. Most wireless billing packages will include features to properly route, accept, process and manage roaming usage, but it is a good idea to specify a requirement for this if your business will handle or allow your subscribers to generate roaming calls.

Currently, taxing jurisdiction is based upon the "home" location of the originating and terminating numbers. However, with subscriber location features being deployed widely, tax jurisdiction determination may shift to physical location of the terminal. International roaming already includes a component of tax "settlements."

3G

3G (third-generation) wireless services have several significant differences from earlier wireless services. First, 3G subscriber terminals are "always on"—that is, maintaining connectivity with the network. This makes simple connection time charges nonsensical—or astronomical! Second, 3G applications are just that—applications that are "content heavy." The value equation in 3G services resides in the content, not the connection. Third, 3G services have significant bandwidth available when required. That bandwidth can support several application functions operating actually or apparently simultaneously. For instance, a subscriber could be using an interactive locator function in conjunction with downloadable "tour guide" information related to the locator information in conjunction with a restaurant reservation application. At the moment, 3G billing algorithms are collecting a transaction record per data packet—making the volumes of usage records in this environment potentially very large.

WLAN

Commercial wireless local area networks (WLANs) are seen as a possible competitor to 3G services for wireless data access to the Internet. However, this is a technology better suited to mobility within a fixed location such as Internet cafes (or McDonald's!) or malls, airports and conference centers. It is unclear yet exactly how such services will be billed. It seems likely that many installations will be prepay only—which requires local authorization and rating, but not bill rendering, delivery or payment management—or credit card usage billing. Others will provide the service as a courtesy. In any case, the uncertainty of the billing context places pay-for-content applications in the position of needing subscriber billing capabilities—which may be yet another application service.

Internet Services

Internet services include the relatively mature services of the Internet market—Internet access, email, Web hosting—as well as the newer voice over IP (VoIP) and application services.

Internet connection

An Internet service provider (ISP) generally provides an Internet connection to the end customer, as well as email service—either on-line or delivered to the end customer's terminating device. The customer is billed based on connect time or a flat fee for connectivity. Currently, only a very few services charge for messages delivered or sent.

Internet host

An Internet presence provider (IPP) or hosting service generally provides Internet accessible Web page hosting capacity—priced either at a recurring flat rate for allocated space or for space dynamically allocated over some base capacity. Such hosts usually provide domain name registration management—which means they act as an agent for their customers to one or more of the domain registrars. They may also provide a variety of software packages, services or plug-ins for use with Web pages or to administer Web sites for a recurring or (rarely) usage-based fee. IPPs often charge the Web site operator for the volume traffic to the site.

Charges for dynamic allocation of space or for usage of software services require a linkage between billing and the server. This is not currently a mature discipline for usage-based billing. Billing for site-specific network traffic requires the capability to capture information about that traffic and interface that information to the billing platform.

The domain name registration agency requires synchronization and reconciliation between payments to the registrar and payments by the subscriber.

Application services

Application service providers (ASPs) are a very new phenomenon. While vendors of tools for ASPs are starting to provide "hooks" for transaction accounting, this is a very immature billing discipline. Most billing features for this type of service will be very leading edge. This means that billing implementations that include support for ASPs should allow both time and expert resources for "getting the kinks out" of the new application.

Management services

Management service providers (MSPs) are a fledgling industry that is positioned to grow quickly. MSPs provide network and server management using remote tools. This allows consolidation of expertise and tools for full-time, highly professional management services. Most MSPs bill based on a negotiated contract, but there is evidence that some MSPs are looking closely at billing models that would contain some usage or professional time charges.

Broadband

Broadband billing has matured in the past several years. Most major billing platforms include adequate support for typical broadband offerings.

Point-to-point

Point-to-point broadband offerings are quite mature and well supported in most telecommunications billing packages.

On demand/configurable

On-demand and configurable broadband services are less mature for usage-based billing, especially when usage is based on packet volume, not connect time.

Broadcast/paging

Simple numeric paging is quickly being replaced by text messaging in many guises. Numeric paging services generally charge a recurring service fee based on the coverage area plus a flat usage charge per message sent to the pager account. Text messaging services also use a recurring charge based on service area, but charge usage based on the number of message characters sent during the billing cycle. Roaming charges may also be assessed, similar to other wireless services.

Cable

Cable companies have moved rapidly from mostly one-way entertainment services with limited on-demand services and charges to interactive Internet access services. The current cable market is mostly consumer and small business and the prevailing charge basis is a flat recurring charge established in relation to the rest of the cable account. If cable providers move into the business and "heavy user" markets, they may be motivated to charge on a usage basis, which would require significant overhaul of billing systems in the cable market or cable interfaces to telecommunications billing systems that already support both packet volume and connect time.

Multiservice Bill Combining

Most modern telecommunications billing systems support combining multiple services on a single bill. However, if you intend to use service-specific billing systems (such as an off-the-shelf wireless billing system) for one service category and if your company wants to offer consolidated billing for all services, then you must specify that the wireless system must provide rated, formatted records to the master billing system and the master billing system must be able to accept and combine other billing information for a complete combined customer bill.

Summary

There may be many unique considerations in developing billing system requirements for your business, not all of them driven by new technology. Sometimes the most complex platforms result from the need to integrate older technologies and rate structures with new services and ways of doing business. If your business has any of the unique considerations identified in this chapter, those considerations should be clearly identified in your RFI, RFP, and vendor contract or contracts.

Chapter

28

Managing and Prioritizing Requirements

In Chapters 21 through 27, we looked at developing system requirements based on information about the business, current account and billing data and business near-term and strategic directions. In this chapter, we consider the management of these requirements and priorities, focusing on the time frame in which a request for information (RFI) is to be published and responses to that RFI are to be assessed.

Appendix B contains a checklist of items for which you should have at least considered developing requirements and evaluation criteria. Not surprisingly, the major categories of requirements correspond to the topics addressed in Chapters 22 through 26, with amplification of some critical topics in Chapters 27 and 28. You may decide that some items have little or no impact for your business. In those instances, you can provide very generic requirements or rely on vendor recommendations. However, you should expect a vendor proposal to address all of the checklist items.

The accompanying CD includes an electronic version of the master checklist outlined in Appendix Bthat contains links to detailed checklists for each line on the master checklist. Note that the master checklist form can be used to evaluate the status of internal requirements and to assess the completeness of vendor submissions.

Checklist Form

The checklist referenced in Appendix B and contained on the CD is a Microsoft Word® form. Of course, it can easily be translated into a Web form or a database input form. However, this is intended as a project control document and probably should not have multiple points of update. It also does not represent very complicated information, so may not justify its own database structure.

Requirements

The first column of the form identifies areas of requirements summarization. These are divided into requirements for the billing system itself, requirements related to customer accounts, requirements related to products and services, requirements related to systems operations and use, requirements related to systems interfaces and requirements related to standards.

% Or ✔

The second column of the form indicates a percent completion of requirements and/or a simple check off that requirements are complete. Percent complete can be used to indicate either the status of requirements preparation or the degree of completeness of information on which requirements are based. Each "cell" in this column contains a text input field and a check box. The check box can be used at a specific milestone in the project to indicate that current requirements are adequate to proceed to the next activity, even though those requirements are not 100 percent complete.

From a project management and tracking perspective, it is useful to identify in comments when the partially complete requirements will be finished and/or the event on which completion is dependent.

Importance

The third column of the requirements checklist is an assessment of the importance or priority of this requirement in evaluating the overall specifications for a solution. Each cell is formatted with a dropdown menu containing the values "high," "medium," and "low." This is often sufficient differentiation of priority. However, if greater granularity is required, the drop down menus can easily be reprogrammed.

In Chapter 14, we discussed some of the ways to determine what is most important to your company from the perspective of business vision and strategy. Whether your company is aiming to be a low-overhead, low-priced commodity provider or a high–value-added, higher-priced service provider will impact the importance of certain requirements areas, for instance.

Other aspects of your business, such as current and anticipated data volumes, can also impact the importance of those requirements. A company with a very large existing or anticipated customer base will not even want to consider smaller platforms that don't completely scale up to the required volumes. A small company may find it appropriate to acquire a platform that is a bit too large if the extra cost is not excessive and the system is a perfect match for all other requirements.

Owner

Ownership of a specific requirements area should be assigned either to an individual member of the billing systems acquisition team or to a specific organization or function. This individual or the designated organization representative is responsible for determining the status of requirements in this area, providing supporting information for the status and, with the concurrence of the team, assigning an importance rating for these requirements. We strongly recommend that an individual—versus an organization or function—be the designee in the interest of accountability and "getting the job done."

The appropriate owner for a specific requirements area will vary depending on your organization and the actual members of the billing requirements management team.

Comments

Comments should be brief, in support of other information on the line. They may be keys to supporting documents and materials. As noted in the discussion of completion status, information relevant to the expected completion or completion dependencies may be entered in this field.

From the Ashes Master Checklist

Figure 28-1 shows the master requirements checklist developed by the From the Ashes billing system planning team just before it prepares a RFI for a new billing system. The From the Ashes team has decided to use the % field to indicate the completeness and level of confidence of supporting information. A ✔ indicates that the requirements are fully documented, based on whatever information is available.

The From the Ashes (FtA) team has decided to use the simple high, medium, and low priority indicators. It is using the initials of the responsible managers to indicate requirements ownership. The FtA chief information officer (CIO) is "kme;" the director of billing operations is "met;" the vice-president (VP) of marketing is "ejv;" and the chief technology officer (CTO) is "jmh." (You may notice some familiar initials here; these roles correspond to positions the authors have held in "real life.") In general, the CIO owns requirements that relate to the system platform or systems operations. The director of billing operations owns requirements that most heavily impact on billing operations. The VP of Marketing owns requirements that are driven by market, customer, or sales needs. The CTO owns all requirements related to overall systems architecture and corporate data standards.

Customer level requirements

Customer level requirements are related to the acquisition, presentation, management and storage of information about customers and accounts. In Section II, we looked at the process of collecting information about current customers and account structures, as well as identifying future needs defined by corporate strategy. In Chapter 22, we developed customer level requirements from the collected data.

Requirements	% or ✔	Importance	Owner	Comments
Customer Level				
Fields/formats	50%	Low	jmh	
Market segments	85% ✔	Medium	ejv	
Media options	90% ✔	High	ejv	Value-added
Billing options	70% ✔	High	met	Value-added
Geographic markets	75% ✔	Medium	ejv	CA, AZ, NV, ?
Account structures	80% ✔	Medium	ejv	Value-added
Product/Service Level				
Service categories	100% ✔	Medium	ejv	Need APP, MSP
Fields/formats	50%	Low	jmh	
Billing basis	70% ✔	Medium	ejv	Need T&M, expense
Volatility	85% ✔	Low	ejv	
Operational & Systems Level				
Functionality	100% ✔	Low	met	
Ease of use	95% ✔	Low	met	
Ease of management	90% ✔	Medium	met	
Exception handling	85% ✔	High	ejv	Need processes
Auditability	90% ✔	High	jmh	
Sizing/volumes	80% ✔	Low	kme	Within all vendor volumes
Billing cycles	100% ✔	Medium	met	Value-added
Platform	25% ✔	Low	kme	
Interface Level				
Types of interfaces	80% ✔	Medium	jmh	
Specifications/tools	65% ✔	Low	jmh	Coordinate with other system teams
Quality tools	75% ✔	High	jmh	Coordinate with other system teams
Capacity	85% ✔	Low	kme	
Standards Level				
Call detail standards	85% ✔	Low	jmh	
Regulatory requirements	95% ✔	Medium	met	
Billing output standards	100% ✔	Medium	met	
Payment processing stds	80% ✔	Medium	met	

Figure 28-1 FtA master requirements checklist.

Fields/formats. Specific requirements can be developed for the fields in a customer record and the formatting of those fields by consolidating all the collected data and determining the mapping of fields from one system to another.

% or ✔. The FtA team has identified all customer and account fields and formats in the broadband billing systems and in the ISP billing system and has mapped those fields and formats to common requirements. However, it does not have complete information on the contents of the old CLEC customer and account records. Since these may prove to be the most complex records, the team has determined that its requirements are only 50 percent complete. The team feels that more information and requirements work will be needed to adequately assess potential solutions.

Importance. FtA feels that the importance of exactly accommodating field and format requirements for customer information is low. There are minimal regulatory strictures on this information and most systems will accommodate fields and formats that will map to the required fields and formats adequately.

Owner. The FtA owner of customer level fields and formats requirements is the CTO.

Market segments. In Chapter 17, we looked at some of the ways a company may choose to segment its target markets. Often these segments have different pricing rules, credit and treatment rules and customer contact rules that are reflected in the billing system. In Chapters 15 and 16, we surveyed collecting and consolidating segmentation information with relationship to products and services.

% or ✔. The FtA team has identified how the new From the Ashes company will segment its markets. They have also gathered customer-specific segmentation information on the existing customers of the two broadband companies, as well as customer profile-based segmentation information on the ISP. The FtA executive council has agreed that all CLEC customers will be slotted into one of the identified market segments during conversion activities. This means that, while data gathering and requirements development in this area are not complete, the requirements are adequate at least through the RFP stage of the acquisition process.

Importance. FtA's marketing strategists believe that the ability to segment markets and "bulk customize" customer interactions—including pricing and billing—is of medium importance to the long-term strategy of being a value-added provider for small and mid-size businesses, while keeping overhead to a minimum.

Owner. The FtA owner of market segments requirements is the VP of Marketing.

Media options. Media options requirements include all desired billing media, specific formats for that media and whether bills are developed and delivered in a single medium or in multiple media for a single account or customer.

% or ✓. The FtA team has firm media information for both broadband companies and for the ISP. However, verbal information from the CLEC indicates that the majority of their billing is conventional paper bills mailed to the customer, with optional CD-based usage information supplied to their largest customers.

Importance. Flexibility in billing media and presentation is very important to the high-value-added strategy of FtA. Not only does FtA plan to offer billing media options to its target business customers, but it also plans to offer some bill analysis capabilities in a number of media by subscription and on demand. Therefore, meeting FtA's media requirements is of high importance.

Owner. The FtA owner of media requirements is the VP of marketing.

Comments. The comments field reflects the rationale for ranking this area of high importance.

Billing options. Billing options cover a range of items including such items as combined billing of services/products, payment options, rounding conventions, the method and extent of treatment options, marketing/sales messages as part of the bill and use of the billing system as a note repository, which can improve the quality of customer interactions.

% or ✓. The FtA team believes it has sufficient information on current billing options to issue an RFI and evaluate responses. It has analyzed current billing information for both broadband carriers. It has an audited statement of billing options as part of the contract documentation for the ISP. For information about billing options in the CLEC, it is relying heavily on tariff and published information.

Importance. Billing options are critical both to meeting regulatory requirements and to the value-added market strategy. Some billing options may also increase available financial controls and checkpoints, which is critical to the company's operational and financial strategy. Therefore, meeting requirements for billing options has been given a high importance.

Owner. The FtA owner of billing options requirements is the director of billing operations.

Comments. The comments field reflects the rationale for ranking this area of high importance.

Geographic markets. Geographic markets defines the service areas and regulatory and tax jurisdictions in which the company is doing or plans to do business.

% or ✔. The FtA team has identified the service areas and jurisdictions for both broadband companies and the CLEC. There is some uncertainty about the extent of the ISP's customer base, although the local access areas are defined.

Importance. FtA plans to offer services in the continental United States only for the next five years. Therefore, any solution must be appropriate to the U.S. market. It must support the local, state and federal regulatory and tax jurisdictions—critical to regulatory approval of acquisition of the CLEC book of business. This has resulted in support for the geographic market requirements being rated of medium importance.

Owner. The FtA owner of geographic markets requirements is the VP of marketing.

Comments. The comments show the known geographic markets and indicate an area of uncertainty.

Account structures. In Chapter 17, we looked at a variety of account structure options. These included relationships of customer master accounts and sub-accounts, relationships of various customer locations and the relationship of accounts and sub-accounts to media options. Typical business accounts are much larger and more complex than the illustrative examples.

% or ✔. The FtA team has reviewed actual account structure information from each of the broadband companies. It has also worked with marketing strategists to define "typical," "complex," and "simple" examples of target accounts per FtA's strategy. It has also used its industry knowledge and some negotiation information provided by the CLEC to hypothesize typical account structures within the CLEC. This information is adequate for the RFI process, but will require refinement before entering into a vendor contract.

Importance. FtA believes that flexible account structures are a significant aspect of the value-added equation for its customers and has rated this area of medium importance.

Owner. The FtA owner of account structures requirements is the VP of marketing.

Comments. The comments field reflects the "value-added" rationale for ranking this area of medium importance.

Product/service level requirements

In Chapter 23, we looked at the development of product and service level business requirements for a new billing platform based on data collection, as outlined in Chapters 15 and 16.

Service categories. Different segments of the communications industry have different billing requirements. Often off-the-shelf billing solutions are designed to support the specific requirements of those industry segments, even though the functions and processes for different offerings are very similar. For instance, there are billing applications designed specifically for the unique rating requirements of the mobile communications industry segment and others specifically tailored to ISPs. We discussed some of these variations in Chapter 23.

Sometimes the same generic software product is designed to support a large range of configurable products and services, but the vendor only supplies preconfigured templates for a specific type of business such as a CLEC.

% or ✓. The FtA team believes that it has identified all relevant service categories for the new billing system. This is based upon actual service records from the broadband companies, due diligence statements from the ISP, advertised service offerings from the CLEC and input from FtA executives and marketing strategists.

Importance. FtA has determined that the ability to accommodate all relevant service categories with a minimum of customization for each service is of Moderate importance. Since FtA will initially be offering conventional telecommunications services and very simple ISP offerings, most telecommunications billing systems will accommodate the initial offerings. However, FtA plans to move into nontraditional value-added offerings, ranging from Web page design and configuration to wireless data interfaces in airports and hotels. Those services are not commonly found in current systems, so customization needs are to be expected.

Owner. The FtA owner of service categories requirements is the VP of marketing.

Comments. The comments field reflects FtA's intention to offer nontraditional services.

Fields/formats. Fields and formats requirements identify the data fields, formats and filters the billing environment must accommodate in order to support current and future product and service offerings.

% or ✓. The FtA team has identified the actual fields and formats used in the broadband billing systems. It has also obtained product and service related fields and formats as part of the due diligence documentation from the ISP. At this point, it has no specific information on the product and service data fields, formats or structures in the CLEC billing environment nor has it been able to work with FtA executive management and product strategists to determine requirements for future products and services.

The team feels that further information will needed to adequately assess RFI responses and to begin preparation of an RFP.

Importance. FtA believes that most telecommunications billing systems will accommodate its data field requirements. There is little customer or regulatory impact if some modifications to field structure and format are required, so fields and formats requirements are rated as low importance.

Owner. The FtA owner of fields and formats requirements is the CTO.

Billing basis. In Chapters 5 and 6, we looked at billing basis for usage-based, one-time and recurring charges. In Chapters 15 and 16, we surveyed collecting and consolidating product and service billing basis information.

% or ✔. The FtA team has determined that it has good requirements for services currently offered by both broadband carriers and the ISP. It has been able to infer billing basis for the CLEC book of business by reviewing published services, but believes that it may not have identified all configurations. In addition, the FtA marketing strategists are still refining their view of future services and their billing basis. The strategists have made clear that any billing system will have to gracefully handle a variety of time, materials and expense billing transactions.

Importance. FtA has determined that support for a variety of billing bases is of medium importance.

Owner. The FtA owner of billing basis requirements is the VP of marketing.

Comments. The comments field reflects the interest in technical services billing on a time, materials and expenses basis.

Volatility. Volatility requirements relate to new services creation or existing services changes or phasing out. This volatility can be in pricing, conditions, discounts, promotions, billing basis and market segment, as well as of new or phased out services. These requirements relate to support for creating or changing products, services and "assemblies" of products and services.

% or ✔. The FtA team has worked with marketing strategists to determine likely product and service volatility over the next three years. It does have volatility history for the broadband products and services and the ISP services, but no volatility history for the CLEC.

Importance. Since volatility requirements generally relate to how easy it is to make changes and FtA plans to locate its service configuration personnel in a low-labor-cost area, FtA has determined that volatility support has a low priority.

Owner. The FtA owner of volatility requirements is the VP of marketing. (In an existing company with some volatility history, ownership of these requirements might be assigned to the director of billing operations.)

Operational and systems level requirements

In Chapter 24, we surveyed the development of operational and systems level requirements. Operational and system level requirements include some requirements related to the business and operational information we introduced in Chapter 1. These requirements also include some requirements related to the technical capabilities and capacity of the new system. They also include requirements related to functionality in support of both system operation and system users.

Functionality. Functionality requirements identify what the system is expected to do. While customer level and product/service level requirements primarily address what information the billing environment must handle and the relationship of that information, functionality requirements identify what actions, and processes must be part of the billing environment. Does the system calculate charges? Does it apply taxes? Does it identify potentially fraudulent situations? Does it provide treatment options? Does it manage printing and mailing and, if so, how or to what extent?

Much of this may seem intuitively obvious. However, many billing products are modular and only provide the functions specified. Some vendors specialize in only one function or in functional add-ons to popular platforms. This makes it very important to clearly identify all required functions.

% or ✔. The FtA team believes it has completely identified all functional requirements for its new platform.

Importance. FtA functional requirements are believed to be typical of industry requirements and, therefore, are not anticipated to be a critical factor in choice of vendor. Functionality is rated as low in importance.

Owner. The FtA owner of functionality requirements is the director of billing operations.

Ease of use. Requirements for ease of use include how user-friendly the system is for establishing accounts, entering and retrieving information, as well as what is automated for the user. This can include average time to accomplish certain tasks, the need for supporting reference documents, the length of training to achieve mastery of each interface, as well as specific user support features.

% or ✔. The FtA team has compiled desired transaction maximum timing, as well as defining user work "positions"—that is, structuring of the user interface. The team believes it has complete requirements in this area.

Importance. FtA has determined that its billing operations center will be located in an area of Arizona with high unemployment levels and low wage rates. This makes user efficiency and work satisfaction of low strategic importance. Thus, ease of use is rated as low importance.

Owner. The FtA owner of ease of use requirements is the director of billing operations.

Ease of management. Requirements for ease of management include such features as management reports and online status, work queue assignment and management and work efficiency and effectiveness tracking. There is significant overlap in the information infrastructure for ease of management and auditability. However, this requirements area deals with requirements for that information to be available and accessible in ways that are appropriate to work management and supervisory concerns.

% or ✔. The FtA team believes it has identified most of the requirements for management support in the new company—based on industry knowledge and strategic business plans.

Importance. FtA, as a post-meltdown company, believes that timely management knowledge and controls are important to the success of the new venture. Therefore, it has determined that meeting requirements for ease of management is of medium importance.

Owner. The FtA owner of ease of management requirements is the director of billing operations.

Exception handling. Exception handling is the ability of the system to identify errors and problems and to support the resolution of those errors and problems.

% or ✔. The FtA team believes it has a good level of understanding of exception handling requirements. Some exception handling requirements will be generated by the choice of architecture, so these requirements are not complete, although they are adequate for evaluation of responses to the RFI.

Importance. FtA places high priority on exception handling requirements, since these are critical to cash flow management and to operational expense.

Owner. The FtA owner of exception handling requirements is the director of marketing.

Comments. The comments field reflects the need to fully identify process requirements associated with exception handling.

Auditability. Auditability requirements relate to capabilities that facilitate tracing transactions through the system, including any transactions against customer account records, pricing structures or other information related to financial calculations, reporting and postings. Auditability requirements may include automated support for database checkpoints and for data and transaction history retention. Auditability requirements may also address the information infrastructure required for ease of management.

% or ✔. The FtA team has developed stringent auditability requirements, based on perceived business needs for financial and operational controls and on regulatory requirements for audit support and retention. As the team obtains access to the CLEC billing system, there may be some need to modify these requirements—especially in relation to legacy data—but the auditability requirements are adequate through the development of the RFP.

Importance. FtA has determined that auditability requirements are of high importance for strategic and regulatory reasons.

Owner. The FtA owner of auditability requirements is the CTO.

Sizing/volumes. FtA has been able to access the billing systems used by both broadband companies and to perform a count of customer accounts and services. The final contract for the purchase of assets and customers from the ISP includes audited counts of customers and services. However, sizing information from the CLEC is an estimate provided as part of the purchase negotiation, not audited information from the current billing system and process.

% or ✔. The FtA team has firm sizing information for both broadband companies and for the ISP. There is some uncertainty concerning the sizing information for the CLEC.

Importance. FtA, even with projected five-year growth, will be at the small end of telecommunications enterprises. Therefore, the ability to accommodate the volume of FtA customers or services is not expected to be a critical acquisition parameter. Of course, appropriate sizing is always desirable.

Owner. The FtA owner of sizing requirements is the corporate CIO.

Comments. The comments field reflects the rationale for ranking this area of low importance.

Billing cycles. Billing cycle requirements include how many billing cycles are supported—usually per month. They also include the ability to specify bill dates for the entire system, for a class of customers and for an individual customer. Billing cycle requirements include the ability to manage critical dates and events within the billing cycle.

% or ✔. The FtA team believes it has adequate but not necessarily complete information on billing cycles used by all four companies. It has actual system parameters for both broadband companies, a billing cycle chart from the ISP and published customer information about the CLEC. It also has gathered strategic requirements for future business directions.

Importance. FtA plans to become a high–value-added supplier to business customers. It has determined that the ability to provide a choice of billing date and customized billing cycles is part of that high–value-added strategy and will be important to customers. However, a customized billing cycle is not anticipated to be as important to customers as other value-added capabilities. Therefore, meeting billing cycle requirements has medium strategic importance.

Owner. The FtA owner of billing cycle requirements is the director of billing operations.

Comments. The comments field reflects the rationale for ranking this area of Medium importance.

Platform. Platform requirements include computer hardware and operating system vendors for servers and for "client" (PCs, PDAs, workstations) systems, DBMSs, network hardware and operating systems, tools and utilities such as reports writers and graphical support and any other third-party resource incorporated into the billing solution.

% or ✔. The FtA team has identified the desktop systems platform currently in use at all companies. The server and networking platforms vary from company to company and details of the CLECs server platform are unknown. Little work has been done on specifying platform elements.

Importance. FtA considers platform of low importance because it is willing to consider all otherwise viable vendors. It has no significant platform investment or expertise at this point in time, other than at the desktop level. Its desktop platform is Windows 2000, which is widely supported.

Owner. The FtA owner of platform requirements is the CIO.

Interface level requirements

Interface level requirements relate to all the external interfaces to the billing environment, including sales, provisioning, call detail generation, external payment processing, printing, general accounting, and management systems.

Types of interfaces. Requirements for types of interfaces identify every expected interface to the billing environment. This includes defining the general type of interface (e.g., order, charges, usage, payment, receivables, payables, regulatory reporting, management reporting, fraud management), whether the interface is inbound or outbound, what information flows over the interface and what kind of system is on the other end of the interface and who controls that system.

% or ✔. The FtA team has identified planned interfaces, as well as existing interfaces for the broadband companies and the ISP. It is lacking complete information on existing interfaces for the CLEC—and many of those interfaces will change when the business is converted to the FtA platform.

Importance. FtA has determined that supporting its types of interface requirements is of medium importance. At the individual interface level, it has determined some interface types that are of high priority, many of medium importance and some that are optional.

Owner. The FtA owner of types of interface requirements is the CTO.

Specifications/tools. Requirements for interface specification and tools include the details of interfacing functions and messages. For a message-oriented interface, these requirements include the message structure, protocols, recovery mechanisms and application-level failure handling. Recovery mechanisms and application-level failure handling should have links into the exception management requirements. For other types of interfaces—such as shared databases or manual inputs or bulk data loads—data fields and relationships are specified, along with control and management functionality. Interface specifications may also include a need for interim storage capacity of interface transactions, if all transactions cannot be processed in real time.

% or ✓. The FtA team does not have complete information on interface specification and tool requirements. Currently, other teams within FtA are developing requirements and beginning the acquisition process for sales and provisioning systems and for a lockbox bank. These activities will need continuous coordination with the billing system requirements and acquisition process.

Importance. FtA believes that it can be flexible in implementing most system interfaces, so they have assessed interface specifications and tools requirements to be low importance.

Owner. The FtA owner of interface specifications and tools requirements is the CTO.

Comments. The comments field reflects the need to coordinate with other systems acquisition teams.

Quality tools. Quality tools requirements include problem detection, notification and resolution capabilities, as well as interface quality analysis and reporting tools.

% or ✓. The FtA team has identified desirable quality tools for all known interfaces. However, these requirements will have to be refined as more information becomes available about the interfaces.

Importance. FtA has determined that meeting quality tools requirements is of high importance, since interface quality can have significant impact on cash flow and on operational overheads.

Owner. The FtA owner of interface quality tools requirements is the CTO.

Comments. The comments field reflects the need to coordinate with other systems acquisition teams.

Capacity. Interface capacity requirements are critical to assessing both the processing platform and the networking platform. These include all traffic between the billing environment and other systems, including message-based transactions, bulk data transfers, error management and correction routines, status and networking overhead messages. Capacity requirements should address average traffic loads, peak traffic loads, total traffic loads by day and week, peak times for each type of interface or transaction, the size of each message or transfer and timeliness parameters for processing the information.

% or ✔. The FtA team believes it has identified the majority of interface capacity requirements with estimates of ±15 percent. If the nature of some major interfaces is not as anticipated (such as a message-based real-time call detail transfer rather than a periodic bulk transfer of call detail), adjustments may be required.

Importance.
FtA has rated interface capacity as low importance in the selection of a billing solution. FtA has lots of excess floor space in its data center location if additional networking equipment and servers are required. It also has excess broadband capacity throughout its area. This means that the networking can grow as needed. Since FtA is fairly small among telecommunications enterprises, it is likely that any solution platform could easily accommodate its volumes.

Owner. The FtA owner of interface capacity requirements is the CIO.

Standards level requirements

Standards level requirements address what external standards must be supported. This includes identification of the standard and the business reason for adherence to the standard, the applicable version, version stability, portion of the standard to be applied and standard update and support requirements. In Chapter 26, we review applicability of standards to specific aspects of the billing system.

Call detail standards. Call detail standards identify the format and content of call or transaction detail records. Generally, these standards are dictated by the network equipment or mediation equipment, the service technology and the geographic market.

% or ✔. The FtA team believes it has solid requirements on call detail standards. The two broadband companies do not currently provide any usage-based services. The ISP has stated that its usage information conforms to IPDR Network Data Management-Usage Specification Version 2.6. The CLEC has not provided any information on call detail standards, but it is known that it is using a Northern Telecom switch that conforms to Telcordia BAF requirements. It has also worked closely with the technology strategists in the CTO organization to identify a need to handle current and upcoming versions of NDM-U (3.1 and beyond), as well as requiring a specifiable usage structure that will accommodate content-oriented usage charges. Business strategists have indicated that some professional services will be treated as usage-based charges, but have suggested no standards for that interface.

Importance. FtA believes that flexibility in specifying usage detail input is more important than support for specific standards and, therefore, has assessed the importance of call detail standards as low.

Owner. The FtA owner of call detail standards requirements is the CTO.

Regulatory requirements. Regulatory requirements include billing information content and format, as well as some auditability and retention requirements.

% or ✔. FtA plans to offer telecommunications service in California, Nevada, and Arizona. It will be subject to the regulatory requirements of those states plus the Federal Communications Commission. These requirements are public record and have been reviewed in developing the system requirements. The small area of uncertainty relates to various proposed rulemaking activities that may come into effect in the life of the billing system.

Importance. FtA has determined that regulatory requirements are of medium importance since demonstrating the ability to meet these requirements is one of the conditions for regulatory approval of the sale of the CLEC.

Owner. The FtA owner of regulatory requirements is the director of billing operations.

Billing output standards. Billing output standards include traditional EDI standards, as well as current and proposed EBPP standards.

% or ✔. The FtA team believes it has complete requirements for billing output standards.

Importance. FtA considers flexible and standard billing output options to be part of its value-added equation and, therefore, of medium requirements.

Owner. The FtA owner of billing output requirements is the director of billing operations.

Payment processing standards. Billing processing standards include new EBPP standards, as well as industry-driven interfaces with credit card companies and clearinghouses.

% or ✔. The FtA team believes it has most of its relevant requirements for payment processing standards. However, it is still in negotiation with some payment processing vendors.

Importance. FtA would like to have flexibility in contracting with payment processing vendors without incurring development costs. Therefore, it has assessed support for formal and de facto payment processing standards to be of medium importance.

Owner. The FtA owner of payment processing standards requirements is the director of billing operations.

Summary

In this chapter, we used a simple requirements checklist to organize, focus and prioritize the business requirements for a new billing system. We looked at an example of the use of that checklist in preparation for developing and publishing to vendors a RFI for the project. In upcoming chapters, we review types of vendors and vendor roles, as well as how to identify potential vendors. Chapter 31 "What is an RFI?" looks at the development of an RFI from these requirements and priorities.

Chapter

29

Types of Products and Vendors

There is a wide variety of options available in software products and services for billing solutions. Some vendors provide an end-to-end billing environment, including mediation, rating, presentation and delivery support. Some vendors provide a narrowly specialized functional module. Others provide custom development and integration services and still others provide functional service bureaus.

Fully Custom Software

Fully custom software is simply software uniquely developed for your business. You get to specify exactly how you want it and you pay all the costs of development and maintenance of the software. At one point, this was the territory of very large companies who could afford to maintain large internal systems development staffs and whose businesses were unique, either due to size or to the nature of their business. In today's telecommunications industry, where support for industry standards can consume significant development resources on an ongoing basis and in which there is generally more than one "player" in any market, fully custom software is found mostly in small, limited-product-line companies who create simple applications for recurring charge billing.

Customizable Software Modules

Many vendors offer customizable software modules or *platforms*. This is software that contains many commonly required functions, but which is designed to have some key areas of format or functionality customized by the vendor's software developers. (For the programmers reading this, these "platforms" are generally high-functionality callable subroutine or object libraries.) This is often a good choice for large and complex enterprises in which the billing environment must gracefully integrate with legacy applications and external partners.

"Off-the-Shelf" Modules

Off-the-shelf software modules contain complete functionality, generally integrated with data-driven configuration capabilities and/or user-defined rule bases. Many of these configurable modules also include both account and service templates for the targeted market or service segment, to make configuration easier for common types of services and companies. These modules could be stand-alone or "plugged together" with other modules from the same or different vendors via vendor defined application programming interfaces (APIs) or common object request broker (CORBA) interactions.

"Off-the-Shelf" Total Solutions

Some vendors provide complete end-to-end billing functionality—often for a specific industry segment—and some vendors combine the available modules of their architecture into a unified total solution.

Service Bureaus

Service bureaus provide specific billing functions on a contract basis. The service bureau owns the software and operations staff and "rents" the service to the telecommunications company. Many service bureaus focus on a single functional area such as bill printing and mailing or payment processing. Some call detail clearinghouses act as rating service bureaus.

Customizing a Product

There are many different ways of customizing a billing product to meet your business needs, short of building a completely custom system. Customizing always incurs some level of risk and has some business impacts. Next, we review these and some possible strategies for optimizing the effectiveness of the different ways of customizing products.

Functionality

Customizing the functionality of a system means actually changing the system's processing logic. This is very tempting if an off-the-shelf module does almost what you want it to do, but not quite. For instance, a telecommunications billing system has very robust support for time-based usage billing, but it expects an input of universal date/time format (UDT) start time and end time. You would like to use these robust features for packet-based billing or character-based billing.

Risks. By asking the vendor to modify an off-the-shelf piece of software, you incur many risks. First, your company will shoulder the entire cost, not only of the new development, but also of any necessary regression testing of the entire module, platform, etc. Second, unless this is a feature that the vendor wants to market to other customers, your company will also be responsible for support, upgrades and regression testing of upgrades to the platform. Third, your company essentially becomes the alpha test site for brand new software, resulting in potential platform instability.

Business impacts. The most obvious business cost is that the custom development, test, support and maintenance may be more costly than buying a platform as is. Other costs include a need for more knowledgeable support of this product within your company, the need to develop unique user and operator training, a longer time to market for changes requiring billing system modification and the ongoing overhead of negotiating continuing support from the vendor.

Benefits include getting the functionality you need without paying for a totally custom development.

Strategies. If you do not find a better-fitting product or combination of products, discuss with the vendor the potential market for the configuration you need—and volunteer to be the beta customer for that configuration. This may result in your getting the appropriate functionality at no additional cost or even at a discount, with extra vendor handholding along the way and standard support, maintenance and upgrades in the future. This is a great strategy if you are the leading edge of a new industry configuration or service class.

If that does not work and you have significant internal systems expertise, negotiate with the vendor for rights to the source code with restrictions on resale of the software. This protects against the vendor's refusing continuing support to some extent. At a minimum, insist on source code escrow—to be released to your company if the vendor fails to support the customized system at a rate negotiated at the time of the original contract.

Customer/account definition

Most modern billing systems include tools for specifying customer and account definition and structure. However, some single category billing systems may have rigid definitions of customer and account structures.

Risks. If customizing the customer and account structure requires software changes, the risks are the same as for functional customization. If customizing requires changes to non-user configurable data tables, the risks are somewhat less, although comprehensive system testing with your parameters is required—generally at your expense.

Business impacts. The business impacts are similar to those resulting from functional customization, although diminished if the changes are data parameters only.

Strategies. Do not purchase any system in which the customer and account structures are hard-coded. In this era, this indicates poor system design and shortsighted architecture.

If the configuration is established in non–user-accessible data or rules bases, discuss with the vendor the possibility of including your desired configurations within the supported product. If that does not work, negotiate for training of your technical staff to manage and maintain the appropriate data.

Product definition

Most modern billing systems include tools for specifying customer and account definition and structure. However, some single-category billing systems may have rigid definitions of customer and account structures.

Risks. If customizing product definition requires software changes, the risks are the same as for functional customization. If customizing requires changes to non–user-configurable data tables, the risks are somewhat less, although comprehensive system testing with your parameters is required—generally at your expense.

Business impacts. The business impacts are similar to those resulting from functional customization, although diminished if the changes are data parameters only.

Strategies. Do not purchase any system in which the product definitions are hard-coded—even when the product definitions are driven by standards. In this era, this indicates poor system design and shortsighted architecture.

If the configuration is established in non–user-accessible data or rules bases, discuss with the vendor the possibility of including your desired configurations within the supported product. If that does not work, negotiate for training of your technical staff to manage and maintain the appropriate data.

Interfaces

In Chapter 25, we discussed interfaces in some detail. Unless you are starting with an entirely new enterprise with no interfaces to any outside entities and all of your company's sales, provisioning, accounting, payment processing, customer care and fraud platforms are supplied by the same vendor who has designed them to work together (not likely), you will need some customization work on or at the interfaces to your billing environment.

Risks. A custom interface is brand new code dependent upon the logic and operations of two unrelated systems that may or may not be well documented. This requires extensive integration testing beyond simple exercising of the new interface.

The biggest risk of customizing interfaces is that the interfacing systems have data that look the same and have the same name and description, but they act differently due to different front-end edits or their relationship to some other data field or key. This can cause the system to act unpredictably in ways that are very difficult to identify and correct.

Business impacts. The full cost of customizing interfaces is generally borne by your company, as is the cost of supporting, maintaining and synchronizing with upgrades to the systems on either side of the interface.

It is virtually impossible to install an effective billing platform without some customized interfaces.

Strategies. To minimize risk and cost, use vendor-defined structured interfaces to each system. Using middleware can improve the management of the interface, but may result in more costly maintenance on the interface.

Develop interfaces where they are needed, but review your architecture for unnecessary interfaces, especially real-time interfaces.

Platform

In Chapter 21, we briefly mentioned issues with major platform customization. Not only will this result in risks and costs similar to those incurred with functional customization, but you risk having development done by developers who do not have expertise in the platform.

Risks. By asking the vendor to modify an off-the-shelf piece of software, you incur many risks. First, your company will shoulder the entire cost, not only of the new development, but also of any necessary regression testing of the entire module, platform, etc. Second, unless this is a feature that the vendor wants to market to other customers, your company will also be responsible for support, upgrades and regression testing of upgrades to the platform. Third, your company essentially becomes the alpha test site for brand new software, resulting in potential platform instability.

Business impacts. The most obvious business cost is that the custom development, test, support and maintenance may be more costly than buying a platform as is. Other costs include a need for more knowledgeable support of this product within your company, the need to develop unique user and operator training and the ongoing overhead of negotiating continuing support from the vendor.

Benefits include getting the platform you desire.

Strategies. The authors cannot recommend any but the most limited platform modification. Limited modification might apply to a different version release of the desktop platform—but that would normally be supported in general release versions by a reliable vendor.

Cosmetic and branding

Most billing system products include the capability for user customization of bill layouts, fonts and other cosmetic features and of bill and bill segment branding. If an otherwise appropriate system is not user customizable in these areas, customization may be necessary and appropriate.

Risks. While all customization has some risks in cost and support, this is generally a minimum risk area.

Business impacts. The business impacts of not formatting a bill to match regulatory requirements or customer expectations can be significant and negative—including fines and loss of customers.

The costs of customization in these areas are usually minimal and with minimal impact on other system modules.

Strategies. Discuss a standard support contract for this type of customization with the vendor or negotiate training and access for your company's development organization to perform and maintain the customization.

Installation Services

In many cases, the software or system vendor will install the billing platform you purchase. However, there are many system contractors and consultants who have a relationship with the software vendor as authorized installation vendors.

Sometimes these vendors can provide a local presence for ongoing support and working with them on installation can build a foundation for an ongoing relationship. At other times, these vendors may have lower overheads than the system vendor and can, therefore, offer equivalent work for a lower cost. Sometimes the most experienced of the vendor's software engineers form their own consulting firms and provide superior expertise.

If your company identifies an authorized installation vendor that is desirable for the above reasons or other good business reasons, there is no reason not to use that vendor. In fact, even if the software vendor does the installation, it is then acting as an installation vendor. It is good practice to have the costs and conditions for installation broken out separately in your contract with the vendor. It is very, very risky to use installation vendors that do not have a relationship with the system vendor and we do not advise doing so.

Configuration Services

As with installation, in many cases, the software or system vendor will configure your billing system. This entails setting up systems and operational parameters for your operation. It may also mean defining products, services and other company-specific data. Again, there are many system contractors and consultants who have a relationship with the software vendor as authorized configuration vendors.

If your company identifies a desirable authorized configuration vendor, there is no reason not to use that vendor. If the software vendor does the configuration, they are then acting as a configuration vendor. It is good practice to have the costs and conditions for configuration broken out separately in your contract with the vendor. It is somewhat less risky to use configuration vendors that do not have a formal relationship with the system vendor since this is usually acting simply as a very knowledgeable user. However, choose such vendors with extreme care.

Conversion Services

Conversion services are at least as much about the old systems as they are about the new platform. In many cases this makes conversion services an ideal venue for a capable independent custom software firm. However, software vendors who make high profit margins on conversion services often do not make bulk load utilities, appropriate for conversions with large numbers of data records, available to customers or other vendors—limiting the load capability that does not void warranties and support agreements to APIs that are designed for single-record-at-a-time upload.

Summary

In this chapter, we look at the many types of vendors that can be involved in establishing a complete billing environment. Your billing system acquisition management team should carefully consider the potential contribution of each type of vendor, as well as whether your company is positioned to act as its own "prime contractor" or if you want a single-vendor solution or a multivendor solution managed by a prime contractor.

Chapter 30

Identifying and Qualifying Vendors

Identifying vendors is not quite as easy as you may think. Often the most venerable vendors have platforms sized to mega-enterprises while newer ventures targeting new and competitive markets may have suffered the fate of many of their customers. Also, in the world of billing, there are many narrowly specialized vendors—focusing on just 2G mobile, just IP usage or just mediation for older technology network equipment. In addition to the primary billing system or service, specialized vendors such as print service bureaus, EBPP vendors and tax module vendors are available to provide a portion of the overall billing environment.

Once you find potential vendors, you want to qualify them quickly—based on both company and product information. There is a lot of overhead associated with each vendor as you move forward in the acquisition process. Continuing to interact with a vendor that does not meet your financial or strategic requirements is a waste of your time and theirs!

Finding Vendors

Finding appropriate and credible vendors is an important part of the procurement process. Existing relationships may not be a good guide to the right vendors in a fast-changing industry where billing is moving from a stodgy backroom operation to a strategic differentiator—and even to service offering status.

Industry sources

If you are about to spend millions of dollars on a new billing system, you should also be spending a few thousand dollars in attending billing trade shows and seminars. There are excellent billing conferences hosted annually by telecommunications trade organizations and by commercial conference organizers. Vendors exhibit at these shows, as well as putting forth some of their "best and brightest" in tutorials and talks. In addition, these are good opportunities to hear from other companies who have recently implemented new billing systems—both in formal case studies and in general conversation at social events.

Trade publications also provide a rich source of information on potential vendors, ranging from development plans to successful and unsuccessful installations to financial and management stability.

Vendor database

In Appendix D, we provide a guide to a database of current vendors of billing products on the CD in an MS Access® database. While we have made a concerted effort to include major players in the billing products and services market and to accurately represent their offerings, we have not exhaustively researched these companies or their products. It does, however, provide a place to start and one perspective on sorting and categorizing vendors.

We look at vendors by functionality, platform and targeted and established markets. Our categorization is based upon vendor self-representation. We do not make any representation about the financial viability of the vendors or the appropriateness of their product to any application.

Advertising for vendors

Advertising in trade publications is a traditional way to stimulate vendors to "self-identify." However, in today's electronic environment, announcing an intended acquisition in any of thousands of contexts on the Internet will get prompt notice.

"Experts"

There are quite a few very expensive reports on the market that provide information on billing trends and vendors. These reports are generally published by major consulting houses and reflect the input of their research staff and their field consultants. They often contain good information and can be quite useful. However, they cannot tell you whether a particular vendor or company is a good fit with your needs and your company.

There are also consultants, the authors among them, who will help you match your requirements to a "short list" of vendors.

Qualifying Vendors

There are a number of things you should "check out" before spending a lot of time with a vendor. First, you want to know that they are financially viable now and in the future. Second, you want to know if they have products and services that are apparently congruent with your billing system needs. Third, you want to know that they are willing to have a financial stake in their performance via monetary performance guarantees and/or penalties. Fourth, you want to know that they have successfully completed similar projects. Finally, you want to know something about the individuals who run the company and who manage the products and services you are considering. We will talk more about filtering out vendors in Chapters 32 and 34!

Even before you send out an RFI, however, you can do some quick checks for relevant information.

Vendor Web sites

Often a quick look at a vendor's Web site will help you determine if the vendor is a serious candidate for all or part of your project. Web sites will usually contain some overview of product information—as well as sales contact information. They often include an archive of press releases and press citations that may help you determine in what type of projects they have been engaged and for what companies. They also may include "investor information" which can provide insights into the financial status of the company and links to recent FEC filings.

Trade press

If the vendor has any history at all, there should be some history of the company and its products in the trade press archives. The Internet has made it much easier to access and search these archives, but billing system decision makers should be making every effort to stay current on billing coverage—both in general and telecommunications specific coverage. Often the trade press will quote billing managers or decision makers in articles on contracts let. Contacting those sources—especially on projects that are nearing completion or are complete—should provide some good insights into vendor performance from sources who are not necessarily selected by the vendor as "references."

Financial press and analysis

If the vendor is publicly held, there should be a wealth of information available about its financial health and stability from the Internet or from major brokerage houses. This, of course, is not infallible information, but it is a good start.

Summary

In this chapter, we reviewed potential sources for identifying vendors and some information sources about the vendors. We have pointed out the utility of the Internet and a telephone in developing a list of credible vendors. In Chapter 32, we will look at qualifying vendors based on RFI responses.

Chapter

31

What Is an RFI?

A request for information (RFI) is just that—a notice to vendors that your company would like information on their products, services and companies. It can be as simple as a published notice that your company is seeking to acquire a new billing platform—which will net you a lot of brochures and sales calls—or it can be a more focused document asking for the information you really want to know.

What Do You Want to Know?

One of the best reasons to have a fairly complete set of business requirements before issuing an RFI is to focus on exactly what is important to you and what you really want to know—about the vendor, their products, services, and support.

Vendor information

There is a lot of information about the vendors that you want to elicit from an RFI. First, you want to clearly identify who they are—the company name, any parent companies or subsidiaries, the corporate location and corporate contact information, as well as the location and contact information for your sales contact. You also want their Dun & Bradstreet number and their corporate history—how long have they been in business and how long have they been in this business. If possible, you would like to know the relationship of the sales contact with the company.

You want to know what their primary geographic market is and what their primary market segment is. Are billing products their primary product line? You also want to know how much they spend each year on research and development. Is the company publicly held? If so, ask for an annual report. If the company is U.S.-based, request a 10-K for the previous year.

Product information

Although this sounds fundamental, you want a description of the product—its functionality, architecture and platform. You want to know about user interfaces, configuration tools and management tools. You want to know about system interfaces and APIs. You want to know about standards supported. You want to know what the capacity of the system is for both current and retained information.

You want to know how old the product is, how much development resource is directed to its enhancement and update each year and what the upgrade path is. You want to know if the product is at the beginning or at the end of its life cycle. Are there any major changes in architecture or platform planned for the next year or two?

You want to know how many of these products have been sold and how many are still in service. You want to know where those products were sold.

References

You want at least five references. Preferably these would be companies in your geographic region—with a similar product mix and target market—using this product. Therefore, you want to know who the references are, location and contact information, when they acquired the product, what version of the product they are running, whether they have been through a version upgrade, as well as a business profile of the reference company.

Sample RFI

Figures 31-1 and 31-2 display the RFI for From the Ashes Communications' new billing system. It is very high level and generic. After all, you are asking open-ended questions in order to elicit answers from the vendor that are not very slanted by your requirements. However, the RFI does ask the vendor to "comment on" several items of key importance to From the Ashes. These vendor comments will give you explicit information, as well as give you insight on communication skills, responsiveness and adherence to the spirit of your RFI.

Title

The title of the RFI clearly states important parameters of the acquisition—that it's a billing system and that it must handle traditional telecommunications and convergence services.

Contacts and background

This section states very briefly what the project is. It then clearly indicates the contact, provides a snail mail and email address and identifies acceptable electronic formats. The lack of a telephone number is not an oversight. FtA has

Request for Information
October 15, 2002

Request for Information:

Telecom and Convergent Services Billing Platform

1. Contacts and Background

From the Ashes Communications, Inc., is interested in acquiring a new billing platform and is requesting that interested vendors respond to:

Ms. Mary Thomas
From the Ashes Communications, Inc.
Oracle, Arizona
met@ashes.com

Electronic responses may be in text, MS Word (.doc) or Adobe (.pdf) format.
Please respond by December 10, 2002.

2. Company Information

Please provide vendor company name, address and contact information, corporate structure (e.g., parent company), company history and Dun and Bradstreet number. If a publicly held corporation, please include your most recent annual report with your submission.
Please comment on the company size and the size of research and development staff in areas related to communications and convergence billing.

3. Product Information

Please describe your billing product or products. Describe functions, features, platform, sizing and interface support.

Comment on applicability to a small value-added convergence venture with some conventional telecom and Internet services billing. When was the product introduced? What is the current version of the product? When was that version generally available?

Please comment on the product's ability to support VoIP, content-based usage billing, packet-based billing and time and materials billing. Please comment on CDR standards supported. Please comment on regulatory and taxation support for U.S. Federal and Arizona, Nevada and California state jurisdictions.

Please comment on the product's support for conventional EDI, Web-enabled EBPP and other billing options. Please comment on the ability to define multiple output streams from the same source data for the same account.

Page 1 of 2

Figure 31-1 FtA RFI, page 1.

determined that it wants all responses in writing to minimize the potential for incomplete information transmission—sometimes characterized as "whispering down the lane syndrome!" The response date is clearly stated.

Company information

This section asks for information that will facilitate the assessment of the financial stability of the vendor, the corporate priorities of the vendor and the resources available for support of the product area.

Request for Information:

Telecom and Convergent Services Billing Platform

4. Services and Support Information

Describe the upgrade history of your current product or products, as well as upgrades currently planned or in development. Describe the typical operational impact of an upgrade effort. Describe your company's typical involvement in a customer upgrade.

Describe implementation and configuration services and staffing. Describe integration and conversion services and staffing. Comment on tools and methodology available for implementation, configuration, integration and conversion.

5. References

Please provide 5 references. Include company name, contact name, title, role in project or projects in which you were the vendor, product or service purchased, date project began, date project completed, address, telephone number and email.

6. Quality

Comment on quality methodology for development of baseline product and for professional services. What portions, if any, of your software development processes are ISO 9001 certified?

7. Pricing Information

Please describe the price range of some typical systems with a description of the configurations represented at the low end and the high end of the range. Describe your company's typical pricing for professional services.

Page 2 of 2

Figure 31-2 FtA RFI, page 2.

Product information

This section makes open-ended requests for information about the product, as well as identifying some specific information related to product maturity and support. It also identifies some areas of particular interest on which information is specifically required. You may notice a direct correspondence between these "special interest" areas and high-level FtA systems requirements.

Services and support information

Since FtA is strategically committed to being a leading-edge, value-added provider, ongoing support for new technologies and new service configurations is important to them. The information requested on upgrade history and plans will assist in the evaluation of a vendor's ability to support leading edge requirements.

FtA expects that the platform vendor or vendors will perform much or all of the implementation, configuration, integration and conversion, as well as ongoing upgrades. Therefore, they express interest in available support and the support environment.

References

This is self-explanatory—and, yes, you do want all that detail! If all the references are in France and Germany, for instance, FtA might question support for U.S. standards and regulations. If all the references are older than three years, it might raise serious questions about the vendor's viability now and in the future.

Quality

Good methodology and auditable processes do not ensure error-free products, but they do indicate a commitment to control and quality that usually results in stable products and tools.

Pricing information

At this point in the acquisition process, vendors are reluctant to give any pricing information at all. However, asking for a range of prices and solutions may elicit useful "ballparks" of cost. It may also help to identify how much of the typical implementation cost is for off-the-shelf items and how much is services related.

Summary

In this chapter, we looked at how to construct an RFI based on business requirements developed in Chapters 22 through 29. In the next chapter, we look at how to evaluate responses to such RFIs.

Chapter

32

Using RFI Responses

In the previous chapter, we surveyed how to construct a request for information (RFI) and looked at a sample RFI. Once you have sent this RFI to the vendors you identified from the sources listed in Chapter 30, including those who self-identify from your advertisements of "intent to purchase," you will probably receive a pile of responses.

At the RFI stage, a typical response will be a cover letter—which may or may not be responsive to your RFI requests—and one or more product brochures with a Web site reference. Some vendors may develop more specific responses, especially if your company is large or visible, if they are eager for your attention. You can, however, do a lot with just the typical amount of information.

Evaluating Product and Vendor Appropriateness

Before you spend a lot of time investigating a company, you should determine whether the respondent actually provides a product or service that is relevant to your high-level business needs.

Initial review

The initial review of responses should be a very quick sorting by type of product or service. Useful categories include "comprehensive billing platforms," "market/technology specific," "function specific," and "services." Somehow, there always seems to be a "none of the above" category as well.

Function specific. If you have decided to outsource your bill print and mailing function, then a vendor of mail processing equipment is probably not relevant to your procurement and you can discard the response. However, if you have decided to bring that printing and mailing function in house as part of a larger billing system procurement and the vendor appears to have equipment that is well suited to your needs, keep that response for further investigation—even if it doesn't meet all of your project needs.

Market or technology specific. Similarly, if your company is a terrestrial-only carrier, you can discard the narrowly targeted turnkey wireless company solutions. However, if your currently terrestrial-only company plans to move into 3G wireless services, you may want to verify that any comprehensive billing platform you consider either currently supports 3G applications or has a committed delivery schedule for 3G features.

Services. What if the respondent does not have a product? This is not a trick question. The respondent may be a systems integrator that would like to act as prime contractor, using other vendors' products. If your company's business or strategy is somewhat unusual, you may want to explore further the capabilities of this type of respondent. On the other hand, if your business is pretty similar to a lot of other companies—varying only by territory, for instance—then using an integrator may be unnecessary overhead.

Other services vendors may include are comprehensive billing services, payment services, integrators for specific interfaces, management consultants and taxation services, among others. Even if your current plans are to perform all functions in house, you may find good economic reasons—when you get actual pricing—for considering some outsourcing. If you do not currently plan to outsource, set these responses aside for possible future reference.

Nonresponsive submissions

Sometimes an RFI will elicit a standard reply that is not responsive to the actual content of the RFI. This may mean that the vendor is simply swamped with RFIs and has decided to handle the volume with form letters. It may mean that they want to create an opportunity for personal contact or it may mean that they are incompetent or not very good salespeople.

If the information you do have indicates that the vendors may be appropriate for all or part of your product, it's worth the effort to check their Web site and/or make a telephone call for further information. If you obtain further information by telephone, it is a really good idea to have the entire acquisition team or their representative on the call to prevent miscommunication. Once you have adequate information, add these vendors to the appropriate sort category.

Compare to requirements

At this point, your team should compare all the "possibles" to the high-level business requirements. (Yes, you had some business requirements in mind when you did the initial sort, but at a stratospheric level.) If the material you have indicates general suitability, but is silent on some specific requirement, do not eliminate these vendors—they may have what you want, but not consider it worth mentioning. Or they may have that feature under development (especially if it is linked to leading-edge technology in one way or another) for the next release of their software—just when you want to start your project.

FtA evaluation of responses

From the Ashes received 20 responses to their RFI. Two were from vendors of bill printing and mailing equipment—a function FtA plans to outsource in a separate project. These two responses were discarded. Three responses were from billing service bureaus specializing in the independent local exchange carrier market. These responses were set aside, since FtA does not plan to outsource core billing functions and the "plain vanilla" service offerings did not meet FtA's value-added criteria. However, since FtA does not yet have good information on their CLEC acquisition, it felt these might become relevant at some point in the future. One response was from a high-functionality billing service bureau that specializes in billing services for the business telecommunications market. This response was held for consideration.

Two responses were from vendors whose products are narrowly focused on the wireless industry. These were discarded, as FtA has no immediate plans to enter the wireless market. Two responses were from a vendor with products narrowly focused on the consumer ISP market. This was set aside, but held in case a need arises as more information becomes available during the ISP acquisition process. Two responses were from vendors in the process of developing mediation equipment for 3G services. These were held for consideration since FtA is interested in providing applications and content services to 3G vendors, although not the wireless infrastructure. Two responses were from tax software vendors—both of whom distribute software and also provide a tax service bureau. These were also held for consideration since they both provide products and services for FtA's service area.

The remaining submissions included three submissions from vendors of fully featured billing platforms and three submissions from systems integrators, two of which specialize in integrating one of the fully featured platforms and one of which specializes in integrating one of the other fully featured platforms. These were all held for further consideration.

FtA began initial evaluation of RFI responses with 20 vendors. After initial review, it has 11 for further consideration, five set aside for possible future investigation and four considered not relevant and discarded. This quick initial sort has eliminated almost half of the respondents—and, hence, the follow-up effort.

It is useful to note that the high-functionality billing service bureau uses one of the fully featured platforms being bid by both the originating company and one of the systems integrators. Thus, among the 11 "surviving" vendors, there are actually three platforms and two tax applications (that work with all three of the platforms) being put forth for consideration.

Evaluating Vendors

Once you have determined that a vendor may have an appropriate product or service, it is time to look more closely at that vendor. You will want to review and update this information before selecting a vendor and signing a contract, but eliminating unstable or uncommitted vendors here will save time and effort in the system acquisition process.

Evaluating vendor stability

In the RFP, you asked for the vendor's Dun and Bradstreet number. This will allow you to run a simple credit check on the vendor, its parent company and any subsidiaries—only the first step in analyzing its stability as a company. If the vendor has little credit history, it may be a new market entrant—possibly a good thing if you are looking for leading edge technology or services support. However, you will want to explore its funding sources and cash flow indicators with some care.

If the vendor is a well-established company that appears to be having cash-flow problems, you should pay particular attention to the size of its customer base for the billing family of products and whether billing systems are a profitable core business for the company. If the billing systems product line or subsidiary is profitable but not a core business, it may be a candidate for selling off to bolster cash reserves. This may cause disruption in product development and support, even for a very successful product line. In Chapter 35, "Negotiating a Billing Solution Contract," we will discuss some ways to mitigate the risks of dealing with such vendors. In this instance, such risk mitigation may be worthwhile and effective.

If billing systems are the company's primary business and it is having cash flow problems, this could be a failing business. Even with risk mitigation, there can be significant "back-end" costs and overheads if the vendor fails as a business.

If the company appears to have a solid credit rating and no cash flow problems, then you are ready to look at its commitment to billing systems, to the product it is identifying to meet your needs and to look at your industry and geographic market.

Evaluating vendor commitment to product and market

How long has the vendor—or its parent company—been in the billing products market? Companies with a solid customer base for a product line generally have a significant commitment to that product area.

How much of the company's research and development (R&D) activities are targeted to the billing products line? Clearly, if a vendor is making a significant commitment of R&D resources to billing-related work such as standards development and billing information models, it has a longer-term interest in the billing market. Of that, how much is relevant to the specific product(s) you are considering? This, of course, indicates commitment to this particular product or product family.

How large is the development staff on this product? Again, sizable resource commitments indicate commitment to the product. When is the next product release and what new features does it contain? Review the product release history. Well planned, feature-rich periodic releases indicate a development process that is under control. Very infrequent releases (less than yearly) indicate a lack of development commitment in a fast-evolving discipline. Overly frequent releases—especially maintenance releases—with lots of "bug" fixes indicate the vendor is having problems with quality control.

To verify information provided by the vendor in the RFI, review industry press information on that vendor for the past year. It is not uncommon for vendors to overlook mentioning that their billing products division is about to be sold off—or to forget to identify maintenance releases of their product!

How much of the vendor's market is relevant to your business? If the vendor is primarily in the business of power company billing systems with an appliqué module for telecommunications billing and your business is mainstream or leading-edge telecommunications, this may not be the right vendor for you. However, if the vendor is interested in moving aggressively into the telecommunications market, it may have significant commitment and provide excellent support. You may want to explore this area further in the request for proposal and during contract negotiations.

Where is the vendor's primary market? If the vendor is located in Europe with the majority of its development organization in India and only sales offices in the United States, you may want to structure the RFP to address the structure for support and services in the United States.

Evaluating vendor performance

At this stage of the systems acquisition process, just check the references provided by the vendor, as well as any of the vendor's clients of which your team may have personal knowledge. Ask the references exactly what the vendor did for them, when and with what products. It is a good idea to ask references to rate the vendor on several dimensions, including product functionality, product stability, product flexibility, operations and controls, as well as installation support, conversion support and ongoing maintenance and upgrade support.

When you are talking to references, they are generally your industry colleagues. Many will be exceptionally forthcoming on all aspects of the vendor's performance. It is not necessarily a bad thing for a reference to evaluate a vendor as "not quite walking on water" if comments indicate solid professional performance. However, any strongly negative comments should be examined carefully, since these are projects the vendor probably considers to be highly successful.

FtA vendor evaluation

In a review of credit ratings of the 11 vendors that remained after the initial screening, one of the systems integrators is shown as having severe cash flow difficulties—slow payment history and significant debt. The FtA determines that this disqualifies the firm as a prime contractor on this project. However, it is a local firm that could provide ongoing support for the platform. FtA will suggest that it subcontract with one of the other companies for integration and migration services.

One of the platform vendors appears to have very limited deployment in North America. Its primary customer base is in Europe. It is apparently in the process of expanding into the Asian and North American markets. Support for U.S. customers is flagged as an issue for further exploration with the vendor.

In checking vendor references, the FtA team determines that one of the other platform vendors has an overall successful record, but did not provide adequate integration support on more than one project. Support for integration is flagged as an issue for further exploration with this vendor and for inclusion in the RFP.

Evaluating Product Information

Product information provided in response to an RFI should be reviewed against the summary requirements used to create the RFI.

Completeness

Has each vendor provided information in response to each area requested in the RFI? If the original response was non-compliant, did the vendor provide adequate information upon subsequent request? If the vendor will not provide high-level feature and functionality information even after personal contact, it probably does not want your business. There are exceptions, so check with industry contacts about the responsiveness of the vendor's sales organization, as opposed to the implementation organization.

One of the authors was once doing a sizable server procurement. There were two candidate vendors, both server manufacturers. One had an active and attentive sales force; the other seemed barely able to answer the telephone. However, industry contacts indicated the first company placed all its support resources in sales and was not responsive for service support. The second company had excellent post-sales support. The second company won the project and there were no regrets in the customer company.

It is not unusual for vendors to be unwilling to disclose pricing at this point. This is not a good reason to disqualify a vendor.

Requirements match

The open-ended language of an RFI generally elicits information about the current and near-term product release functions and features. Identify where these appear to be a close match or an exact match to your requirements. If any of the vendor platforms appear not to support major critical requirements, you may decide not to pursue that vendor. You may also want to just flag the gaps and carefully review how the vendor deals with those requirements in their proposal.

FtA product evaluation

FtA found that both of the tax software vendors cover their intended markets and are capable of supporting FtA's services. They both provide adequate reporting and control functions and both are adequate for FtA's projected volumes of tax transactions.

FtA determined that one of the platform vendors does not currently support either IPDR or TIPHON standards for usage data interfaces as a preconfigured input type, although it has a flexible interface that would allow the definition of such an interface. Another of the vendors, with a primarily European market base, does not have predefined billing output formats to meet regulatory standards in FtA's market. Otherwise, they all appear to be capable of supporting FtA's high-level requirements. Since these deficiencies can be eliminated with custom configuration or development for all the platforms, they are not deemed reasons to eliminate these vendors.

Summary

Once RFI responses have been evaluated, your company has better information on the "state of the art" in available billing platforms and modules. This may result in some minor modifications in your requirements. For instance, if you require TIPHON support, but none of the vendors provides that in standard products, you may want to amend requirements that indicate the need for general product support of the TIPHON standard to indicate a requirement for support of the TIPHON standard, whether general product or custom interface support.

These amended business requirements will play an important role in the development and content of the request for proposal (RFP). The next chapter, "Writing a Good RFP," provides guidance on format, content, and use of RFPs.

Chapter

33

Writing a Good RFP

Writing a good Request for Proposal (RFP) is a critical factor in the success of a billing information project. It is the key to getting vendor proposals that reflect your company's requirements, which will allow the vendor to do a good job on your project.

RFPs serve two purposes. They let the vendor know exactly what you want to know and they provide the vendor with enough information to be able to tell you what you want to know. If you don't provide the vendor with accurate information concerning requirements or sizing, you are doing both your company and the vendor a disservice.

The included CD contains a format for an RFP that you may modify for your project. Later in this chapter, we look at a sample RFP that used this format as a basis.

What Does the Vendor Want to Know?

The vendor wants to know the scope of the project, your company's requirements, any limits on how it will meet those requirements, when you need to meet those requirements, what resources you will provide to support the project, where the work will take place, what role your staff will play in the project, whether they will be coordinating with other vendors and whether your company will be an ongoing customer.

Project scope

Exactly what does the project include? Clearly identify all functional and operational modules. Does the project include mediation equipment for some or all types of usage-based information? Does it include platform support for pre-pay and real-time rating? Does it include tax software? Does it include the data processing platform (computers, OS, DBMS), networking platform and/or desktop platform?

Does the project include installation and configuration services? If so, for what system configuration and in what location? Does it include integration services? How many interfaces will require integration and how complex are they? Service creation and configuration? How many services have been identified and how complex are they? Data conversion? How many data sources will there be and how many unique source formats? How many instances of products and services will be converted? Provide some evaluation of the quality of the source data.

Scheduling and logistics

When does your company plan to begin the project? Are there any "hard" dates for completing the project—that is, dates that cannot be moved for any reason? Such dates usually relate to legal compliance, although the critical Y2K dates were an unusual example of hard dates for converting from non-compliant platforms. Are there any "extremely desirable" dates for project completion? These are usually driven by financial or strategic considerations, such as the expiration of a service contract or the entry into a new market—or seeking a new round of financing.

In what locations will work be performed? This usually involves the locations where your servers are located, the locations in which users of the system will be located and the locations of your subject matter experts and management team. If the vendor can perform much of the work remotely, is that acceptable to your company? If integration work requires coordination with other vendors, who are they and where are they located?

Are there any special conditions, restrictions, rates or arrangements for travel that should be factored into estimates of project expenses? Does your company require contractors to use a specific travel provider or booking agency?

Resources

What resources will be available to the vendor's team? This includes all the documentation of data collection and requirements you have been doing, access to management, and subject matter experts, as well as physical resources such as workspace, computers and telephones.

What Do You Want to Know?

Your company wants to know if the vendor can meet your company's business requirements, how it will meet those requirements, when it will meet those requirements, exactly what the vendor will deliver, what performance surety the vendor will offer, any flexible terms and how much all that will cost.

Beyond documents

An RFP may require the demonstration of a product or a particular product capability. This may be a vendor "demo" system, set up for customer access. It may be a formal proof of capability in which your staff gets to poke and prod, not just at the user interface, but at the supporting platform and configuration and to engage in full capability exercises and performance reviews. Also of great value are opportunities to visit existing installations and review the platform in a fully implemented production environment. This has the added advantage of giving you additional access to folks who are actively using the platform.

Pricing

Your company wants to know how much this project is going to cost. The management of the project needs to know the basis for that project pricing. Often a vendor will enter a low bid on a basic product and then inflate professional services prices that can create a larger margin. That does not mean that a fixed price for a platform and some time and expense prices for configuration, customization, integration and conversion is not an appropriate pricing structure—it is very appropriate.

If you request and get the basis for project price estimation, you can then truly compare the various proposals you receive. Otherwise, you will be guessing about the variables—and perhaps doing the "apples and oranges" thing.

Performance surety

The RFP should indicate if your company requires a performance bond for the project and what other performance surety and warranties are required. This will be reviewed and perhaps amended in the contract negotiation, but it should be information available to the vendor when estimating the project.

Relationship to Requirements

The RFP is an opportunity to communicate your business and technical requirements to the vendor. This will allow the vendor to accurately determine how closely its platform meets those requirements and whether it can meet the requirements at all.

Business requirements

It is generally a good idea to have a signed nondisclosure with each of your "short list" vendors before releasing an RFP to them. If the business requirements are good requirements, they will reveal a great deal about your business and your business strategy.

Comprehensive business requirements should be a lengthy appendix to the RFP. Be very explicit in these requirements—spelling out and clearly identifying all terminology and references. As we have said elsewhere about requirements, good requirements are exactly as specific as necessary. Therefore, if the current version of standard x is version 2 and some function you require was first addressed in version 2, your requirement is to have a feature "compliant with version 2 or beyond of standard *x*." Do not say either "compliant with standard *x*" or "compliant with standard *x*, version 2." The first statement may get you a product compliant with version 1, while the second statement might constrain a vendor who is developing that feature to be compliant with proposed version 3 of the standard.

How requirements are met

Good requirements will leave a great deal up to the vendor. The vendor may choose any platform elements not specified in the requirements, such as development environment, hardware and operating system, documentation tools, delivery media, appearance of user interfaces for billing operations and third-party software. The RFP should indicate that you want to know these things.

The vendor may be using other vendors' products to satisfy your requirements. The RFP should reflect that you want to know what these are, how they will be supported, who will support them, who will resolve any issues between the third-party product and the development done by the primary vendor. This is particularly important if the primary vendor is an integrator or consulting house using mostly third-party elements for principle functionality. However, the maintenance and coordinated upgrade of such small platform elements as reports writers or major platform elements such as database management systems can be an ongoing nightmare if these relationships are not clearly understood and the integrated product is not well supported.

You want to know if the vendor is using off-the-shelf products to meet your needs or doing new development or customization of an existing product for you.

When requirements cannot be met

Sometimes vendors cannot meet your requirements. The language of the RFP should invite them to offer alternative solutions for the business issue or to suggest a phased project structure that would meet the requirements in a later phase.

Example RFP

The FtA team has developed an RFP based on its business requirements and business planning. We look at sample pages of that RFP next.

Background and contacts

Figure 33-1 shows the contacts background section of the FtA RFP. Note that the background is considerably more detailed than that provided in the sam-

RFP for Billing System
January 20, 2003

1. Contacts and Background

From the Ashes Communications, Inc., is a new communications and information company that will focus on high value-added services for small and mid-range business customers. From the Ashes Communications, Inc., has purchased telecommunications and Internet business from several sources. This business is currently being billed in a variety of systems on a variety of platforms.

From the Ashes Communications, Inc., is soliciting proposals for a new billing platform to support its convergent telecommunications and Internet business. Vendors should submit proposals to:

Ms. Mary Thomas
From the Ashes Communications, Inc.
Oracle, Arizona 85623
met@ashes.com
1-800-555-0987

Electronic responses may be in text, MS Word (.doc) or Adobe (.pdf) format.

Please submit a final proposal by March 15, 2003.

Questions and clarifications about the billing operations environment at From the Ashes should be directed to Ms. Thomas using the contact information listed above.

All other questions and clarifications about the business, technical or strategic environment at From the Ashes should be directed to:

Ms. Johanna Henry
From the Ashes Communications, Inc.
P.O. Box 1984
Reno, Nevada 89523
jmh@ashes.com
1-800-555-1234

This document and all attachments and supporting data are proprietary to From the Ashes Communications, Inc., and are being supplied only to vendors that have returned properly executed non-disclosure agreements. This information is to be used only for the purpose of developing a proposal for the acquisition described herein. If in the course of developing such a proposal a vendor determines a need to disclose any of this information to a potential supplier or subcontractor, the vendor must apply to Ms. Thomas for permission to do so.

Page 1 of 43

Figure 33-1 FtA RFP background and contacts section.

ple RFI and that the contact information is much more extensive. The contact information encourages personal contact with FtA personnel.

Project scope

Figure 33-2 displays the project scope section of the main FtA RFP for a billing platform. Figure 33-3 displays a similar section from the FtA RFP for tax software. Note that the main RFP also includes tax software. Often major platform vendors can obtain preferential pricing from the tax software vendors if they do a lot of business with the vendor.

RFP for Billing System
January 20, 2003

2. Scope

From the Ashes Communications, Inc., is seeking proposals for a comprehensive telecommunications and Internet billing platform. The proposal should include software for customer account creation and maintenance, for processing service orders into customer account records, for usage collection and rating, for bill calculation, for tax calculation, and for bill presentation in a number of formats. In addition, the proposal should include software supporting payment, collections, and adjustments processing.

This proposal should also include the implementation of interfaces and integration to the following FtA or third-party systems:

- Provisioning system (for all account and service order information)
- Sales system (for rate tables and product and service definitions)
- Usage data interfaces in FAF, IDPR, and TIPHON formats (at least five different sources)
- Lockbox bank payment processing
- Credit card charge and payment systems
- Bill print system (third party)
- Fraud management system
- Credit bureau system
- Corporate accounting (tax reporting, accounts payable, revenue accounting, revenue analysis)

This proposal should also include the conversion of existing account and service records from four separate sources. FtA estimates the number of existing accounts to be 300,000 ± 100,000 and including 1.2M ± 200,000 products and services. These numbers are subject to change.

FtA intends to provide the server hardware, network, and desktop platform for this billing system. Those items should not be included in the pricing of this project. However, the proposal should include detailed specifications for the required platform and infrastructure.

Page 2 of 43

Figure 33-2 FtA billing platform RFP scope section.

RFP for Billing System
January 20, 2003

2. Scope

From the Ashes Communications, Inc., is seeking proposals for a telecommunications tax system and database. The proposal should include software for tax calculation and for all required telecommunications and sales tax reports. In addition, the proposal should include tax and reporting databases for U.S. federal taxes and Arizona, California, and Nevada state and local taxes.

This proposal should also include the implementation of interfaces and integration to the following FtA or third-party systems:

- Billing
- Corporate accounting (tax reporting, accounts payable, revenue accounting, revenue analysis)

This proposal should also include support for both real-time and batch tax application transactions.

FtA intends to provide the server hardware, network and desktop platform for this billing system. Those items should not be included in the pricing of this project. However, the proposal should include detailed specifications for the required platform and infrastructure.

Page 2 of 43

Figure 33-3 FtA tax software RFP scope section.

Project structure

Figure 33-4 is the project structure section of the main FtA billing platform RFP. Note that this section is rather open ended, requesting the vendor to propose a structure, rather than imposing one. If your company has some specific project structure requirements, such as the use of a particular subcontractor, this would be the place to detail those requirements.

RFP for Billing System
January 20, 2003

3. Project Structure
Identify the structure you propose to use for this project. Please indicate organization, staffing, and key personnel. If you intend to subcontract any portion of the project, identify all subcontractors, their qualifications, and the work you intend to subcontract.

Page 4 of 43

Figure 33-4 FtA project structure RFP section.

Business requirements

FtA has chosen to append its detailed business requirements to the RFP and refer to those requirements in the body of the RFP. Figure 33-5 shows the first page of the business requirements section of the RFP, while Figure 35-6 illustrates a page of the referenced requirements.

Note that the RFP section provides a framework of required functionality and types of services, while the business requirements appendix contains detailed business requirements that will map directly to system and configuration requirements.

Other requirements

FtA has determined that it wants to see a demonstration of the proposed platform that would include:

- IP usage processing load equivalent to 10 million packets per hour
- Billing operations management interface
- Audit and reconciliation capabilities
- Billing operations user interface
- Customer care user interface

This demonstration is to take place prior to contract signing.

Project schedule

In the project schedule section of the FtA RFP, FtA indicated a desire to begin its billing environment project by April 15, 2003. It also asked vendors to

RFP for Billing System
January 20, 2003

4. Business Requirements

Detailed business requirements are available in Appendix A of this document.

The billing system should provide the following functionality:

- Real-time and batch rating
- Bill compilation and calculation
- Tax determination and reporting or standard interface to third-party tax determination and reporting software
- Flexible bill formatting for multiple media
- Account inquiry and customer care support
- Payment processing and account maintenance support
- Robust treatment and collections capabilities
- Support for external payment processing
- Robust process controls and process management
- Financial reporting
- Sales reporting

FtA intends to offer combined Internet and telecommunications services, primarily to the business market. The billing system will be required to accommodate the following types of service:

- POTS
- 800 and other terminating end charged services
- 900 and other content related services
- Centrex
- Point-to-point private line including local access
- Fixed broadband capacity

Page 5 of 43

Figure 35-5 Sample of FtA business requirements RFP section.

RFP for Billing System
January 20, 2003

Index

Page 11 of 43

Figure 35-6 Sample of business requirements appendix to FtA RFP.

assess the feasibility of completing implementation by September 30, 2003 and conversion by December 15, 2003. These dates would allow FtA to terminate current billing arrangements by the end of calendar year 2003.

Resources available

In the resources section of the RFP, FtA indicates that it will provide conditioned space and power, servers that meet vendor specifications, as well as operating system and data base management system software for the billing environment. It will also provide the data network and user desktop computers.

FtA indicates that a full-time project coordinator from FtA will be available to facilitate interaction with FtA resources. Members of the billing system management team are committed to 25 percent availability to this project and subject matter experts will be made available as needed with limited notice to the project coordinator.

FtA will provide two administrator workstations co-located with the servers, as well as 10 private work areas ("cubicles") with network connections to the FtA network and data center, Internet access and dial-up access to off-net locations. All work areas will be equipped with telephones. FtA will also provide a conference room with a teleconference phone, network and dial-up connections. If required, FtA will also provide private office space for a vendor on-site project manager. FtA expects the vendor to provide personal computers for its professional staff. FtA will provide a training room for 20, with access to the new system, as required.

FtA indicates the desirability of the vendor's performing as much work as possible remotely, to minimize travel expenses and to ensure depth of technical support.

Performance surety and quality

FtA requires a performance bond for the project.

It also has identified multiple "quality gates" within the project. These quality gates are points at which FtA will actively review and accept or reject the work to that point, according to quality standards (generally verification or testing) to be agreed upon by FtA and the vendor. Payment will be contingent upon meeting these quality gates. These agreements will be referenced in the final contract. The quality gates indicated in the RFP include:

1. Platform installation
2. Interface integration and configuration
3. Platform configuration
4. Migration of current accounts and services
5. Turn-up to service

If the vendor is not fully ISO 9001:2000 certified for all processes involving this product, FtA requires review of product quality management processes and standards. For all vendors, FtA requires information on their product support terms, their process for managing and resolving maintenance requests and all product and service warranties.

Terms, payments, and pricing

Figure 33-7 displays the terms, payment and pricing section of the FtA RFP. Note that FtA leaves the actual pricing structure open to vendor proposal, but requires information on the basis of all pricing.

RFP for Billing System
January 20, 2003

9. Terms, Payments and Pricing

Proof of financial responsibility will be required. Proof of compliance with applicable labor laws and regulations will be required. Proof of liability insurance in the amount of $1 million and of errors and omissions insurance in the amount of $1 million—or equal amounts held in escrow for the purpose of indemnifying against general liability and project errors and omissions—will be required.

Provide an estimate for the entire scope of the project. Identify the basis of all estimated charges (e.g., fixed price, time and materials, per module). Indicate maximum charges ("caps").

Provide a payment schedule. All payments should be tied to tangible deliverables. Indicate payment terms.

Provide all payment details including contract contact, payment address, payment method and bank information if requesting EFT.

Page 40 of 43

Figure 33-7 FtA RFP: terms, payments, and pricing.

Summary

In this chapter, we have looked at the process of developing a RFP for a billing system platform. This RFP is then sent to the "short list" of vendors developed as a result of reviewing RFI responses. In Chapter 34, we review the process of evaluating the proposals you receive in response to the RFP. There normally is, and should be, a significant amount of interaction with the vendors as they prepare their proposals.

Chapter

34

Evaluating Proposals

After sending out RFPs to specific vendors, discussing the details of requirements and procurement with them and perhaps participating in demonstrations and presentations of products, your company has received a stack of proposals for your billing system project. This chapter reviews ways to sort through and analyze all this information comparatively and in relation to your company's needs, requirements and budget.

Evaluation Against Business Requirements

Each vendor response should be evaluated for responsiveness to each identified business requirement. On the first pass of evaluation, the "fit" with requirements will be evaluated on a numeric scale. The scale for "fit" should have room for gradations of fit. A 10-point scale works well. Previously, each requirement or class of requirements will have been assigned a level of importance, as discussed in Chapter 28, "Managing and Prioritizing Requirements."

Key project team members should review the results of this quantitative evaluation to ensure overall fit. Sometimes numbers do lie—or at least mislead.

If a requirement is of significant strategic and operational importance, you may want to give some weight to having that requirement met within the vendor's core product rather than by custom development. If the requirement is met within the core product, it is more likely to have comprehensive and ongoing support from the vendor.

Table 34-1 shows a small portion of the FtA proposal evaluation summary for one vendor.

Typically, teams of subject matter experts in each requirements area will review such evaluations across all potential vendors. Representatives from each team will then meet to compare assessments and rank vendors overall on requirements.

TABLE 34-1 Sample requirements evaluation.

Requirements	1–10	Importance	Comments
Customer Level			
Fields/formats	7	Low	Address validation not adequate; no language indicator
Market segments	10	Medium	Flexible; reassignable by algorithm
Media options	3	High	No standard EBPP
Billing options	8	High	
Geographic markets	10	Medium	
Account structures	9	Medium	Needs better support for restructuring subaccounts within a master account
Product/Service Level			
Service categories	7	Medium	Weak on content-based and professional services
Fields/formats	6	Low	Does not have contract info
Billing basis	7	Medium	Not fully and flexibly configurable
Volatility	10	Low	Can easily handle volumes at minimum configuration
Operational and Systems Level			
Functionality	9	Low	Good operational functions
Ease of use	10	Low	Great graphical interface!
Ease of management	7	Low	Needs better queue and work management
Exception handling	5	Medium	Doesn't support online resolution
Auditability	3	High	Logs difficult to read and analyze; some transactions not fully tracked
Sizing/volumes	10	Low	
Billing cycles	10	Medium	
Platform	8	Low	Runs on same servers used in ISP; uses Sybase, not Oracle which is standard in data center
Interface Level			
Types of interfaces	5	Medium	Would require custom development for credit bureau; marginal provisioning API
Specifications/tools	8	Low	
Quality tools	4	High	No real-time metrics; other metrics require unique query
Capacity	10	Low	
Standards Level			
Call detail standards	5	Low	No support for IPDR or TIPHON
Regulatory requirements	10	Medium	
Billing output standards	10	Medium	
Payment processing standards	5	Medium	Supports OFX but not IFX

Evaluation of Vendor and Guarantees

In Chapter 32 we discussed a preliminary review of vendors, their financial stability and their commitment to the product and product line. At this point in the procurement process, all of those dimensions should be revisited with current and more thoroughly researched information.

Revisiting vendor stability and performance

In addition to formal credit reports and vendor-supplied references, search the trade and financial press for news of the vendor firm and product. Identify customers who were not given as references and contact them. Identify any management or ownership changes that might impact stability or commitment to the product.

If your research identifies any troubling information, determine what risk mitigation strategies would be appropriate to the perceived risk. If the risk is primarily financial, financial mitigation is generally manageable. If the risk is strategic, legal, regulatory or operational, you need to carefully consider whether the risks can be mitigated. These then become part of your contract negotiation strategy, as we discuss in Chapter 35, "Negotiating a Billing Solution Contract."

Financial and quality guarantees

What financial and quality guarantees have the vendors offered in their proposal? Will they post a performance bond? Do they carry errors and omissions insurance? What is the financial value of such coverage?

What are the vendors' commitments regarding correction of problems with their software? With any third-party products they may supply? Generally, vendors have a hierarchy of trouble resolution commitments, ranging from immediate resolution of service affecting problems and errors to resolution in weeks, months or the next generic for essentially cosmetic problems or minor operational problems. Your company must ensure that this hierarchy meets your business needs. For instance, something the vendor might consider to be cosmetic may have regulatory importance and have a major impact on your business.

Vendor's quality process

In the FtA RFP, we saw a requirement for ISO 9001:2000 certification or equivalent presentation of process controls and quality management. The authors recommend that your company evaluate the quality processes of a vendor as an important procurement parameter. Typically, platform vendors will have attained ISO certification or be in the process of doing so. However, many integrators have not done so. It is important to a stable environment that the vendor have control processes in place for custom work, as well as for core platform elements.

Evaluation of Support and Maintenance Commitment

What ongoing support and maintenance does the vendor offer? For how long? How much will it cost? What surety will the vendor provide for ongoing support?

Does the vendor routinely update its core product to incorporate new technologies and standards? Does it make these updates easy to install and backwards compatible with existing databases and configurations? Can these updates be performed independently of any proposed customization? This is particularly important to the long-term viability of the platform for your business.

Evaluation of Proposed Project Structure

Does the vendor's proposal incorporate elements that your company can supply economically? For instance, if your company already has a favorable site license and contract for some third-party software such as a DBMS, you may choose to provide that element yourself. If another proposal utilizes software for which you have no other agreement, you may want the procurement to include that software.

Is the vendor proposing to be the ongoing source of support and upgrade for third-party hardware and software or does it merely plan to act as a purchasing agent? If your company, like our fictitious FtA, plans to have extensive in-house technical support and expertise, you may want to deal directly with the third-party vendors. If you have minimal in-house technical resources to support this type of hardware or software, you may determine that "turnkey" support and maintenance responsibility is appropriate.

This is often subject to negotiation between the vendor's initial proposal and a final statement of work for contractual purposes.

Evaluation of Project Schedule

How well does the vendor's proposed project schedule meet your realistic business needs? Does it provide enough time and support for extensive operational testing, user acceptance testing and validation, and personnel training. In Section IV, we look at realistic project schedules and the factors that can create a wide variance in project duration. Does the vendor's schedule appear to conform to these general guidelines?

Often experienced vendors build time in a project schedule to accommodate the lack of timely availability of people and information. If you can assure your vendor that such delays will be held to a minimum on your project, it may be willing to commit to a more aggressive schedule—or at least a more aggressive schedule contingent on resource availability. Certainly, well documented and controlled data collection and requirements will go a long way in reassuring a skeptical vendor.

Evaluation of Project Pricing

Evaluating project pricing is more complicated than you might expect. While some vendors will provide an estimated bottom line based on whatever data were provided in the RFP, more typically they will provide some fixed-price items and some per-item and/or time-based items. It is rare that these cover the same territory in the same way for different vendors.

One of the authors has found that graphical representation of the various pieces of the system, overlaid with notations of what is core product and what is custom development, along with each vendor's pricing structure, can help to sort out "apples and oranges" in pricing. If there are unknowns in your environment (such as number of services, sources or accounts), set up a spreadsheet where you can examine each bid for the best-case and worst-case numbers for those unknowns. Try to make the range fairly wide. In one project in which the authors participated, the client company estimated the potential number of service instances for conversion to be between .25 million and .5 million. When we worked with the legacy systems organization, the actual number was determined to be 1.5 million!

Summary

In this chapter, we have looked at the process of analyzing vendor proposals for a billing systems project. The results of this analysis are used in choosing a potential vendor and in negotiating contract terms. In the upcoming chapter, we walk through that negotiation process and the contents of a contract for a billing systems project.

Chapter

35

Negotiating a Billing Solution Contract

Once you have evaluated the RFP responses, you may have identified a "clear winner" that meets all of your needs or you may have a very small number of "finalists." Even if you have a clear choice, there are still important areas of negotiation before you are ready to sign a contract.

Identifying Negotiation Priorities

If the proposed solution meets all of your business needs, including pricing that is affordable and in line with industry benchmarks, you still have some negotiating to do and we will get to those items in a moment. First let us look at the more common situation in which a vendor has a proposal that gives you close to what you need for 25 percent more than your desired price point. Let us assume that the proposed price does not exceed the maximum feasible amount your company is capable of paying for this project.

"Close to..."

First, identify the ways in which the vendor's proposal does not match your company's requirements. Are any of these strategically impacting? Are they strategically impacting right now?

For instance, your company may have plans to offer wireless location-based services beginning in approximately 24 months. The vendor does not currently have service templates or detail processing for this type of service and has not offered those templates in this proposal.

Find out if the vendor has location-dependent templates and processing under development. See if the vendor will include the enhanced capabilities in the "deal"—perhaps supporting your company as a beta customer—at no additional charge.

What if the vendor has no plans for supporting location-based services and no interest in doing so? Can the vendor help you identify a workaround? Perhaps you can locate other companies that use this vendor who are also contemplating entering the location-based market. Then you can work together with the vendor to fund the development of the relevant capabilities.

If the failures to match requirements are operational rather than strategic, estimate the financial and intangible impacts of the mismatch on your business. If workarounds are required, can the vendor suggest or support the workarounds? Are these acceptable business risks? You can always use the mismatch as leverage to reduce that 25 percent difference from your desired price point.

"More than we want to pay..."

Okay, we all know some vendors put some "wiggle room" into their initial estimates. However, if the vendor wants to do business with you, there will not be much "fat" in its pricing. There may be more flexibility in payment terms, as well as the possibility of the vendor's financing the project at little or no interest, although this is actually more difficult to obtain now than it was a few years ago.

Sometimes restructuring portions of the project can achieve a more acceptable bottom line. For instance, if a server vendor wants to do business with you on other projects, it may give you a more favorable price on servers than you would get if the billing system vendor provides the servers. Recently, companies have been getting great prices on servers by acquiring the assets of failed dot-com ventures at pennies on the dollar. The billing system vendor may be locked into a contract with a server manufacturer for "factory-new" systems only.

Your company may want to defer implementing some functionality or interfaces until they are actually needed, rather than incorporating them into a single large implementation. While you will eventually have to pay for them, your company has the use of that money for a longer period.

You may also want to review what you want the vendor to do and what your own staff will do. Often vendors are charged with extensive data gathering that could be done better and faster by your own staff. Of course, if you have followed the guidelines in Section II for data gathering, you already have a head start on the whole data gathering arena—and have saved yourself significant money in implementation and conversion effort. Be certain that potential vendors are aware of this when you are negotiating price. Vendors will assume a poorly defined and documented environment unless you tell them differently.

Realistic dates

What is the driving force behind your project dates? If vendors indicate that they cannot perform a conversion of over a million legacy records and run two parallel bill cycles in three months, listen to them! Unrealistic project dates have led to more failed projects than any other negotiating criteria. If you force a vendor to commit to doing six months of work in three months, the project will either be late, of poor quality or both.

Help your vendors understand the business drivers for a particular date—they may have a solution that resolves that particular issue without endangering the entire project. If the vendors are not willing to work with you on these priorities, then they may not be the right vendors for you.

Pricing Structures and Options

There are many different ways to structure the pricing of a billing systems project, as discussed in Chapter 34, "Evaluating Proposals." You may want to use one or several of those pricing structures for various aspects of the project. However, your company will at all times want current tracking on costs, as well as the ability to predict and control costs.

A note of caution is appropriate here. We all like to think of ourselves as great negotiators and it is important to ensure that you get the best value for your company's money. However, driving the price of a project below the vendor's cost and some reasonable margin often results in inferior work, inadequate quality control, the use of less-experienced professional resources and/or a financially unstable vendor.

Fixed price

A fixed price to cover standard billing modules and routine professional services, such as software loads, is entirely appropriate. After all, this should be an entirely known quantity to the vendor.

If the vendor is to provide integration with other systems and you have provided it with both good interface requirements and specifications for the "other side" of the interface, the vendor should also be able to provide a fixed price for each interface. Obviously, interfaces that the vendor considers "standard" should be less expensive than any unique or totally custom interfaces.

Per unit

Where some information is unknown or only partially known, a per-unit pricing structure may be very appropriate. It is advisable to identify a maximum approved cost above which the vendor must seek approval and a contract amendment at some minimum time before commencing work beyond that maximum. This is useful for project cost control and for budgeting purposes and improves vendor accountability.

If the total number of billable elements is not identified at the time of contract and the vendor will be providing configuration services that include establishing a database of products and services, then a rate for configuration per product, service or billable feature is appropriate. This is often particularly appropriate for conversions. Conversion costs may be estimated per source, per account converted, per service or product converted—or some combination of all of those parameters.

Time-based

Professional services related to very incomplete work definitions or uncertain work conditions may appropriately be based on vendor time expended. In one way, this is the most open-ended of pricing schemes, since total costs are generally in a fairly broad range with occasional reauthorization.

On the other hand, time-based charges are usually related to specific project resources. You may negotiate for approval of the specific resource and their billing rate. Also, if a resource is not committed fulltime to your project, this may be a way to be charged only for the time spent on your company's project.

When being charged on the basis of vendor time expended, it is extremely important to ensure that the resources a vendor requires from your company are available to the vendor as needed. If a vendor professional resource has to spend three days on your site to get an hour with each of four groups of subject matter experts, you will be charged for three days, not four hours. The exception to this may be some resources who can work remotely on other projects while waiting for your company to get its act together.

Expenses

The contract should contain formal expense reporting points, as well as an estimated "not to exceed" expense figure. The contract should also specify the process triggered when expenses are running higher or lower than estimates at the reporting points, as well as identifying what happens when the "not to exceed" figure is approached.

This portion of the contract should also include any specific conditions for expense items including preferred vendors, maximum rates or classes of service, items allowable in travel expenses and any other conditions.

Preparing for the Unexpected

Major systems projects are rarely free of unexpected or unknown-at-the-time-of-contract circumstances. Over the course of a lengthy and complex project, new information emerges and conditions change on both the vendor and the customer side and in the business environment. The best of system plans cannot anticipate everything that may occur during the project. However, the contract can include strategies that will mitigate any negative effects of unexpected conditions.

Performance guarantees, remedies, and penalties

Requiring specific performance guarantees, identifying remedies for inadequate performance and penalties for failure to perform can mitigate the business impact of a vendor failure. It is important to assess the value of the vendor activity and the related business risk. For instance, if the failure to meet a specified date generates the need to extend service bureau contracts and makes it impossible to enter hot new markets in a timely manner, a value must be placed on both those conditions and addressed by one or more remedies.

Acceptance criteria. Clearly defined and strongly enforced acceptance criteria are the key to enforceable performance guarantees, remedies and penalties. These criteria may be defined fully in the contract or the contract may include a process to establish some of those criteria during the project by mutual agreement.

At a minimum, the contract should identify the points in the project—or "quality gates"—for which acceptance criteria exist or will be established. It should also identify the basis for those acceptance criteria.

Escrows. Anything can be placed in escrow—money, software, and documents—even hardware. There are vendors that specialize in software and data escrow, refreshing media and verifying readability of media periodically.

Financial escrows. The vendor may require that your company escrow part or all of the value of the contract—especially if your company is a new enterprise with little or no credit history. To protect your company's interests, you may choose to require escrow of interim payments until the project reaches a successful completion.

Typically, client companies do not require escrow of vendor funds, as there are other ways of ensuring vendor financial responsibility such as performance bonds.

Source code escrow. Unless the vendor is willing to provide source code for its core product at little or no additional cost, escrow of source code is a very good idea. Requiring escrow of source code can provide some protection for ongoing support and maintenance if the vendor no longer exists or no longer supports the product. If there are other companies using the same core product (do verify release and version!), you may be able to share escrow costs.

Ideally, you will require escrow of all related documentation and specifications as well. Requiring an independent audit of code completeness and readability can provide an additional level of protection.

Performance bonds. The performance bond guarantees that the vendor will complete the contract in accordance with the terms, conditions and specifications of the contract. Many public acquisition laws and regulations require a vendor to post a performance bond—which is basically insurance, paid for by the vendor, against the vendor's failure to perform. Since the bond would be paid by a third party, it provides a guarantee against financial failure of the vendor.

Financial incentives. If there are specific goals of significant value to your company, but which would require extraordinary effort—or luck—on the part of the vendor, you may wish to create financial incentives in the contract for meeting those goals. This is more conducive to good vendor relations than using financial penalties to enforce "stretch" goals.

Financial penalties. Financial penalties are most appropriately brought into play when reasonable performance parameters are not met. For instance, if the vendor exceeds its own completion estimate by 50 percent of the total task or project time, this may be an appropriate condition for assessing strong penalties.

Dependencies in your company

There may be project dependencies within the control of your company, such as the completion of an acquisition or obtaining additional venture financing. These should be identified in the contract, along with potential project impacts and related management processes. For instance, the start of a phase of conversion may be dependent upon the completion of an acquisition. A process should be outlined in the contract for amending the project schedule, payment schedule and other impacted areas of the project.

Vendor dependencies

Similarly, the vendor may be dependent upon an official standards release or some third-party software or hardware availability. These should be identified in the contract—again along with potential project impacts and related management processes. For instance, the start of a phase of conversion may be dependent upon the completion of an acquisition. A process should be outlined in the contract for amending the project schedule, payment schedule and other impacted areas of the project.

Spending caps

In the discussion of pricing options, we looked at applying spending caps. Spending caps should be specified for all variable pricing.

Mutual agreement

The contract should define the conditions and process for establishing "mutual agreement" for project and contract responses to unexpected circumstances.

What Belongs in a Contract?

The basic rule of thumb in contracts is that they should incorporate everything you have agreed upon, including preparations for the unexpected and they should protect your business interests and priorities.

Legal stuff

The authors are not lawyers and will not provide guidance or direction on contract legal boilerplate, phrasing or terms and conditions. For that you need good contract lawyers. If your company and the vendor company are in different legal jurisdictions, engaging a contract lawyer in each jurisdiction, at least for contract review, is very valuable—even when it appears that the relationship of the jurisdictions for contracts is quite clear, as between states in the United States. When, as is often the case, your vendor is headquartered in a different country, this is an absolute imperative.

It is not wise to use a vendor's contract boilerplate. Vendors' lawyers exist to protect their clients and vendor boilerplate often severely restricts client flexibility, enforcement and redress.

Statement of work

The statement of work should state clearly what the vendor is going to do and when, what your company is going to do and when and how they are related. For instance, the statement of work may say, "Vendor X will install the Vendor X Basic Billing Module, version 2.3 on two Computer Company Q HotServers, model A.09, running the Qunix operating system version 3.4. The HotServers will be supplied, installed and configured by From the Ashes Communications, Inc., by May 15, 2003, according to specifications to be supplied by Vendor X by April 1, 2003."

What and when. The statement of work should clearly indicate exactly what is to be done and what the quality measurements for acceptance will be. Often in negotiating quality and acceptance, both the vendor and the company develop a clearer understanding of expectations. If some external reviews will be applied, this is the place to identify the standards to be used in the review. Your company should also identify any procedures for tracking and mitigation of deficiencies and the penalties for unresolved deficiencies.

"When" certain work is to be started and completed can be stated as calendar dates or as a time in relation to some other event. For instance, platform installation could be specified to begin within 10 business days after verification of the servers and operating systems.

Relationships and dependencies. To prevent misunderstandings, illustrating dependencies for each section or paragraph of the statement of work within the statement of work is very useful. Sometimes contract negotiators are unwilling to include such "project management" detail, but the lack of clarity at this level often results in time-wasting and costly contract disputes that do not benefit either party.

Unknowns. Sometimes there are things neither you nor the vendor know at the time you write the statement of work. This is often the case when establishing a contract for conversions from systems not under your control or when moving to a new technology or beta product. It is wise to acknowledge these unknowns and how they will be handled when further information is available. Some possible ways to handle unknowns include provisions for updating the statement of work and other contract terms, the negotiation of an ancillary contract or establishing payment terms (such as pure time and expenses billed monthly) and information-dependent schedules subject to review by both the vendor and your company to encompass these areas of uncertainty.

Schedule

Generally, a contract should contain a high-level project schedule that identifies the significant quality gates in the project. Uncertainties and dependencies should be identified and acknowledged here. A process should be clearly established for review and amendment of this schedule.

Payment schedule

The payment schedule should clearly identify what the conditions are for each payment and how quickly after the conditions are met the payment must be made. The payment schedule section also usually includes information about how the payment will be made, including payee, destination, bank details and other relevant information.

Contract administration

Contract administration responsibilities and contacts should be clearly spelled out for both the vendor and the client company. If there are multiple administration or management contacts, their roles should be clearly identified and responsibilities detailed.

Summary

In this section, we have reviewed all the steps in specifying and procuring a new billing system for a communications enterprise. In Section IV, we turn our attention to managing the billing system implementation and conversion projects. These activities cannot exist in totally separate spheres. Project planning activity will impact the project schedules incorporated into any vendor contracts for part or all of the billing system. Likewise, requirements and specifications will impact the project plan, as will final agreements with the vendor. Maintaining good communications between project management and the procurement teams throughout the entire project—starting with the definition of project scope and direction and extending through the early stages of billing production—is very important to project success.

Section

IV

Implementing a Billing System

Section IV takes a look at a billing system project from the project management perspective. We look at scheduling, resource planning, tracking and quality assurance for all aspects of project preparation, implementation, conversion and integration of a billing environment.

There are obvious overlaps to information covered in previous sections. However, the earlier sections were targeted at business analysts and decision makers. The chapters in this section take the project management perspective. Where processes and activities have been defined in Sections II and III, this section will simply identify the project management parameters associated with those processes and activities. Where processes and activities appear in Section IV for the first time, the processes and activities will be reviewed in more depth. For the most part, precontract activities are well defined in earlier chapters, while postcontract activities will require some detailing in this section.

Section IV is keyed to two sets of linked project plans. One set is designed as a template for you to modify for your own project. The task lists for these project plans are provided in this chapter and a corresponding Microsoft Project® project file is on the included CD. A second set, based on the template set, contains project plans for the billing system implementation, conversion and integration for the fictitious company we introduced in Section III, From the Ashes Communications, Inc.

Each set is comprised of five project plan files. One file covers activities that are common to implementation, conversion and integration. A second covers implementation-specific activities, while a third covers migration-specific activities and the fourth focuses on integration activities. In addition, there is an overall project file that provides external links to each of the subprojects.

Many billing systems projects will involve all of these activities and both the template and the sample project plans incorporate links between the project plan files as appropriate for such a comprehensive project.

If you are implementing a billing system for a new enterprise that has no existing services, you may need only the common, implementation and integration files. If you are planning a migration of current customers from multiple systems to one due to a merger, you may want to utilize the common and migration templates only. If your company has implemented a stand-alone billing system and now wants to integrate that system with other corporate systems, you may need only the common and integration plans.

While each portion of the overall project plan includes the appropriate testing and validation activities, we will look at testing as a separate topic. Testing is critical to the billing environment throughout the system life cycle. Good testing practices in the billing environment can be essential to your company's long-term financial health.

The sample project plans for From the Ashes Communications (FtA) make use of the effort-based scheduling feature of MS Project® to develop the project schedule. This illustrates the real-world scheduling constraints of staffing and expertise.

Decision makers and project planners should note that the FtA project plan was developed as if initial data gathering began in June 2002. The contract for a new platform is signed in Spring 2003 and final conversion activity concludes almost exactly two years after the project began, in June 2004.

Chapter

36

Overall Project Plan

The overall project plan is little more than a connection point between the individual project plans for preparation, implementation, migration and interfacing. It allows a view of an entire project from the perspective of overall duration and facilitates linkages between the subprojects.

The overall project plan may be more elaborate, containing more information about linkage points and shared resources. In this simple form, it provides a view of total project duration and dependencies.

Template for Overall Project

Table 36-1 shows the tasks in the template for an overall project. "Precontract activities" is an external start-to-start link to the common project plan; "interface implementation project" is an external link to the interface project plan; "implementation project" is an external link to the implementation project plan; and "conversion project" is an external link to the conversion project plan.

TABLE 36-1 Overall Project Tasks

Task Name	Duration
Overall billing project	283d
Start project	0d
Precontract activities	30d
Interface implementation project	149.5d
Implementation project	253d
Conversion project	240d
Complete overall billing project	0d

FtA Overall Project Plan

Table 36-2 shows the overall project plan for the FtA billing system project. Note that the Conversion project extends beyond the Implementation project. This is due to the decision to delay conversion of the CLEC services until after system turn-up to service.

TABLE 36-2 FtA Overall Project Task List

Task Name	Duration
Overall billing project	415d
Start project	0d
Precontract activities	90d
Interface implementation project	203d
Implementation project	304d
Conversion project	325d
Complete overall billing project	0d

Planning with Resource Constraints

Throughout this section, we will be examining the impact of resource constraints on project schedules. The duration variance between the template project plan and the FtA overall project is primarily the result of resource constraints, although a few tasks have been added or lengthened.

Most vendors of comprehensive billing platforms provide optimistic implementation project plans similar to the template. In planning your billing project, it is critical to identify resource constraints and their impact on your project plan. In the FtA plan, we have assumed that the vendor is unconstrained for resources, but that is also never a good assumption in the real world. One of the authors once worked on a project where the platform installer's immigration difficulties delayed the entire project by a week!

Developing a workable project plan requires a solid understanding of your business, the system requirements, available expertise and resources, as well as a good grasp of the complexity of your project and environment. The "science" of project planning is in mastering and incorporating all the pertinent details in your project plan.

There is, however, an "art" side to project planning. Intangible information is often as important as or more important than quantitative and calendar information. Often the organization culture has a significant impact on project schedules. How does your organization make decisions? Is there a single decision maker who receives input directly and makes decisions immediately? Is there a single decision maker who receives recommendations from layers of subordinates and consultants? Is decision making a group process of seeking consensus?

Once you have developed a project view from the perspective of your company, it is important to integrate that with the vendor's project view. You may find some unexpected constraints and scheduling issues!

Summary

In this chapter, we took a quick look at a project plan whose only purpose is to provide an overview of and linkage between other project plans. This is occasionally a useful technique for developing an overview of end-to-end project duration, but of little other practical value. However, the same technique can be used to connect project plans at points where resources must be shared and coordinated.

In the upcoming chapters, we will look at the contents of the "common" preparation phase project plan, the billing system implementation project plan, the interface implementation project plan and the conversion project plan.

Chapter

37

Preparation Planning

All of the preparation tasks should take place prior to contracting for a new billing system. In the event that this has not happened—perhaps your management did not read this book—these tasks will have to be incorporated into the project plan and schedule. This may impact the actual start date for the vendor and will certainly impact subsequent project dates.

Table 37-1 shows the common preparation tasks for a billing system project as they appear in the MS Project® project template. The tasks are listed under "task name." The duration shown is based on totally unconstrained resources and minimal coordination requirements. The task number of the predecessor task in the "Pred." column indicates task dependencies.

In this table, "*Pre-Contract Activities" is a summary task—its duration is "rolled up" from the duration of subtending tasks and their dependencies. Both Task 2 "gather information" and Task 9 "develop business requirements" are summary tasks rolled up from the subtending tasks listed below the summary tasks. When duration rollup is performed—if resources are not considered—subtask duration and dependencies are considered. Since "develop business requirements" is dependent upon the completion of "gather information," the five days required for "gather information" is added to the 25 days for "develop business requirements." If the dependency did not exist, the duration for "*precontract activities" would be 25 days.

As we look at each of these tasks, we will be considering individual units of work that have a fixed duration with a specific assignment of resources. The duration should not be considered subdividable in most cases. For instance, if a task requires one analyst and one subject matter expert (SME) for (the same) two days talking to each other, adding additional analysts or SMEs will not shorten the task duration—and in the real world, may add significant overhead for information handoffs. (The classic project management example of this is scheduling nine women to produce a baby in one month when the fixed normal duration of human gestation is nine months.)

TABLE 37-1 Preparation Phase Task List

#	Task Name	Duration	Pred.
1	* Pre-Contract Activities	30 days	
2	Gather Information	5 days	
3	Gather Product and Service Information	2 days	
4	Consolidate and Analyze Product and Service Information	2.25 days	3
5	Gather Customer and Account Information	3 days	
6	Gather Tax and Regulatory Information	2 days	
7	Gather Interface Information	3 days	
8	Gather Business Process Information	5 days	
9	Develop Business Requirements	25 days	2
10	Develop Customer-Related Requirements	10 days	5
11	Develop Product/Service Level Requirements	10 days	3
12	Develop Operational and Systems Level Requirements	15 days	2
13	Develop Data Interfaces Requirements	20 days	7
14	Develop Standards Requirements	5 days	10, 11, 12, 13

Gather Information

Section II of this book provides extensive guidance on the data gathering process and supporting tools. The time required for information gathering is determined by the time required to perform the subtending tasks.

Gather product and service information

Under ideal circumstances, gathering information on a single product or service should require two duration days. During those two duration days, at least one business analyst will be involved full time. In addition, several SMEs will be involved part time—totaling to two full effort days for SME resources. The SMEs will be drawn primarily from marketing, sales and billing operations, although input from provisioning and network engineering may also be required. Also, any current products or services will require some systems analysis and programming support, totaling perhaps a half day's effort, to extract information from the old billing systems.

In a world of unconstrained resources, then, all product and service data gathering would be completed in two calendar days using 2.25 (one analyst, one SME, and .25 systems expert) person-days per day for those two days—or a total of 4.5 person-days per product or service. In the real world, of course, resources are not unconstrained. Your project may have five analysts available for product and service data gathering and have 40 services to research and document. Just based on analyst availability, this would extend your product and service data gathering to 16 days.

Initially, you may not have a good estimate of the number of products and services and may have to adjust this information in response to preliminary data gathering.

Consolidate and analyze product and service information

Consolidating and analyzing product and service information is a function of the volume of that information. However, as the number of products and services that share some characteristics increases, the time per service decreases. Analysts will perform most of the consolidation and analysis of product and service information, but appropriate subject matter management—from product management, billing management, network engineering and interfacing systems—will be required to review and validate the analysts' output.

As an estimating rule of thumb, plan for one day of effort per old system plus one day of effort per old line of business plus .25 day for each service. Assume that one analyst and one equivalent subject matter manager are required for all effort. In the real world, increasing the number of analysts may incrementally reduce the total calendar time, since they are working with multiple subject matter managers. However, if your company is small, the availability of subject matter managers may become a critical path issue for your project.

Gather customer and account information

Customer and account information gathering effort is driven by the number of old billing systems, number of market segments and by corporate strategy for customers and accounts. Assume three days of calendar time for collecting information on corporate strategy and three days each for the old billing systems. Add one day for each market segment. If market segments will be realigned during conversion, count each current market segment and each planned market segment for one day each. As in most data gathering, assume that an analyst and an equivalent SME will be required full time for the customer and account information gathering activities.

Gather tax and regulatory information

If your company has well-organized tax and regulatory functions, plan for one day per tax jurisdiction with one day of effort on the part of an analyst and one day of effort from a tax SME. Also plan one day per regulatory jurisdiction with one day of effort on the part of an analyst and one day of effort from a regulatory SME.

Gather interface information

Plan for three days of calendar time for an analyst and an equivalent SME for each major interface function or system. This three days should not be subdivided among multiple analysts.

Gather business process information

If your company has well-defined and documented business processes, including cash flow management processes, this becomes a matter of accessing those documents and understanding the impacts on the billing process and system. This can be accomplished in five business days utilizing an analyst and one equivalent process manager (SME) resource.

If your company does not have well-defined and documented business processes, this may require the development of such processes and documentation. Since this would require a significant effort as well as executive review, assume at least 10 business days—each requiring the full-time effort of a process analyst and an equivalent SME resource. If the process analyst is not also a systems business analyst for the billing system, add an additional business day for knowledge transfer involving the process analyst and the billing system business analyst full-time.

Develop Business Requirements

Chapters 21 through 28 of this book provide guidance and tools for developing business requirements for a new billing system. Again, the duration for developing business requirements is driven by the time required for and dependencies of the subtending tasks.

Develop customer-related requirements

Chapter 22 provides information on customer-related requirements. In order to estimate scheduling for the development of customer-related requirements, identify one business day per account structure type, three days for all billing option requirements—although more may be required in a complex billing environment, one day per new market segment and one day per media option. For U.S. state markets, calculate one day for the U.S. market and .25 days for each state. For international markets, calculate one business day per market. Each business day should be calculated utilizing one analyst and one equivalent functional executive from marketing, customer care and/or billing operations. Ten days is a minimum duration of this activity for a very simple customer environment.

Develop product/service level requirements

Chapter 23 provides extensive guidance on the content and development of product and service level requirements. In order to estimate scheduling for the development of product and service level requirements, assume six business days of baseline effort plus three days for each product/service category and one day for each billing basis. Each of the days of effort will require one analyst and two SME equivalents from a variety of disciplines.

Develop operational and systems level requirements

In Chapter 24, operational and systems level requirements are detailed. Scheduling for these requirements in a moderately complex business environment should be approximately 15 days, utilizing one analyst and one equivalent SME, primarily from Billing Operations. If your environment is unusually simple, 10 days may be adequate. More complex enterprise environments may require additional time and resources for additional review and approval.

Develop data interfaces requirements

Information on the development of data interfaces requirements can be found in Chapter 25. Business requirements for data interfaces are complex and "business requirements" for integration of data interfaces require some technical specifications, as well. Planning for the development of data interface requirements should include two days for the general interface environment utilizing an analyst and one SME equivalent, usually from technical support or technical architecture, and 10 business days for each interface utilizing one analyst, one SME resource from the interfacing system or function and .25 SME resource from process management per day. The minimum time for such activities in a very, very simple environment will be 20 business days.

Develop standards requirements

Standards requirements are derived from other requirements areas. Plan for a total of five business days utilizing .5 analyst and .5 standards SME per day to identify current standards and contents and develop and review formal requirements. If your company does not maintain a technical standards organization, this activity may take longer.

FtA Common Preparation Tasks

In this preparation phase, FtA has assigned five business analysts to the billing system project. In addition, a number of subject matter experts have been assigned to support the project as required, with a commitment of either 50 or 100 percent of their time when needed. Table 37-2 displays a table of these resources.

TABLE 37-2 FtA Resources for Preparation Phase

Resource Name	Initials	Group	Maximum Effort
Analyst 1	SD	CTO	100%
Analyst 2	RD	CTO	100%
Analyst 3	CD	CTO	100%
Analyst 4	WD	CTO	100%
Analyst 5	MD	CTO	100%
Business services	AP	MKTG	50%
Consumer services	CP	MKTG	50%
Internet services	IP	MKTG	100%
Professional services	PP	MKTG	100%
Consumer sales	CS	SALES	100%
Business sales	BS	SALES	50%
Sales systems	SS	CIO	100%
Billing operations 1	BA	BO	100%
Billing operations 2	BB	BO	50%
Old billing systems 1	OA	CIO	100%
Old billing systems 2	OI	ISP	100%
Old billing systems 3	OC	CLEC	100%
Provisioning 1	PA	OPS	100%
Network engineering 1	NEA	ENG	100%
Tax	T	TAX	100%
Regulatory	R	REG	100%
Sales and provisioning process	SP	EXEC	100%
Financial processes	FP	EXEC	100%
Billing process	BP	EXEC	100%
Technical architecture	TA	CTO	100%
Provisioning systems	PS	CIO	100%
Financial and fraud systems	FS	CIO	100%
Standards	SD	CTO	100%
Executive	EV	EXEC	50%

As we look at the FtA Preparation Phase project plan, we identify how resource constraints affect a project schedule.

We also look at other real-world impacts on a project. What if complete information simply is not available to the project at this time? FtA is a company just being formed. Some decisions have not been made. Negotiations for the CLEC have barely begun and the CLEC is not releasing any information not already in the public domain. This may require deferring some activities or portions of activities until a later project phase.

Also, sometimes a knowledgeable project manager may determine that a particular instance of an activity or service configuration is so common in the industry or is completely specified by a formal industry standard that less effort is required for that item than the "rule of thumb."

Except where otherwise indicated, tasks in the FtA project plan are initially "fixed-duration" task types in MS Project®. Table 37-3 shows the FtA Preparation Phase plan without resource consideration. While there is additional detail specific to FtA's business, the three days' deviation in overall duration from the template is due to a project planning shortcut used in formulating tasks 36 through 40 that made assumptions about resources.

TABLE 37-3 FtA Preparation Phase without Resources

#	Task Name	Duration	Pred.
1	* Precontract activities	33d	
2	Gather information	11d	
3	Gather product and service information	2d	
4	Gather info on BB I service 1	2d	
5	Gather info on BB I service 2	2d	
6	Gather info on BB II service 1	2d	
7	Gather info on BB II service 2	2d	
8	Gather info on ISP service 1	2d	
9	Gather info on ISP service 2	2d	
10	Gather info on ISP service 3	2d	
11	Gather info on CLEC service 1	1d	
12	Gather info on CLEC service 2	1d	
13	Gather info on CLEC service 3	1d	
14	Gather info on CLEC service 4	1d	
15	Gather info on CLEC service 5	1d	
16	Gather info on CLEC service 6	1d	
17	Gather info on CLEC service 7	1d	
18	Gather info on CLEC service 8	1d	
19	Consolidate and analyze product and service information	4.75d	
20	Analysis—services	3.75d	3
21	Analysis—BB I system	1d	4, 5
22	Analysis—BB II system	1d	6, 7
23	Analysis—ISP system	1d	8, 9, 10

continued on next page

TABLE 37-3 FtA Preparation Phase without Resources (continued)

#	Task Name	Duration	Pred.
24	Analysis—CLEC system	1d	11, 12, 13, 14, 15, 16, 17, 18
25	Analysis—LOB BB	1d	4, 5, 6, 7
26	Analysis—LOB ISP	1d	8, 9, 10
27	Analysis—LOB CLEC 1	1d	11, 12
28	Analysis—LOB CLEC 2	1d	13, 14, 15
29	Analysis—LOB CLEC 3	1d	16, 17, 18
30	Gather customer and account information	6d	
31	Gather corporate strategy information	3d	
32	Gather account info—BB I system	3d	
33	Gather account info—BB II system	3d	
34	Gather account info—ISP system	3d	
35	Gather account info—CLEC system	2d	
36	Old market segments	6d	
37	New market segments	5d	
38	Gather tax and regulatory information	4d	
39	Gather tax information	4d	
40	Gather regulatory information	4d	
41	Gather interface information	11d	
42	Gather business process information	5d	
43	Gather product management process info	2d	
44	Gather provisioning process info	3d	
45	Gather billing process info	5d	
46	Gather payment and collections process info	2d	
47	Gather accounting process info	3d	
48	Develop business requirements	22d	2
49	Develop customer-related requirements	5d	30
50	Account structure requirements	5d	32,3 3,3 4, 35
51	Billing option requirements	3d	31, 45
52	Market segment requirements	5d	36, 37
53	Media option requirements	4d	31, 45
54	Market requirements	1.75d	31
55	Develop product/service level requirements	6d	3
56	Baseline P/S requirements	6d	
57	P/S type requirements—LEC	3d	

continued on next page

TABLE 37-3 FtA Preparation Phase without Resources (continued)

#	Task Name	Duration	Pred.
58	P/S type requirements—IXC	3d	
59	P/S type requirements—ISP	3d	
60	P/S type requirements—T&M	3d	
61	P/S type requirements—APP	3d	
62	P/S type requirements—MSP	3d	
63	P/S type requirements—IPP	3d	
64	Billing basis	3d	
65	Develop operational and systems level requirements	15d	30, 42
66	Develop data interfaces requirements	10d	41
67	Develop standards requirements	7d	49, 55, 65, 66

Gather information

FtA has decided to gather, document and put under change control as much information as possible as early as possible in its billing systems project. There are some pieces of information that do not exist at this point in the project. For instance, many of the interfacing systems have not yet been selected, although their functionality and the need for an interface has been identified. Other pieces of information, such as some of the CLEC proprietary information, have not yet been made available to FtA.

Gather product and service information. FtA's project planner has expanded the summary "gather product and service information" task, represented by a single task line on the template project plan, into subtasks. This will facilitate resource assignment and identify resource constraints on the schedule.

FtA knows that each of the broadband companies provides fixed bandwidth point-to-point services for a recurring monthly charge, as well as technical services (engineering, coordinated testing, etc.) on demand for time and materials charges. Hence, the four tasks (tasks 4 through 7) are configured for services of the two broadband companies. In its due diligence with the ISP, FtA has learned that it offers a basic consumer service and two levels of business service—driving the 3 ISP product and service information tasks (tasks 8through 10).

From published sources concerning the CLEC—including service brochures and the company Web site, FtA has been able to identify seven unique service offerings in the business and consumer markets. Through industry contacts, FtA is also aware that much of the CLECs business is based on unique customer contracts, resulting in eight tasks (tasks 11 through 18) for CLEC product and service data gathering. Since this information will be high level until the CLEC due diligence is complete, the FtA project planner has allotted only one day to do data collection for the CLEC services. This will drive a need for additional data gathering in later project phases.

Consolidate and analyze product and service information. FtA will have four "source" systems for current products and services—illustrated by four separate tasks in the FtA project plan—tasks 21 through 24. The FtA project planner is aware that the BB I company has a very simple billing system and a very small number of service records, while the CLEC has a much more complex system and many more services and service records. However, FtA has all requisite information to analyze on the BB I system and only limited information on the CLEC system, so this seems adequate for planning purposes at the moment.

The lines of business in the BB I and BB II companies are similar enough to be treated as a single line of business. The ISP considers all of its services to be a single line of business, while the CLEC apparently has consumer, business switched and business private line lines of business. There are five tasks for line of business analysis: tasks 25 through 29.

Gather customer and account information. The FtA project planner has established a task (task 31) for gathering corporate strategy information at the recommended duration. She has also established individual tasks (32 through 34) for gathering customer and account information from each of the old billing systems, also at the recommended duration.

Due to the lack of access to CLEC information, the FtA project planner has determined that only two days can be spent productively gathering customer and account information for the CLEC (task 35), instead of the "rule of thumb" three days.

The FtA planner has also taken a shortcut in establishing market segment-related tasks. She knows that the same personnel will be involved in gathering old market segment information for all segments, so she has created a single task—task 36—for that. Likewise, she has created a single task—task 37—for gathering information on new market segments. This can be a bit risky in an environment with shifting resources and priorities, but it can also simplify the project plan for discussion and presentation, as well as for tracking.

Gather tax and regulatory information. Again, the project planner has taken the shortcut of combining all the tax and regulatory tasks into one tax task, task 39, and one regulatory task, task 40. Once original resource assignments are made to this task, she will designate the task as being of the "fixed work" type, so that the availability of additional resources may modify the total duration.

Gather interface information. Since FtA is a newly forming enterprise, many decisions about interfacing systems were not made at the time this preparatory work was to occur. Therefore, FtA scheduled only one day per interface type at this time, with the caveat that further data gathering will be necessary later in the project. This is task 41.

In a mature environment where the new billing system is to be integrated to existing interfacing systems, all of the interface data gathering could take place in the preparation phase. This is somewhat more efficient than splitting the task, minimizing the need for refamiliarization and knowledge transfer.

Gather business process information. FtA has established a process development organization that is charged with working with functional managers to establish business processes and metrics. This group resides in the chief technology officer's organization. While these processes have not been completed or approved, initial process work will provide guidance for initial requirements developments. Tasks 43 through 47 cover the process development activities.

Develop business requirements

FtA plans to develop, document and place under change control all business requirements for the new billing system as early as possible in the overall project.

Develop customer-related requirements. The FtA project planner has allocated five days to account structure requirements, based loosely on the idea that there will be simple consumer accounts, complex consumer accounts, simple business accounts, complex business accounts and "new concept" accounts. Note that the project manager has again taken the shortcut of "lumping" this work into one task, task 50. Three days have been scheduled—task 51—to develop billing option requirements based on an assessment that FtAs requirements will be typical in complexity. FtA will have five market segments going forward and requirements will be developed in task 52. It plans to bill via traditional paper bills, via credit card, via Internet-based EBPP and via CD; requirements will be developed in task 53. FtA has three state markets, all in the United States; requirements will be developed in task 54.

Develop product/service level requirements. The acquisitions completed or planned by FtA include LEC, IXC, ISP, and technical services. In addition, FtA plans to offer Internet hosting (IPP), applications services (APP), and management services (MSP). FtA has usage-based services, fixed recurring charge services and as-incurred time and expenses offerings. These requirements are developed in tasks 56 through 64.

Develop operational and systems level requirements. The FtA project planner has determined that the proposed FtA business environment will be moderately complex, especially in relation to operational and systems controls. She has, therefore, scheduled this task, task 65, for a 15-day duration.

Develop data interfaces requirements. While developing comprehensive data interfaces requirements for the 11 interfaces identified by FtA would require two days of general architecture and management requirements and 11 tasks of 10 days' duration each, not enough information is yet available on interfacing systems to utilize that many resources at this time. An overall task of 10 days will establish the general architecture and management requirements and will allow confirmation of plans for interfaces and a high-level description of the information flow between systems. This is task 66.

Develop standards requirements. FtA has hired an expert in telecommunications standards and an expert in systems standards to be the core of a permanent standards group within the CTO organization. They will provide support to the standards definition task, task 67.

FtA task schedule (resource balanced)

Table 37-4 shows the high-level tasks in the FtA Preparation Phase plan with resource balancing. Note that the unconstrained 30 to 33 days has increased to 90 days with realistic constraints of resources and expertise.

TABLE 37-4 FtA Preparation Phase Resource Leveled

Task Name	Duration
Precontract activities	90d
Gather information	43d
Gather product and service information	15d
Consolidate and analyze product and service information	34d
Gather customer and account information	34d
Gather tax and regulatory information	8d
Gather interface information	11d
Gather business process information	35d
Develop business requirements	47d
Develop customer-related requirements	28d
Develop product/service level requirements	40d
Develop operational and systems level requirements	15d
Develop data interfaces requirements	10d
Develop standards requirements	7d

Summary

In this chapter, we looked at the data gathering and business requirements development tasks that are common to all billing systems projects. In the next three chapters, we take a look at what comes next—billing system implementation in Chapter 38, interface implementation in Chapter 39 or migration in Chapter 40.

Chapter

38

Planning Your Implementation

In this chapter, we look at planning a billing system implementation from the development of an RFI to having a tested, working billing environment with all services defined in the system and all billing output configured for the appropriate delivery media.

Precontract Activities

Implementation precontract activities are those activities leading directly to a contract for billing system implementation. These activities are dependent upon the completion of the common precontract activities discussed in Chapter 37. In the MS Project® file for the implementation plan, an external link has been established between the first task in the implementation plan and the completion of business requirements.

Table 38-1 shows the tasks in the template billing system implementation plan, with their associated durations and predecessor tasks.

Request for information process

The development of an RFI and assessment of responses is only slightly impacted by resources. Therefore, these activities are treated as fixed duration with variable effort and resources within the project plan.

Develop and distribute RFI. The development of an RFI is reviewed in Chapter 31. With good requirements, the actual development of an RFI requires only one to two days. The rest of the duration for developing and distributing an RFI involves some time for executive approval, time for vendor identification and time for physical distribution and/or publication of the RFI to vendors.

TABLE 38-1 Tasks from Implementation Project Template

#	Task Name	Duration	Pred.
1	Precontract activities	171d	
2	Develop business requirements	25d	
3	Request for information process	51d	Common\10
4	Develop and distribute RFI	15d	
5	Analyze RFI responses	20d	4
6	Proposal process	109d	3
7	Develop RFP	25d	
8	Obtain vendor nondisclosures	2d	
9	Distribute RFP	2d	7, 8
10	Analyze proposals	30d	9FS+45d
11	Review vendor demos, etc.	5d	10
12	Select vendor pending contract negotiations	2d	11
13	Develop initial project plan	4d	6
14	Identify internal resources and availability	2d	
15	Identify third-party resources and availability	2d	
16	Identify scheduling constraints	2d	14, 15
17	Integrate resources, constraints, and vendor schedule	4d	
18	Negotiate contract	10d	12
19	Sign contract	1d	13, 18
20	Implementation phase	82d	19
21	Familiarization	5d	
22	Develop issue and risk management process	2d	19
23	Develop status reporting process	3d	19
24	Finalize high-level project plan	5d	13
25	Kickoff meetings	2d	21
26	Platform implementation	26d	
27	Platform deployment	18d	
28	Specify platform elements	1d	19
29	Procure platform elements	15d	28
30	Install platform elements	2d	29
31	Software installation—core software	8d	27
32	Install required infrastructure software	2d	27
33	Install core billing software	2d	32
34	Configure infrastructure and core billing software	2d	32
35	Install user interface software	2d	33

continued on next page

TABLE 38-1 Tasks from Implementation Project Template (continued)

#	Task Name	Duration	Pred.
36	Load vendor-configured data	2d	33
37	Installation/configuration knowledge transfer	1d	34, 35, 36
38	Software install verification, review, and sign-off	1d	37
39	Develop functional specification	10d	
40	Communicate product, business, and network data	7d	
41	Provide product and business process documents	1d	
42	Provide equipment and network inventory documentation	1d	41
43	Review data gathering documents with vendor	5d	42
44	Communicate billing requirements	3d	
45	Provide billing requirements	1d	
46	Review billing requirements with vendor	1d	45
47	Develop configuration specification	10d	
48	Match requirements with system capabilities	1d	43SS, 44FF
49	Define product configuration	8d	
50	Define customer and account configuration	2d	
51	Define system level configuration	6d	
52	Define user interface configuration	3d	
53	Define bill output configuration	3d	
54	Define reports configuration	1d	
55	Define security configuration	4d	
56	Review and approve functional specification	2d	49, 50, 51, 52, 53, 54, 55
57	Implement functional specification	13d	47
58	Implement product configuration	13d	
59	Implement customer and account configuration	1d	
60	Implement system level configuration	5d	
61	Implement user interface configuration	4d	
62	Implement bill output configuration	4d	
63	Implement reports configuration	1d	
64	Implement security configuration	2d	
65	Perform vendor functional validation	6d	57
66	Update functional specification	1d	57, 65
67	Configuration knowledge transfer	1d	66
68	Training	33d	

continued on next page

TABLE 38-1 Tasks from Implementation Project Template (continued)

#	Task Name	Duration	Pred.
69	Develop training plan	3d	
70	Train the trainers	5d	69
71	Train key users	5d	70
72	Train all end users	20d	71
73	Acceptance test	33d	65
74	Define scope of acceptance test	1d	48
75	Specify test environment	5d	74
76	Create test environment	5d	75, 56
77	Define test scenarios	2d	47, 74
78	Define test cases	10d	77
79	Perform functional acceptance test	15d	78, 65
80	Perform user acceptance test	5d	71, 79
81	Operational test	42d	
82	Define scope of operational test	1d	74
83	Specify test environment	2d	75, 82
84	Create test environment	2d	76, 83
85	Define test scenarios	3d	77, 84
86	Define test cases	10d	78, 85
87	Perform operational test	10d	80, 86
88	Configure for production	10d	87
89	Turn system up for live service	1d	

Analyze RFI responses. Analysis of RFI responses is detailed in Chapter 32. Again, this is not a resource-constrained task, but the duration may be impacted by the quantity of valid responses.

Proposal process

Typically, the proposal process is dependent on the RFI process and is itself the predecessor task of the contract signing.

Develop RFP. The contents of a billing system RFP are presented in Chapter 33. The duration of this task is impacted by the complexity of the RFP, as well as by the quality of RFI responses. It may also be impacted by the complexity of your team structure and management structure, as these can affect review and approval time.

Obtain vendor non-disclosures. Frequently, good RFPs contain critical strategic and operational information about your company. RFPs should only be distributed to vendors that have appropriately executed non-disclosure agreements with your company.

Distribute RFP. Since an RFP is generally considered sensitive proprietary information, it is typically distributed in controlled hardcopy, which requires at least one to two days. If your company has incorporated a formal question period for potential vendors, this task should incorporate subtasks for delivery of the RFP, for receiving vendor questions (typically between 10 and 30 days depending on the complexity and uniqueness of the project) and for replying to vendor questions. Ten days is usually sufficient for this activity. Often both the receipt of questions and the replies are handled via email or secure Web site.

Analyze proposals. This plan allows a 45-day interval from distribution of the RFP to the start of proposal analysis. Thirty days is allocated for the actual review of proposals. Chapter 34 provided a look at the proposal review process.

Review vendor demos, etc. Often RFPs will incorporate a request for a comprehensive demonstration of the vendor's platform or a "proof of concept" demonstration for some custom feature. This often entails off-site travel or lengthy on-site meetings. This can be impacted by the number of vendors. The template plan assumes three serious vendors.

Select vendor pending contract negotiations. The vendor selection process itself usually requires team meetings and executive approval. The allocated two days assumes a streamlined team structure with executive participation. If your team and management structure is more extensive or not co-located, this activity will require additional time.

Develop initial project plan. Once the vendor is selected, your team should work with the vendor to develop a preliminary project plan. The agreed-upon plan will be incorporated in the contract. The four days allocated to this activity assumes that the vendor has a standard project plan template and that your team has access to information about your company's resources and availability. Typically, the vendor will be able to supply information on its proposed activities and anticipated need for resources from your company.

Identify internal resources and availability. The two days allocated for this activity assume that key resources have previously been identified and availability determined. This then becomes the process of matching proposed schedule to resources. Note that resources may include nonpersonnel resources such as financing, shared computing and network resources and space.

Identify third-party resources and availability. If your project involves third-party resources to be provided by your company, they should be identified and tentatively scheduled in this task.

Identify scheduling constraints. Are there scheduling conflicts or overloads? Your project management system can help identify these if you provide it with scheduling and resource information.

Integrate resources, constraints, and vendor schedule. This requires meeting with the vendor project planner, combining information and negotiating a mutually agreeable schedule.

Negotiate contract

The process of contract negotiation is outlined in Chapter 35. The template project plan allows 10 days for contract negotiation. If you have complex or unique requirements or if the vendor is new to the product or the market, this may take longer.

Sign contract

While signing a contract may seem a matter of moments, it generally involves a final review of terms by the legal departments of both companies, as well as some internal review and recommendation before the contract is actually signed.

Implementation Phase

The implementation phase includes all the tasks to deploy, configure, test and turn up a billing system to service. These activities are dependent upon contract signing since they involve vendor resources. This plan assumes a configurable, off-the-shelf vendor system with no custom software development other than through a preconfigured reports writer. If your implementation includes custom features and unique development, you will need to add development life cycle activities and vendor estimates of development time and effort to your plan in the "develop functional specification" and "implement functional specification" activities.

Familiarization

In any system implementation, the vendor's organization and the contracting organization must establish communications and management processes for the project. These processes define roles, information flow, document content and format and review points. In a typical project, this may take five to ten days.

Develop issue and risk management process. Typically, issue and risk management are the daily responsibility of the project management team, with provisions for a formal escalation process. If your company or the vendor company has existing processes for issue and risk management or a template for intercompany issue and risk management, these may be quickly tailored to incorporate the appropriate contacts and schedules for your project. That is the assumption for the two-day duration for this task. If neither party has existing tools and processes, then this activity will take more time. Resources for this task include decision makers and program and project managers from both companies. The task primarily involves meetings to negotiate responsibilities, deliverables and quality standards.

Develop status reporting process. If your company or the vendor company has existing processes for status reporting or a template for status reporting, these may be quickly tailored to incorporate the appropriate contacts and schedules for your project. That is the assumption for the three-day duration for this task. If neither party has existing tools and processes, then this activity will take more time. Resources for this task include decision makers and program and project managers from both companies. The task primarily involves meetings to negotiate responsibilities, deliverables and quality standards.

Finalize high-level project plan. This activity involves the entire project management team, including your company's project management staff and the vendor's project management resources. Starting with the work plan incorporated in the project, the team now adjusts dates, identifies resources and makes any other adjustments required. Typically, this project plan is reviewed with executive management and approved prior to the kickoff meetings.

One work week is generally sufficient for this activity. The resources involved depend upon the size of the project management team and the complexity of the review process.

Kickoff meetings. Kickoff meetings are forums for introducing vendor personnel and your company's personnel, as well as for reviewing the project plan, policies and procedures to the entire project team. Typically, there will be a series of meetings held sequentially—an executive meeting, a team leader meeting and one or more team meetings involving all of the project personnel or their direct supervision. However, a very small company or project may hold a single meeting. More complex project structures may involve longer sessions to cover project plans.

Platform implementation

Platform implementation includes the installation of system hardware, software, third-party software and the billing vendor's off-the-shelf core software. Your company's system administrators may perform these activities or the vendor may perform some or all of the activities. This will have been specified in the contract, but should also be reflected in the resource assignment in your project plan.

Platform deployment. In this era of "just-in-time" server deployment, 18 days for specifying, procuring and installing servers, operating systems and network connections is aggressive but possible. If your planning horizon allows a somewhat more leisurely approach to server acquisition, there may be some benefit in both platform vendor support and pricing. Also, a longer interval allows your data center staff to coordinate this activity with their ongoing responsibilities.

Software installation—core software. The template project plan assumes that the billing system vendor will perform this software installation—hence the knowledge transfer and installation verification and sign off activities. If your system administration staff is performing the installation, your contract should include both vendor support for the installation and detailed standards for installation documentation. You may want to add a task for an acceptance review of installation documentation before the infrastructure software installation task.

Most vendors have significant experience with installation of their own platforms and software and can provide a realistic schedule for these activities upon request.

Transfer business information

Back in the preparation phase, your company gathered lots of information about products and services, customers and accounts, business processes and metrics and about business operations. You also developed detailed business requirements for your new billing system. In this activity, you make that information available to the billing system vendor personnel and spend some time reviewing the material, clarifying information and establishing a mutual understanding of your business and what the system is expected to do. The duration for these tasks is estimated based on good documentation under change control and co-location of vendor and company personnel for these activities.

The resources for this activity are dependent upon the size and roles of the project team. However, the duration is based on the availability and quality of documentation in all areas. If information or related documentation is missing or poor in one or more areas, this activity will take longer.

Develop configuration specification

This activity relates your business information and requirements to the functions and data tables in the vendor platform. Often vendor project plans for this activity will be structured in terms of the structure of the vendor system. If your company's project management is using a version of the project plan that relates to your information and business functions and the vendor's project management is using a version of the project plan that relates to system structure, verify that the relevant schedules and resources match.

Match requirements with system capabilities. In this activity, your company's personnel and the vendor personnel establish a common vocabulary and understanding of what business requirement relates to what system capability. This activity generally involves the vendor team with support from your business analysis team equivalent to two analysts.

Define product configuration. The effort required for this activity is dependent upon the number of service types to be implemented. Generally, eight days is sufficient for a single type such as LEC, legacy wireless, or IXC. Defining product configuration requires a vendor analyst/system engineer and support from your business analysis team equivalent to one full-time analyst.

Define customer and account configuration. The effort required for this activity is dependent upon the number of customer and account configurations to be implemented. Generally, two days is sufficient for a single market segment. Defining customer and account configuration requires a vendor analyst/system engineer and support from your business analysis team equivalent to one full-time analyst.

Define system level configuration. The effort required for this activity is relatively constant. Defining system level configuration requires a vendor analyst/system engineer and support from your business analysis team equivalent to one full-time analyst.

Define user interface configuration. The effort required for this activity is somewhat dependent upon the number and complexity of user interface types to be implemented. However, three days is a useful rule of thumb for the typical system in which the vendor has predefined user interface configurations. Defining the user interface configuration requires a vendor analyst/system engineer and support from your business analysis team equivalent to one full-time analyst.

Define bill output configuration. The effort required for this activity is dependent upon the number of bill output or invoice types to be implemented. Generally, three days is sufficient for a single type such as print bill for consumer local accounts. Defining bill output configuration requires a vendor analyst/system engineer and support from your business analysis team equivalent to one full-time analyst.

Define reports configuration. The effort required for this activity is somewhat dependent upon the number of reports types to be implemented. Generally, one day is sufficient for configuring the vendor "canned" reports options. However, if your company requires unique reports or reporting formats, allow a day of effort for each of those unique reports. Defining reports configuration requires a vendor analyst/system engineer and support from your business analysis team equivalent to one full-time analyst.

Define security configuration. The effort required for this activity is dependent upon the number of systems users and the number of user classes to be implemented. Generally, four days is sufficient for a moderate size organization with typical security requirements. Defining security configuration requires a vendor analyst/system engineer and support from your business analysis team equivalent to one full-time analyst.

Review and approve functional specification. Relevant SMEs and functional managers should review and approve the functional specification before the team proceeds to implementation.

Implement functional specification

This activity implements the configuration defined in the previous activity. The template project plan assumes that vendor personnel will perform this implementation. If your company decides that it wants to enhance internal expertise in system configuration by having your personnel perform the configuration, this activity becomes dependent upon training those personnel. Also, given relative unfamiliarity of your staff with the platform, it is advisable to double the time and effort for each implementation activity. Thus, an activity requiring one vendor resource two days would require one internal resource four days. If one person is scheduled to perform significant amounts of the same type of configuration (such as product/service configuration), the effort could be dropped to vendor levels after the first five. All configuration implementation activities are performed by a single resource.

Implement product configuration. Typically, billing system platforms provide the ability to create a template or example for a class or type of products and services and then use that template or example as a basis for similar product or service configurations. If your vendor has a different schema, it should be able to provide guidance on planning product configuration.

In the template project, we have assumed eight days for configuring the template or example and .5 days for each unique product or service based upon that template. Often the only difference in unique product or services will be the service designation, market code and price.

Implement customer and account configuration. For each unique customer type assume .5 day of configuration with an associated .5 day of account configuration for each generic account configuration.

Implement system level configuration. Typical system level configuration requires five days. If your company requires unusually complex system and operational configurations, this duration could increase by a day or two. If the vendor platform does not have friendly interfaces for system level configuration, this duration could double. The vendor should be able to provide reliable estimates of this activity.

Implement user interface configuration. The four days duration in the template plan is based upon the assumption of typical user interface definitions and reasonably "friendly" configuration tools.

Implement bill output configuration. Bill output configuration, with reasonably good configuration tools, should require four days effort (one person for four days) for each bill output configuration. If the tools are less than optimal or the output configuration is unusually complex, this may take longer.

Implement reports configuration. One day is generally sufficient for configuring an option on "canned" reports. Any custom reports development will have to be estimated by the vendor.

Implement security configuration. Like the specifications for this activity, the effort for this activity is related to the number of users and the number of user classes. The two days estimated in the template is based on a moderate-size organization with typical security requirements.

Perform vendor functional validation

The vendor verifies the accuracy, completeness and operational readiness of the configuration activities in this task. Most vendors have predefined validation plans. Your company's contract should specify that your company receive documentation of the validation results. The duration of this activity is controlled by the vendor and should not be truncated, as this is an important quality gate.

Update functional specification

Often, validation will identify some issues with the functional specification. The specification should be updated to reflect the resolution of these issues and the changes reviewed with and approved by your company's project team.

Configuration knowledge transfer

If the vendor performed the configuration activities, the knowledge of that configuration and its implications for future configuration activities and for systems functions in general must be shared with the personnel who will support future configuration and with the project team.

Training

Training of system administrators, applications administrators and systems users, as well as of any ongoing training personnel, is a critical activity in any system implementation.

Develop training plan. Developing a training plan requires involvement by the vendor, by your company's project manager and by the manager responsible for training delivery. The training delivery manager may be the functional manager of your company's training department—if there is one—or the functional manager of one or more of the organizations containing personnel requiring training.

Typically, this effort consists of meeting to understand what the specific training needs are and to determine the options for and availability of specific training. This is followed by the vendor and training manager's determining specific resource availability, followed by establishing a detailed training schedule.

Train the trainers. If your company has an internal training function, typically some of your training personnel will attend vendor "train-the-trainers" sessions. These are often scheduled for a business week at a vendor location. There will often be a week-long course for user functions and another for application administration and configuration functions. There may be an additional course targeted to system administrators.

Train key users. Certain users may be designated to perform user acceptance testing and to act as "lead" users within their work group. Typically, they will be trained in advance of training all end-users. Again, typical training modules are designed for a business week.

Train all end users. Unless your user base is very small and none of them has to be "minding the store" while others are in training, it will typically take at least four sequential training sessions to cover the user base, hence the 20-day duration.

Acceptance test

Acceptance testing allows your company to determine if the vendor has actually delivered the system and configuration that you specified and your company needed. Successful completion of acceptance testing is a primary quality gate for implementation and may be tied to the payment schedule.

Define scope of acceptance test. The acceptance test should include representative product, services, customers, accounts, billing output media and billing cycles. It should also include typical functions such as the application of payments, the application of credits and the handling of overdue accounts. SMEs, business analysts and testers are generally all involved in an acceptance test scope meeting.

Specify test environment. Testers and vendor personnel are usually involved in specifying the configuration of the test environment, including any test tools if required. Other personnel, such as system administration, may have some involvement.

Create test environment. The effort involved in creating the test environment is dependent upon the complexity of the specifications, but five days accommodate the creation of a typical test environment (often a test configuration on the system being prepared for production).

Define test scenarios. Test scenarios describe the business transaction or conditions to be tested. The effort involved in developing test scenarios is dependent upon the complexity of the implementation. Two days is sufficient time for the development of a fairly simple set of test scenarios. This time generally involves meetings involving SMEs, business analysts and testers.

Define test cases. Test cases identify in detail what transactions will take place and what the anticipated result will be. This task involves the development of the test cases—obviously involving effort proportional to the number of test cases—by testing personnel and a brief review and approval by the relevant SMEs. For estimating purposes, assume that a test case designer can complete two test cases per day. A very simple test plan would include 20 test cases.

Perform functional acceptance test. The functional acceptance test verifies that all system functionality performs according to contract requirements or subsequent supplementary agreements. In a typical environment, this requires a minimum of 15 days, since some functionality testing is sequential. For instance, create a customer account, establish service in an account and then generate a bill and apply payment. Functional acceptance testing is performed by the testing group and reviewed by your project manager.

Perform user acceptance test. User acceptance testing is performed by a subset of the end-user and administrator community. This testing verifies that user interfaces perform as required and that user documentation and training is accurate and usable. User testing verifies that system interfaces meet ease of use and efficiency requirements.

Operational test

Operational testing creates a subset of the production environment and performs transactions similar to those performed in normal billing cycles. Typically, the operational test covers two complete billing cycles in order to properly test application of payment against amounts generated in the first billing cycle. Often the duration of operational testing is compressed by "fooling" the application by manipulating system dates. While this would allow the template duration of 10 test days, it does create some risks in not truly mirroring a production environment.

Operational test activities are similar to acceptance test activities and have much the same resource requirements. The shorter durations for specifying and creating the test environment assume some overlap with the acceptance test environment activities. The operational test tasks are:

- Define scope of operational test
- Specify test environment
- Create test environment
- Define test scenarios
- Define test cases
- Perform operational test

Configure for production

Once testing is complete, the system must be configured to receive "live" transactions and to produce actual customer bills. This may entail minimal administrative effort or backing out all the test transactions, enabling production feeds and other extensive configuration and database work. Ten days is a minimum interval for this work. Typically, this effort involves vendor personnel plus your company's ongoing system administration and application administration personnel.

Link to conversion project

If your overall project includes data conversion from an old billing environment and the conversion is scheduled to occur with system turn up, conversion should be complete prior to turning the system up for service.

Link to interface project

If your overall project includes integration of the billing system to interfacing systems, the interface implementation should be complete before turning the system up for service.

System turn-up

This activity involves turning the system up for live service. The activity may range from simply administratively turning the system over to real users or it may involve running predefined manual or automated procedures.

Scheduling Considerations/Issues

Your system implementation plan should take into consideration the impact of the system turn up date on billing cycles and related operational dates. This is even more critical if your project includes conversion from old systems, especially if billing cycles are defined differently in the new platform.

FtA Implementation Project Plan

The FtA implementation plan is, for the most part, typical in size and structure to the template plan. However, there were several tasks that were partially deferred from the preparation phase, due to the lack of information about the CLEC business and old billing system. The FtA project also includes significant conversion and interface projects that require coordination with the billing implementation project.

FtA project tasks—unconstrained view

Table 38-2 shows the tasks in the FtA implementation project with no resource constraints.

TABLE 38-2 FtA Implementation Project—No Constraints

#	Task Name	Duration	Pred.
1	Precontract activities	179 days	
2	Develop business requirements	47 days	
3	Request for information process	59 days	Common FtA/53
4	Develop and distribute RFI	15 days	
5	Analyze RFI responses	20 days	4
6	Proposal process	109 days	3
7	Develop RFP	25 days	
8	Obtain vendor nondisclosures	2 days	
9	Distribute RFP	2 days	7, 8
10	Analyze proposals	30 days	9FS+45 days
11	Review vendor demos, etc.	5 days	10
12	Select vendor pending contract negotiations	2 days	11
13	Develop initial project plan	8 days	6
14	Identify internal resources and availability	2 days	
15	Identify third-party resources and availability	2 days	
16	Identify scheduling constraints	2 days	14, 15
17	Integrate resources, constraints, and vendor schedule	4 days	16
18	Negotiate contract	10 days	12
19	Sign contract	1 day	13, 18

continued on next page

TABLE 38-2 FtA Implementation Project—No Constraints (continued)

#	Task Name	Duration	Pred.
20	CLEC data gathering	4.5 days	
21	Gather info on CLEC service 1	1.5 days	
22	Gather info on CLEC service 2	1.5 days	
23	Gather info on CLEC service 3	1.5 days	
24	Gather info on CLEC service 4	1.5 days	
25	Gather info on CLEC service 5	1.5 days	
26	Gather info on CLEC service 6	1.5 days	
27	Gather info on CLEC service 7	1.5 days	
28	Gather info on CLEC service 8	1.5 days	
29	Analyze and consolidate CLEC P/S information	3 days	21, 22, 23, 24, 25, 26, 27, 28
30	Gather customer and account details	2 days	
31	Update requirements to reflect improved CLEC information	5 days	Common FtA/53
32	Implementation phase	86 days	19
33	Familiarization	5 days	
34	Develop issue and risk management process	2 days	19
35	Develop status reporting process	4 days	19
36	Finalize high-level project plan	5 days	13
37	Kickoff meetings	2 days	33
38	Platform implementation	27 days	
39	Platform deployment	18 days	
40	Specify platform elements	1 day	19
41	Procure platform elements	15 days	40
42	Install platform elements	2 days	41
43	Software installation—core software	9 days	39
44	Installation/configuration knowledge transfer	1 day	
45	Install required infrastructure software	2 days	39
46	Install core billing software	3 days	45, 44
47	Configure infrastructure and core billing software	4 days	45
48	Install user interface software	4 days	46
49	Load vendor-configured data	4 days	46
50	Software install verification review and sign off	1 day	
51	Develop functional specification	8 days	
52	Gather information	43 days	

continued on next page

TABLE 38-2 FtA Implementation Project—No Constraints (continued)

#	Task Name	Duration	Pred.
53	Communicate product, business, and network data	7 days	Common FtA/3
54	Provide product and business process documents	1 day	
55	Provide equipment and network inventory documentation	1 day	54
56	Review data gathering documents with vendor	5 days	55
57	Communicate billing requirements	3 days	Common FtA /53
58	Provide billing requirements	1 day	
59	Review billing requirements with vendor	1 day	58
60	Develop configuration specification	8 days	
61	Match requirements with system capabilities	1 day	56SS, 57FF
62	Define product configuration LEC switched	8 days	
63	Define product configuration LEC PL	8 days	
64	Define product configuration IXC	8 days	
65	Define product configuration ISP	8 days	
66	Define product configuration IPP	8 days	
67	Define product configuration APP	8 days	
68	Define product configuration MSP	8 days	
69	Define product configuration professional services	8 days	
70	Define customer and account configuration	5 days	
71	Define system level configuration	6 days	
72	Define user interface configuration	3 days	
73	Define bill output configuration—print	3 days	
74	Define bill output configuration—Web EBPP	3 days	
75	Define bill output configuration—credit card	3 days	
76	Define bill output configuration—CD	3 days	
77	Define reports configuration	1 day	
78	Define unique reports	3 days	
79	Define security configuration	4 days	
80	Review and approve runctional specification	2 days	70, 71, 72, 76, 77, 89
81	Implement functional specification	13 days	60, 95
82	Implement product configuration	13 days	
83	Implement customer and account configuration	1 day	
84	Implement system level configuration	5 days	

continued on next page

TABLE 38-2 FtA Implementation Project—No Constraints (continued)

#	Task Name	Duration	Pred.
85	Implement user interface configuration	4 days	
86	Implement bill output configuration	4 days	
87	Implement reports configuration	1 day	
88	Implement security configuration	2 days	
89	Perform vendor functional validation	6 days	81
90	Update functional specification	1 day	81, 89
91	Configuration knowledge transfer	1 day	90
92	Training	33 days	
93	Develop training plan	3 days	
94	Train the trainers	5 days	93
95	Train key users	5 days	94
96	Train all end users	20 days	95
97	Acceptance test	33 days	89
98	Define scope of acceptance test	1 day	61
99	Specify test environment	5 days	98
100	Create test environment	5 days	99, 80
101	Define test scenarios	2 days	60, 98
102	Define test cases	10 days	101
103	Perform functional acceptance test	15 days	102, 89
104	Perform user acceptance test	5 days	95, 103
105	Complete interface implementation	1 day	
106	Operational test	75 days	
107	Define scope of operational test	1 day	98
108	Specify test environment	2 days	99, 107
109	Create test environment	2 days	100, 108
110	Define test scenarios	3 days	101, 109
111	Define test cases	10 days	102, 110
112	Perform operational test	10 days	104, 111
113	Complete preturn-up conversion	1 day	
114	Configure for production	10 days	112
115	Turn system up for live service	1 day	114

The FtA project planner has made several changes to the template implementation project. Some of these changes are in response to FtA's unique business situation. Others are simply expansions of the template with detailed subtasks. Some relate to specific implementation choices made by FtA.

FtA intends to close its acquisition of the CLEC on April 1, 2003, allowing the remaining CLEC data gathering to commence on April 15, 2003. These tasks have been added to the FtA project plan with the constraint of starting no earlier than April 15. A task has also been added to update the existing requirements based on the CLEC data gathering. The added tasks related to CLEC information are tasks 20 through 31.

Both FtA's selected vendor and FtA have implemented automated status tracking systems. Both companies deemed it desirable to develop an electronic interface between the two systems. In order to accommodate this mini-development, task 35 has been extended to four days.

FtA has chosen to have its own system administrators perform platform implementation tasks. As a result, the knowledge transfer task has been moved before the software installation tasks and the duration of the installation and configuration of any vendor software tasks (tasks 46 through 49) have been increased to accommodate lack of familiarity with the platform.

FtA has identified tasks to define product configurations for each type of service it intends to offer, using the standard eight-day duration. It has assigned a duration of ten days to defining the customer and account configuration based on five market segments and an estimated five types of account structures.

The FtA project planner has created a definition task for each of the required bill media types. As FtA works with the vendor and the vendor product, it may identify additional output definitions. For instance, it may be that different print formats will be required for residence and business customers.

In discussions with the vendor, FtA has determined that it will require three unique reports that can be configured through a reports-writer utility. A task has been added to reflect the related definition task.

FtA has determined that it wants to have certain key users participate in the configuration implementation activities. Therefore, it has made key user training (task 95) a predecessor of "implement functional specification"—task 81.

FtA also has significant interface implementation activity, as well as migration from multiple old billing systems within the scope of its billing project. These activities are shown as external dependencies in the FtA plan (tasks 105 and 114).

FtA implementation project with FtA constraints

The FtA project planner has made the initial assumption that vendor resources other than the vendor project manager are essentially unconstrained. Also, FtA has retained a consulting firm to lead and perform the majority of testing tasks that do not require end users. The firm will be able to make available sufficient testers for testers to be considered an unconstrained resource as well. Therefore, resource constraints on implementation are primarily related to FtA SMEs and users.

In Table 38-3, it is apparent that even these very limited resource constraints have a significant impact on the project end date—pushing that date out by more than two months.

TABLE 38-3 FtA Implementation Project with Resource Constraints

#	Task Name	Duration	Pred.
1	Precontract activities	189 days	
2	Develop business requirements	47 days	
3	Request for information process	69 days	Common FtA\53
4	Develop and distribute RFI	15 days	
5	Analyze RFI responses	20 days	4
6	Proposal process	109 days	3
7	Develop RFP	25 days	
8	Obtain vendor non-disclosures	2 days	
9	Distribute RFP	2 days	7,8
10	Analyze proposals	30 days	9FS+45 days
11	Review vendor demos, etc.	5 days	10
12	Select vendor pending contract negotiations	2 days	11
13	Develop initial project plan	8 days	6
14	Identify internal resources and availability	2 days	
15	Identify third-party resources and availability	2 days	
16	Identify scheduling constraints	2 days	14, 15
17	Integrate resources, constraints, and vendor schedule	4 days	16
18	Negotiate contract	10 days	12
19	Sign contract	1 day	13, 18
20	CLEC data gathering	4.5 days	
21	Gather info on CLEC service 1	1.5 days	
22	Gather info on CLEC service 2	1.5 days	
23	Gather info on CLEC service 3	1.5 days	
24	Gather info on CLEC service 4	1.5 days	
25	Gather info on CLEC service 5	1.5 days	
26	Gather info on CLEC service 6	1.5 days	
27	Gather info on CLEC service 7	1.5 days	
28	Gather info on CLEC service 8	1.5 days	
29	Analyze and consolidate CLEC P/S information	3 days	21, 22, 23, 24, 25, 26, 27, 28

continued on next page

TABLE 38-3 FtA Implementation Project—No Constraints (continued)

#	Task Name	Duration	Pred.
30	Gather customer and account details	2 days	
31	Update requirements to reflect improved CLEC information	5 days	20, Common FtA\53
32	Implementation phase	133 days	19
33	Platform implementation	2 days	35
35	Familiarization	5 days	
36	Develop issue and risk management process	2 days	19
37	Develop status reporting process	4 days	19
38	Finalize high-level project plan	5 days	13
39	Kickoff meetings	2 days	35
40	Platform implementation (from FtA Common)	4 days	41
41	Platform implementation	27 days	
42	Platform deployment	18 days	
43	Specify platform elements	1 day	19
44	Procure platform elements	15 days	43
45	Install platform elements	2 days	44
46	Software installation—core software	9 days	42
47	Installation/configuration knowledge transfer	1 day	
48	Install required infrastructure software	2 days	42
49	Install core billing software	3 days	48, 47
50	Configure infrastructure and core billing software	4 days	48
51	Install user interface software	4 days	49
52	Load vendor-configured data	4 days	49
53	Software install verification, review, and signoff	1 day	
54	Develop functional specification	55 days	
55	Gather information	43 days	
56	Communicate product, business, and network data	7 days	Common FtA\3
57	Provide product and business process documents	1 day	
58	Provide equipment and network inventory documentation	1 day	57
59	Review data gathering documents with vendor	5 days	58
60	Communicate billing requirements	2 days	Common FtA\53
61	Provide billing requirements	1 day	
62	Review billing requirements with vendor	1 day	61

continued on next page

TABLE 38-3 FtA Implementation Project—No Constraints (continued)

#	Task Name	Duration	Pred.
63	Develop configuration specification	55 days	
64	Match requirements with system capabilities	1 day	59SS, 60FF
65	Define product configuration LEC switched	8 days	
66	Define product configuration LEC PL	8 days	
67	Define product configuration IXC	8 days	
68	Define product configuration ISP	8 days	
69	Define product configuration IPP	8 days	
70	Define product configuration APP	8 days	
71	Define product configuration MSP	8 days	
72	Define product configuration professional svcs	8 days	
73	Define customer and account configuration	5 days	
74	Define system level configuration	6 days	
75	Define user interface configuration	3 days	
76	Define bill output configuration—print	3 days	
77	Define bill output configuration—Web EBPP	3 days	
78	Define bill output configuration—credit card	3 days	
79	Define bill output configuration—CD	3 days	
80	Define reports configuration	1 day	
81	Define unique reports	3 days	
82	Ddefine security configuration	4 days	
83	Review and approve functional specification	2 days	73, 74, 75, 79, 80, 82
84	Implement functional specification	13 days	63, 98
85	Implement product configuration	13 days	
86	Implement customer and account configuration	1 day	
87	Implement system level configuration	5 days	
88	Implement user interface configuration	4 days	
89	Implement bill output configuration	4 days	
90	Implement reports configuration	1 day	
91	Implement security configuration	2 days	
92	Perform vendor functional validation	6 days	84
93	Update functional specification	1 day	84, 92
94	Configuration knowledge transfer	1 day	93
95	Training	33 days	
96	Develop training plan	3 days	

continued on next page

TABLE 38-3 FtA Implementation Project—No Constraints (continued)

#	Task Name	Duration	Pred.
97	Train the trainers	5 days	96
98	Train key users	5 days	97
99	Train all end users	20 days	98
100	Acceptance test	33 days	92
101	Define scope of acceptance test	1 day	64
102	Specify test environment	5 days	101
103	Create test environment	5 days	102, 83
104	Define test scenarios	2 days	63, 101
105	Define test cases	10 days	104
106	Perform functional acceptance test	15 days	105, 92
107	Perform user acceptance test	5 days	98, 106
108	Complete interface implementation	100 days	
109	Operational test	122 days	
110	Define scope of operational test	2 days	101
111	Specify test environment	2 days	102, 110
112	Create test environment	2 days	103, 111
113	Define test scenarios	3 days	104, 112
114	Define test cases	10 days	105, 113
115	Perform operational test	10 days	107, 114
116	Complete preturn-up conversion	100 days	
117	Cconfigure for production	10 days	115
118	Production load	87 days	117
119	Turn system up for live service	1 day	117

Summary

Implementation projects can stand alone. More commonly, they are related to interface implementations and migrations and require coordination of planning. If you are installing a brand new billing system in a brand new company, at a minimum you will need to be able to receive customer account and order information very early in the life cycle of the billing system. More typically, even in a start-up mode, you will have some existing customers and services that need to be migrated into the new system.

FtA, of course, has both interface and migration activity that must be coordinated with the implementation project. Both interface and migration are dependent upon the selection and implementation of the billing platform.

Chapter

39

Planning Integration

Integration is often the least predictable part of a billing system project. Integration involves the specification, design and development of system-to-system interfaces. However, it also entails verifying that beginning-to-end transactions that use the interfaces work properly under all circumstances.

Common Interfaces to Billing Systems

In Chapter 25, we looked at typical interfaces to billing systems. Planning the implementation of these types of interfaces is the subject of this chapter. However, interfaces involve more than the two systems and the interface. Interfaces are a critical point in processes and data flows that may originate several steps upstream and cause actions several steps downstream. When implementing an interface, the project plan (and planner) must consider the potential impact of that interface on end-to-end data flows and processes.

Integrating with sales and provisioning systems

Customer account information and service order information originate in sales order entry systems—although some of the information may originate in and be transferred from sales inquiry systems. Often some customer and order information is passed unedited through provisioning systems—which may be the point of interface with billing. If a change is made in the sales system environment or if there is more than one sales order system, such as one per market segment, this can affect the functioning of the interface between the provisioning system and the billing system, even though neither of those systems or the interface changes.

When performing integration testing for billing systems interfaces, it is important to include the systems and users that originate transactions in the real world. How do the upstream systems know that the billing system has encountered a problem? What is the problem resolution process?

Integrating with bill production

When integrating with bill production, it is important to understand bill production processing. Does production simply take a bill image and print it? Does it collect customer address information for mail processing purposes?

For electronic billing, what does bill production do to the bill information the billing system provides? Does it add formatting? Callable software? How does this work in the production environment or in a test partition with the information provided via the billing system interface?

How does billing get notified if there is a downstream problem? Is there a reconciliation and problem resolution process?

Integrating with corporate accounting

Fundamentally, billing is about money. All the calculating and processing and managing is about getting revenues for your products and services. That means linkages to the corporate accounting environment are key to billing systems fulfilling their business purpose.

Accounts receivable. Typically, billing systems interface at some point in the accounts receivable process, often interfacing directly to the receivables journal. How billed amounts (expected revenues) and adjustments are handled and how they are reconciled to cash accounting is key to planning integration and integration testing for billing-generated receivables information. When are transactions posted to the accounts receivable ledger and when do transactions post to the general ledger? How much additional information, such as specific products/services associated with each amount, is provided? How are adjustments handled and synchronized?

In today's accounting-sensitive environment, it is even more critical than in the past that interface implementers understand the end-to-end process and the impact of interface implementation on that process. What end-to-end controls exist to ensure that all transactions are properly booked and fully auditable?

Accounts payable. Typically, billing systems provide "accounts payable advice" via a system-to-system interface. Since this represents money flowing out of the company, it is critical to understand and document how the interface functions in the context of the payables process, including controls and auditability. Typically, such accounts payable advice includes payments to taxing authorities for billed taxes and payments to customers or former customers for account overpayments, return of deposits, etc.

Integrating with third-party vendors. Billing systems may interface with a number of additional third parties and these interfaces must be included in the implementation planning. Third-party vendors include such things as usage data collection systems, tax data processors, lockbox banks and credit bureaus. Third-party interface implementers must have a clear understanding of where the vendor adds value to the end-to-end process and the impact of interface implementation on that process.

Template Interface Project Plan

The template interface project plan contains all activities required to find a vendor and contract for a billing system interface. With the preparation phase project plan, it can be used as a guide for planning stand-alone interface projects or for contracting with an integrator to develop interfaces in the context of a more extensive billing system project. Much of the acquisition activity is identical to that in the implementation project, so we will not review it here. Since a single interface is perhaps less complex than an entire billing platform, the RFP development time has been somewhat shortened. However, a project that implements custom interfaces—single or multiple—may be more complex than implementation of a standard billing solution and the time should be adjusted accordingly.

Table 39-1 displays the task list for the template project plan.

TABLE 39-1 Interface Implementation Project Template Tasks

#	Task Name	Duration	Pred.
1	Complete overall billing project	0 days	2
2	Interface implementation project	149.5 days	
3	Gather information	5 days	
4	Request for information process	65 days	Common\5
5	Develop and distribute RFI	15 days	
6	Analyze RFI responses	20 days	5FS+30 days
7	Proposal process	36 days	4
8	Develop RFP	10 days	
9	Obtain vendor nondisclosures	2 days	8
10	Distribute RFP	2 days	9
11	Analyze proposals	15 days	10
12	Review vendor demos, etc.	5 days	11
13	Select vendor pending contract negotiations	2 days	12
14	Develop initial project plan	8 days	7
15	Identify internal resources and availability	2 days	
16	Identify third-party resources and availability	2 days	

continued on next page

TABLE 39-1 Interface Implementation Project Template Tasks (continued)

#	Task Name	Duration	Pred.
17	Identify scheduling constraints	2 days	16
18	Integrate resources, constraints, and vendor schedule	4 days	17, 15, 16
19	Negotiate contract	10 days	7
20	Sign contract	1 day	19
21	Implementation phase	37.5 days	20
22	Familiarization	5 days	
23	Develop issue and risk management process	2 days	
24	Develop status reporting process	3 days	
25	Finalize high-level project plan	5 days	
26	Kickoff meetings	2 days	25, 23, 24
27	Platform implementation	2 days	
28	Determine platform requirements	2 days	
29	Allocate platform resources	2 days	
30	Analysis/software requirements	29.5 days	
31	Review business requirements for interfaces	5 days	
32	Identify interface tools in remote system	2 days	31
33	Identify interface tools in billing system	1 day	32
34	Determine data mappings, mismatches and discrepancies	5 days	33
35	Determine flow controls	5 days	34
36	Determine error protocols	5 days	35
37	Draft preliminary software specifications	3 days	36
38	Develop preliminary budget	2 days	37
39	Review software specifications/budget with team	0.5 days	38
40	Incorporate feedback on software specifications	1 day	39
41	Design 4 days	30	
42	Develop functional specifications	3 days	
43	Review functional specifications	1 day	42
44	Development	10 days	
45	Develop code	10 days	
46	Modular testing (dummy interfaces)	5 days	45FS-75%
47	Testing 19 days		
48	Develop interface test plans	3 days	
49	Develop integration test plans	4 days	
50	Interface testing	7 days	

continued on next page

TABLE 39-1 Interface Implementation Project Template Tasks (continued)

#	Task Name	Duration	Pred.
51	Review interface code	1 day	48
52	Test component modules to specifications	1 day	51
53	Identify anomalies to product specifications	1 day	52
54	Modify code	2 days	53
55	Retest modified code	1 day	54
56	Load test interface	1 day	55
57	Integration testing	9 days	
58	Test module integration	5 days	56
59	Identify anomalies to specifications	1 day	58
60	Modify code	2 days	59
61	Retest modified code	1 day	60
62	Documentation	4 days	41
63	Document interface flow	2 days	
64	Document interface data	2 days	
65	Document interface error conditions and protocol	4 days	
66	Deployment	3 days	57
67	Train support staff	1 day	
68	Deploy software	2 days	67
69	Turn up for service	1 day	

Acquisition and familiarization tasks

Acquisition tasks include the RFI process, the proposal process and negotiating and signing a contract. Familiarization includes all the activities required to organize a team comprised of resources from different organizations within your company and vendor resources. For more detail on acquisition and familiarization tasks, refer to Section III and to Chapter 38 of this book. For the interface project template, the duration for development of the RFP is shown as 10 days and the duration for analysis of proposals is shown as 15 days, both considerably shorter than for a major platform implementation. However, if the project includes more than one custom interface, these tasks may require significantly more time.

Platform

Although system architects are fond of showing interfaces as lines between systems, in fact interfaces consist of software that has to run somewhere as well as network resources—which may be "smart" (self-analyzing and reporting) or "dumb" (pure transport).

Analysis/software requirements

With the exception of interfaces in which each system has interface handlers that are fully compliant with fully specified standards, interfaces are generally "mini" custom software development projects. Note the phrase "fully compliant with fully specified standards." Often standards are not fully specified. That is, they allow certain fields, controls or transactions to be vendor-specified. This means that even if you have two fully compliant systems, it may require custom development to get them to fully understand each other. Also, many standards specify only information format, not control or error management.

Review business requirements for interfaces. Interface business requirements should have been developed to support acquisition. This is the point to review and clarify these requirements for the systems analysts and designers who will be participating in interface implementation. The duration of this task is fixed for a single interface, although the resources will include the analyst who developed the business requirements and however many vendor resources will be part of the implementation.

Identify interface tools in remote system. Does the interfacing system have an existing interface for this function? A formally specified API or CORBA interface? What are the characteristics of these interface tools?

Identify interface tools in billing system. Does the billing system have an existing interface for this function? A formally specified API or CORBA interface? What are the characteristics of these interface tools?

Determine data mappings, mismatches, and discrepancies. What data does the "sending" system have? What does it look like? What edits have been performed on it? What data does the "receiving" system expect? What does it look like? What edits does it expect to have been performed? What data problems will cause a transaction to fail in the receiving system? Where will it fail (e.g., at the point of interface, downstream in the system or in subsequent systems)?

Determine flow controls. How will flow be controlled over the interface? How does the sending system (on any transaction including reply transaction) know that the receiving system is ready to receive and is receiving successfully? What should the sending system do if the receiving system is not ready to receive or is not receiving successfully?

Determine error protocols. How are errors at the interface handled and tracked? How are downstream errors handled and tracked? What constitutes an error? Do all errored transactions fail or are there "soft" or "nonfatal" errors that generate notification, but continue to process?

Draft preliminary software specifications. Based on requirements, tools, flow controls and error protocols, a preliminary software specification can be generated by the design team.

Develop preliminary budget. Since this is custom design, a resource estimate is valuable for planning, even on a fixed-price project. On a time and expenses-based contract, the preliminary budget should become the reference for a limit for project spending without additional approval. The limit may not be the estimate itself, but the estimate plus or minus some specified percent.

Review software specifications/budget with team. If analysts developed the specifications and budget and development resources are joining the team at this point, reviewing software specifications becomes an important familiarization activity. It is also an opportunity for your company's business analysts and SMEs to review the specifications. Again, this activity is of fixed duration and may involve a variable number of resources.

Incorporate feedback on software specifications. Review meetings may produce improvements on the specifications, as well as serving an educational purpose. This task supports the incorporation of these improvements into the specifications.

Design

This is the activity in which software designers figure out how to implement technically the detailed specifications developed in the preceding task. Depending on the size and composition of your team, this may be "rolled up" into the previous activity.

Development

Think of development as the stuff that geeks go off in a corner to do after you have told them what to do. This includes writing and validating code and performing stand-alone tests that generally use a "dummy" piece of code for each end of the interface—something that "looks" like the interfacing system from the "outside." If the developer departs from the specification with regard to error conditions, appearance of control reports and other necessary information for testers and system administrators, relevant documentation should be created and made available for user documentation and training development.

The duration of this activity is dependent upon the clarity and complexity of the specifications, as well as upon the tools available in each system.

Testing

Interfaces are the most likely place for catastrophic and unrecoverable system problems. Testing billing systems interfaces extremely thoroughly is an important investment in your company's future. In Chapter 41, we look more closely at testing issues for billing systems.

Develop interface test plans. The complexity of interface test plans is proportional to the complexity of the interface and the variability of the data handled by the interface. If the interface handles only one type of transaction in one format from one original source, the interface test plans will be much simpler and require less time and effort to develop than if the interface handles 20 different types of transactions in five formats from eight original sources. The three days in the template project plan will support a typical small record interface with several types of transactions (e.g., new, change, disconnect, cancel plus replies) in consistent formats with very little variability in data structure or content.

Develop integration test plans. The complexity of the business process and downstream and upstream environments is the critical factor in determining the effort required to develop good integration test plans. Again, the duration in the project plan relates to a typical small record interface with a single upstream and a single downstream system.

Interface testing

Since this is custom development, interface testing includes formal code inspection, testing and activities designed to correct problems found. The smaller the portion of custom code in the interface, the less time may be required for "resolving anomalies"—i.e., fixing problems. Load is a potential issue for interfaces; all interfaces should be tested with an offered peak and average load that mimics a production load.

Integration testing

The effort involved in integration testing is dependent more on the complexity and variability of the process and the end-to-end environment than it is on the complexity or uniqueness of the interface itself. Also, if most of the environment is newly deployed, additional time may be required, due to the inherent instability of newly deployed platforms.

The task durations in the template are based on a mature environment with well defined processes and process controls.

Documentation

Documentation of interfaces for systems and network operations personnel is absolutely essential to the successful deployment of a new interface.

Document interface flow. Obviously, the effort involved in documenting the interface information flow and flow control is directly related to the complexity of that flow and inversely related to the clarity of specifications for the flow. The time allocated to this documentation task is appropriate to a fairly simple interface with clear specifications.

Document interface data. The effort involved in documenting the data that pass over the interface and how they are handled by the interface is related to the complexity of the data (number and type of fields), the variability of the data, the number of edits or transformations performed on the data as part of the interface and the clarity of data interface specifications. The time allocated to this task is appropriate for a typical small record interface with several types of transactions (e.g., new, change, cancel plus replies) in consistent formats with very little variability in data structure or content and edits for data type and format only.

Document interface error conditions and protocol. Documenting interface error conditions and the protocols associated with those error conditions is the most critical piece of operations documentation associated with an interface. This document should be complete, accurate, clear and carefully validated. The allocated four days is a minimum time to produce an adequate document for a simple interface with a limited number of error conditions.

Deployment

The deployment of billing systems interfaces may have to be coordinated with system implementation and migration activities if the interface is part of a larger project.

Train support staff. Operations and administration staff must be fully trained in a new interface before that interface is placed in service. If documentation is good, actual interface training can usually be accomplished in one day. Relevant staff are usually limited in number and co-located, making sequential training sessions unnecessary. Obviously, if there are multiple locations or a larger staff or shift scheduling issues, the training duration may have to be extended.

Deploy software. Typically interfaces are fairly self-contained—not requiring major database rebuilds or other intrusive system activity. This tends to make deployment fairly rapid.

Turn up for service. If the interface is replacing another interface, this activity must be coordinated with the termination of the old interface. Sometimes this coordination takes the form of dual feeding the same information across two different interfaces.

FtA Interface Implementation Plan

In the FtA billing system project, interface implementation is part of the main billing system contract, so FtA has removed all the acquisition and familiarization tasks from their Interface project plan and created an external link to the acquisition and familiarization tasks in the Implementation project plan.

Table 39-2 shows the FtA Interface Implementation tasks, along with the resource-constrained duration.

TABLE 39-2 FtA Interface Implementation Tasks

#	Task Name	Duration	Pred.
1	Interface implementation project	123.5 days	
2	Familiarization	5 days	
3	Platform implementation	2 days	Imple-mentation—FtA\35
4	Determine platform requirements	2 days	
5	Allocate platform resources	2 days	
6	Implement provisioning interface	81 days	3
7	Analysis/software requirements	51.5 days	
8	Review business requirements for interfaces	7 days	
9	Identify interface tools in remote system	2 days	8
10	Identify interface tools in billing system	1 day	9
11	Determine data mappings, mismatches and discrepancies	10 days	10
12	Determine flow controls	5 days	11
13	Determine error protocols	5 days	12
14	Draft preliminary software specifications	4 days	13
15	Develop preliminary budget	2 days	14
16	Review software specifications/budget with team	2 days	15
17	Incorporate feedback on software specifications	1 day	16
18	Design	11 days	7
19	Develop functional specifications	3 days	
20	Review functional specifications	1 day	19
21	Development	10 days	
22	Develop code	10 days	
23	Modular testing (dummy interfaces)	5 days	22FS-75%
24	Testing	58 days	
25	Develop interface test plans	5 days	

continued on next page

TABLE 39-2 FtA Interface Implementation Tasks (continued)

#	Task Name	Duration	Pred.
26	Develop integration test plans	4 days	
27	Interface testing	9 days	
28	Review interface code	1 day	25
29	Test component modules to specifications	3 days	28
30	Identify anomalies to product specifications	1 day	29
31	Modify code	2 days	30
32	Retest modified code	1 day	31
33	Load test interface	1 day	32
34	Integration testing	20.5 days	
35	Test module integration	10 days	33
36	Identify anomalies to specifications	1 day	35
37	Modify code	2 days	36
38	Retest modified code	2 days	37
39	Documentation	4 days	18
40	Document interface flow	2 days	
41	Document interface data	2 days	
42	Document interface error conditions and protocol	4 days	
43	Implement usage data interfaces	67 days	
44	Analysis/software requirements	62 days	
45	Review business requirements for interfaces	5 days	
46	Identify interface tools in remote system	5 days	45
47	Identify interface tools in billing system	1 day	46
48	Determine data mappings, mismatches, and discrepancies	10 days	47
49	Determine flow controls	5 days	48
50	Determine error protocols	5 days	49
51	Draft preliminary software specifications	3 days	50
52	Develop preliminary budget	2 days	51
53	Review software specifications/budget with team	0.5 days	52
54	Incorporate feedback on software specifications	1 day	53
55	Design	3 days	44
56	Develop functional specifications	2 days	
57	Review functional specifications	1 day	56
58	Development	5.5 days	
59	Develop code	2 days	

continued on next page

TABLE 39-2 FtA Interface Implementation Tasks (continued)

#	Task Name	Duration	Pred.
60	Modular testing (dummy interfaces)	5 days	59FS-75%
61	Testing	51.75 days	
62	Develop interface test plans	3 days	
63	Develop integration test plans	3 days	
64	Interface Testing	5.5 days	
65	Review interface code	1 day	62
66	Test component modules to specifications	1 day	65
67	Identify anomalies to product specifications	1 day	66
68	Modify code	0.5 days	67
69	Retest modified code	1 day	68
70	Load test interface	1 day	69
71	Integration Testing	5.25 days	
72	Test module integration	4 days	70
73	Identify anomalies to specifications	0.25 days	72
74	Modify code	0.25 days	73
75	Retest modified code	0.25 days	74
76	Documentation	2 days	55
77	Document interface flow	1 day	
78	Document interface data	1 day	
79	Document interface error conditions and protocol	2 days	
80	Implement Tax Interface	111.5 days	
81	Analysis/Software Requirements	16.25 days	
82	Review business requirements for interfaces	1 day	
83	Identify interface tools in remote system	0.5 days	82
84	Identify interface tools in billing system	0.5 days	83
85	Determine data mappings, mismatches, and discrepancies	0.5 days	84
86	Determine flow controls	1 day	85
87	Determine error protocols	1 day	86
88	Draft preliminary software specifications	0 days	87
89	Develop preliminary budget	2 days	88
90	Review software specifications/budget with team	0.5 days	89
91	Incorporate feedback on software specifications	0.25 days	90
92	Design	48.25 days	81
93	Develop functional specifications	1 day	

continued on next page

TABLE 39-2 FtA Interface Implementation Tasks (continued)

#	Task Name	Duration	Pred.
94	Review functional specifications	1 day	93
95	Development	1 day	
96	Develop code	1 day	
97	Modular testing (dummy interfaces)	0 days	96FS-75%
98	Testing	59 days	
99	Develop interface test plans	2 days	
100	Develop integration test plans	4 days	
101	Interface Testing	1.5 days	
102	Review interface code	0.5 days	99
103	Test component modules to specifications	0 days	102
104	Identify anomalies to product specifications	0 days	103
105	Modify code	0 days	104
106	Retest modified code	0 days	105
107	Load test interface	1 day	106
108	Integration Testing	54 days	
109	Test module integration	5 days	107
110	Identify anomalies to specifications	1 day	109
111	Modify code	1 day	110
112	Retest modified code	1 day	111
113	Documentation	2 days	92
114	Document interface flow	1 day	
115	Document interface data	1 day	
116	Document interface error conditions and protocol	2 days	
117	Implement product definition interface	87 days	
118	Analysis/software requirements	74 days	
119	Review business requirements for interfaces	5 days	
120	Identify interface tools in remote system	2 days	119
121	Identify interface tools in billing system	1 day	120
122	Determine data mappings, mismatches, and discrepancies	5 days	121
123	Determine flow controls	5 days	122
124	Determine error protocols	5 days	123
125	Draft preliminary software specifications	3 days	124
126	Develop preliminary budget	2 days	125
127	Review software specifications/budget with team	0.5 days	126

continued on next page

TABLE 39-2 FtA Interface Implementation Tasks (continued)

#	Task Name	Duration	Pred.
128	Incorporate feedback on software specifications	1 day	127
129	Design	4 days	118
130	Develop functional specifications	3 days	
131	Review functional specifications	1 day	130
132	Development	10 days	
133	Develop code	10 days	
134	Modular testing (dummy interfaces)	5 days	133FS-75%
135	Testing	42 days	
136	Develop interface test plans	3 days	
137	Develop integration test plans	4 days	
138	Interface testing	6 days	
139	Review interface code	1 day	136
140	Test component modules to specifications	1 day	139
141	Identify anomalies to product specifications	1 day	140
142	Modify code	2 days	141
143	Retest modified code	1 day	142
144	Load test interface	0 days	143
145	Integration testing	9.5 days	
146	Test module integration	5 days	144
147	Identify anomalies to specifications	1 day	146
148	Modify code	2 days	147
149	Retest modified code	1 day	148
150	Documentation	4 days	129
151	Document interface flow	2 days	
152	Document interface data	2 days	
153	Document interface error conditions and protocol	4 days	
154	Implement bill print interface	109 days	
155	Analysis/software requirements	93.5 days	
156	Review business requirements for interfaces	5 days	
157	Identify interface tools in remote system	2 days	156
158	Identify interface tools in billing system	1 day	157
159	Determine data mappings, mismatches, and discrepancies	5 days	158
160	Determine flow controls	5 days	159
161	Determine error protocols	5 days	160

continued on next page

TABLE 39-2 FtA Interface Implementation Tasks (continued)

#	Task Name	Duration	Pred.
162	Draft preliminary software specifications	2 days	161
163	Develop preliminary budget	2 days	162
164	Review software specifications/budget with team	0.5 days	163
165	Incorporate feedback on software specifications	1 day	164
166	Design	6.5 days	155
167	Develop functional specifications	1 day	
168	Review functional specifications	1 day	167
169	Development	1.25 days	
170	Develop code	1 day	
171	Modular testing (dummy interfaces)	1 day	170FS-75%
172	Testing	53 days	
173	Develop interface test plans	3 days	
174	Develop integration test plans	4 days	
175	Interface testing	7 days	
176	Review interface code	1 day	173
177	Test component modules to specifications	1 day	176
178	Identify anomalies to product specifications	1 day	177
179	Modify code	2 days	178
180	Retest modified code	1 day	179
181	Load test interface	1 day	180
182	Integration Testing	21 days	
183	Test module integration	5 days	181
184	Identify anomalies to specifications	1 day	183
185	Modify code	2 days	184
186	Retest modified code	1 day	185
187	Documentation	4 days	166
188	Document interface flow	2 days	
189	Document interface data	2 days	
190	Document interface error conditions and protocol	4 days	
191	Implement lockbox bank interface	112.5 days	
192	Analysis/software requirements	67 days	
193	Review business requirements for interfaces	3 days	
194	Identify interface tools in remote system	1 day	193
195	Identify interface tools in billing system	1 day	194
196	Determine data mappings, mismatches, and discrepancies	3 days	195

continued on next page

TABLE 39-2 FtA Interface Implementation Tasks (continued)

#	Task Name	Duration	Pred.
197	Determine flow controls	3 days	196
198	Determine error protocols	3 days	197
199	Draft preliminary software specifications	1 day	198
200	Develop preliminary budget	2 days	199
201	Review software specifications/budget with team	0.5 days	200
202	Incorporate feedback on software specifications	1 day	201
203	Design	8.5 days	192
204	Develop functional specifications	1 day	
205	Review functional specifications	1 day	204
206	Development	2 days	
207	Develop code	2 days	
208	Modular testing (dummy interfaces)	1 day	207FS-75%
209	Testing	59.5 days	
210	Develop interface test plans	3 days	
211	Develop integration test plans	4 days	
212	Interface Testing	7 days	
213	Review interface code	1 day	210
214	Test component modules to specifications	1 day	213
215	Identify anomalies to product specifications	1 day	214
216	Modify code	2 days	215
217	Retest modified code	1 day	216
218	Load test interface	1 day	217
219	Integration Testing	17 days	
220	Test module integration	5 days	218
221	Identify anomalies to specifications	1 day	220
222	Modify code	2 days	221
223	Retest modified code	1 day	222
224	Documentation	4 days	203
225	Document interface flow	1 day	
226	Document interface data	1 day	
227	Document interface error conditions and protocol	4 days	
228	Implement revenue accounting interface	116.5 days	
229	Analysis/software requirements	98 days	
230	Review business requirements for interfaces	5 days	
231	Identify interface tools in remote system	2 days	230

continued on next page

TABLE 39-2 FtA Interface Implementation Tasks (continued)

#	Task Name	Duration	Pred.
232	Identify interface tools in billing system	1 day	231
233	Determine data mappings, mismatches, and discrepancies	5 days	232
234	Determine flow controls	5 days	233
235	Determine error protocols	5 days	234
236	Draft preliminary software specifications	3 days	235
237	Develop preliminary budget	2 days	236
238	Review software specifications/budget with team	0.5 days	237
239	Incorporate feedback on software specifications	1 day	238
240	Design	8.5 days	229
241	Develop functional specifications	3 days	
242	Review functional specifications	1 day	241
243	Development	6.25 days	
244	Develop code	5 days	
245	Modular testing (dummy interfaces)	5 days	244FS-75%
246	Testing	60 days	
247	Develop interface test plans	3 days	
248	Develop integration test plans	4 days	
249	Interface Testing	7 days	
250	Review interface code	1 day	247
251	Test component modules to specifications	1 day	250
252	Identify anomalies to product specifications	1 day	251
253	Modify code	2 days	252
254	Retest modified code	1 day	253
255	Load test interface	1 day	254
256	Integration Testing	17 days	
257	Test module integration	5 days	255
258	Identify anomalies to specifications	1 day	257
259	Modify code	2 days	258
260	Retest modified code	1 day	259
261	Documentation	4 days	240
262	Document interface flow	2 days	
263	Document interface data	2 days	
264	Document interface error conditions and protocol	4 days	
265	Implement revenue analysis interface	119.5 days	

continued on next page

TABLE 39-2 FtA Interface Implementation Tasks (continued)

#	Task Name	Duration	Pred.
266	Analysis/software requirements	84.5 days	
267	Review business requirements for interfaces	4 days	
268	Identify interface tools in remote system	2 days	267
269	Identify interface tools in billing system	1 day	268
270	Determine data mappings, mismatches, and discrepancies	4 days	269
271	Determine flow controls	2 days	270
272	Determine error protocols	2 days	271
273	Draft preliminary software specifications	3 days	272
274	Develop preliminary budget	2 days	273
275	Review software specifications/budget with team	0.5 days	274
276	Incorporate feedback on software specifications	1 day	275
277	Design	8 days	266
278	Develop functional specifications	3 days	
279	Review functional specifications	1 day	278
280	Development	5 days	
281	Develop code	4 days	
282	Modular testing (dummy interfaces)	4 days	281FS-75%
283	Testing	32.5 days	
284	Develop interface test plans	3 days	
285	Develop integration test plans	3 days	
286	Interface testing	5 days	
287	Review interface code	1 day	284
288	Test component modules to specifications	1 day	287
289	Identify anomalies to product specifications	1 day	288
290	Modify code	1 day	289
291	Retest modified code	1 day	290
292	Load test interface	0 days	291
293	Integration testing	6 days	
294	Test module integration	3 days	292
295	Identify anomalies to specifications	1 day	294
296	Modify code	1 day	295
297	Retest modified code	1 day	296
298	Documentation	2 days	277
299	Document interface flow	1 day	

continued on next page

TABLE 39-2 FtA Interface Implementation Tasks (continued)

#	Task Name	Duration	Pred.
300	Document interface data	2 days	
301	Document interface error conditions and protocol	2 days	
302	Implement Credit Bureau Interface	111.5 days	
303	Analysis/Software Requirements	84.5 days	
304	Review business requirements for interfaces	3 days	
305	Identify interface tools in remote system	2 days	304
306	Identify interface tools in billing system	1 day	305
307	Determine data mappings, mismatches and discrepancies	2 days	306
308	Determine flow controls	2 days	307
309	Determine error protocols	2 days	308
310	Draft preliminary software specifications	2 days	309
311	Develop preliminary budget	1 day	310
312	Review software specifications/budget with team	0.5 days	311
313	Incorporate feedback on software specifications	1 day	312
314	Design	5 days	303
315	Develop functional specifications	3 days	
316	Review functional specifications	1 day	315
317	Development	2 days	
318	Develop code	2 days	
319	Modular testing (dummy interfaces)	1 day	318FS-75%
320	Testing	38.5 days	
321	Develop interface test plans	3 days	
322	Develop integration test plans	2 days	
323	Interface testing	6 days	
324	Review interface code	1 day	321
325	Test component modules to specifications	1 day	324
326	Identify anomalies to product specifications	1 day	325
327	Modify code	2 days	326
328	Retest modified code	1 day	327
329	Load test interface	0 days	328
330	Integration Testing	13 days	
331	Test module integration	2 days	329
332	Identify anomalies to specifications	1 day	331
333	Modify code	1 day	332

continued on next page

TABLE 39-2 FtA Interface Implementation Tasks (continued)

#	Task Name	Duration	Pred.
334	Retest modified code	1 day	333
335	Documentation	2 days	314
336	Document interface flow	1 day	
337	Document interface data	1 day	
338	Document interface error conditions and protocol	2 days	
339	Deployment	4 days	6, 43, 80, 117, 154, 191, 228, 265, 302
340	Train support staff	1 day	
341	Deploy software	2 days	340
342	Turn up for service	1 day	

FtA interfaces

The FtA project planner has identified separate activities for all but one of the interfaces identified for the FtA billing system. The exception is the fraud system interface that FtA has decided to defer for the moment. The fraud group will be receiving periodic preconfigured reports via email initially.

Provisioning interfaces. The FtA provisioning system will be implemented in parallel with the FtA billing system. This is a large and complex interface with large numbers of transaction messages, a large number of possible message formats and significant variability among the messages.

Both the billing system and the provisioning system have well defined APIs for the provisioning-billing interface. They also both have moderate support for flow control and error management and allow some flexibility in specifying responses to detected errors.

The complexity of the interface has prompted increasing the duration of requirements and design activities. However, the availability of good interface tools has kept that increase to a minimum.

The duration of most testing activities has been increased significantly over the template values. The complexity of the provisioning environment has driven an even greater increase in integration testing duration than for interface testing.

Usage data collection interfaces. All of the usage data collection systems that will be providing data to the FtA billing system utilize standards-compliant interface formats (BAF and IPDR initially). The vendor billing platform includes interfaces for these formats. However, there are options in implementation that make clear specifications and careful configuration and testing imperative for these interfaces. Since there will be little or no unique software development for these interfaces, the durations for those tasks have been reduced significantly, while specification and testing activities reflect significant effort.

Tax data interfaces. FtA has chosen one of the major tax software and data vendors. The billing system vendor has standard interfaces to this tax platform, so there is little effort required for design or development. However, it is important to verify the end-to-end configuration for tax data, from determination of jurisdiction through the creation of payable advice to accounts payable. Therefore, significant effort remains in integration testing.

Product definition interface. This interface is unique to FtA. FtA plans to develop "product catalogs" in a custom system for product management and then distribute the information to sales, provisioning and billing systems. This is a low-volume interface and will feed the product/service creation APIs in the billing system. The effort estimated for this interface is very similar to the template effort. Load testing has been eliminated.

Bill printing interface. The billing system platform includes the capability to format bills for interface with bill printing vendors. The bill printing vendor will develop/configure the bill print side of the interface. The effort here is more configuration than development, so development tasks are minimized, but the need for testing remains intact. It should be noted that testing of the print side of the interface will be done as part of the operational and user acceptance testing activities.

Lockbox bank interface. FtA has chosen a major national bank to provide lockbox services. The bank has a well-defined and stable interface module for relaying payment information to billing systems. The billing system vendor has implemented numerous interfaces from its platform to the lockbox platform, so no custom development is required. Again, the need for comprehensive testing remains.

Revenue accounting interface. The revenue accounting interface will be built around some configurable interface capabilities within the billing platform and tailored to the FtA accounting systems. This is anticipated to involve minimal unique code, primarily to meet the stringent control, management and error tracking requirements in the FtA environment.

Revenue analysis interface. This interface is based on standard configurable queries within the billing platform. The vendor has implemented numerous similar interfaces. This is a periodic, rather than a real-time interface. Volumes are predictable. Therefore, flow control is minimal and error recovery is a simple request for re-send. This reduces the development and testing required.

Credit bureau interface. The credit bureau interface is supported within the billing system platform, although the FtA configuration will require specification and testing. Test support from the credit bureau is limited, so test time is minimal.

Resource constraint impacts

The FtA interface implementation project without resource constraints has a duration of 52 days following the completion of the billing project familiarization activity. With resource constraints, the interface project requires 123.5 days, finishing just six days before the billing platform turn-up to service.

Summary

In this chapter, we looked at interface implementation activities that ranged from routine to unique and complex. Most billing systems today are required to be fully integrated with a variety of business and network systems. This makes the planning and management of interface implementation key to a successful billing system project.

In the upcoming chapter, we scrutinize another key factor in many billing system implementations—data conversion from legacy and/or external systems.

Chapter

40

Planning Conversion/ Migration from Legacy Billers

Billing data conversions are frequently associated with new billing system implementations, as the records in the old system must be moved to the new platform. However, conversions can also be associated with the acquisition of a company or a book of business in order to move the new acquisition to your existing billing platform.

Conversion of legacy data and migration to new billing platforms has unique challenges. This is almost always custom development. Often information about the legacy platform such as data validations and transformations is unavailable for the legacy system or only available by tracking transactions through murky code. Frequently, too, the legacy environment may not have enforced standards of consistency—internally or in relation to other systems and databases. It is common—and good—practice to include data cleanup activities in a legacy conversion schedule, especially if the legacy environment is not under your control.

Template Conversion Project Plan

The template project plan assumes that your data conversion involves customer and account records and their associated service instances. It also assumes that product and service definitions—including rate tables—will be created manually within the new system and are not part of the conversion. However, "override" or custom pricing information associated with the customer account record is considered part of the conversion.

As was the case in the interface template, the template conversion project plan contains all activities required to find a vendor and contract for a billing system conversion. With the preparation phase project plan, it can be used as a guide for planning stand-alone conversion projects or for contracting with an integrator to develop conversion support in the context of a more extensive billing system project. Much of the acquisition activity is identical to that in the implementation project, so we will not review it here. Since a single conversion is perhaps less complex than an entire billing platform, the RFP development time has been somewhat shortened. However, a project to implement multiple conversions may be more complex than implementation of a standard billing solution and the time should be adjusted accordingly.

Table 40-1 shows the task list for the template conversion project plan.

TABLE 40-1 Conversion Project Task List

#	Task Name	Duration	Pred.
1	Migrate legacy billing data to new platform	187.5 days	
2	Pre-contract activities	82 days	
3	Develop business requirements	25 days	
4	Request for information process (conversion)	35 days	Common\15
5	Develop and distribute RFI	15 days	
6	Analyze RFI responses	20 days	5
7	Proposal process	36 days	4
8	Develop RFP	10 days	
9	Obtain vendor nondisclosures	2 days	8
10	Distribute RFP	2 days	9
11	Analyze proposals	15 days	10
12	Review vendor demos, etc.	5 days	11
13	Select vendor pending contract negotiations	2 days	12
14	Develop initial project plan	8 days	7
15	Identify internal resources and availability	2 days	
16	Identify third-party resources and availability	2 days	
17	Identify scheduling constraints	2 days	16
18	Integrate resources, constraints, and vendor schedule	4 days	15, 16, 17
19	Negotiate contract	10 days	7
20	Sign contract	1 day	19
21	Implementation phase	99 days	20
22	Familiarization	7 days	
23	Develop issue and risk management process	2 days	
24	Develop status reporting process	3 days	
25	Finalize high-level project plan	5 days	
26	Kickoff meetings	2 days	23, 24, 25

continued on next page

TABLE 40-1 Conversion Project Task List (continued)

#	Task Name	Duration	Pred.
27	Platform implementation	4 days	
28	Determine platform requirements	2 days	
29	Allocate platform resources	2 days	28
30	Develop migration strategy	16 days	
31	Review customer and account data	10 days	
32	Review product/service data	10 days	
33	Review customer and account data structures and tools in new system	2 days	31
34	Review product/service data structures and tools in new system	2 days	32
35	Document migration strategy	4 days	31, 32, 33, 34
36	New system data load software	26 days	33, 34
37	Design data load software	15 days	
38	Develop data load software	6 days	37
39	Synchronize data load software with migration strategy	3 days	38, 35
40	Test data load software	2 days	39, 27
41	Develop test plan	5 days	30
42	Define scope of data migration software test	5 days	
43	Define scope of parallel processing	1 day	
44	Define testing activities	1 day	
45	Establish testing standards	2 days	
46	Parallel processing preparation	18 days	43
47	Develop parallel data feed requirements	5 days	
48	Develop functional specifications for parallel processing support from old system	2 days	47
49	Implement functional specifications for parallel processing support from old system	5 days	48
50	Develop functional specifications for accepting parallel data in new system	2 days	47
51	Implement functional specifications for accepting parallel data into new system	2 days	50
52	Test parallel processing	13 days	
53	Define test scope	1 day	47
54	Define test scenarios	2 days	53
55	Define test cases	3 days	54
56	Establish test parallel feed	4 days	49, 51
57	Perform test	2 days	55, 56

continued on next page

TABLE 40-1 Conversion Project Task List (continued)

#	Task Name	Duration	Pred.
58	Data extraction	32 days	
59	Draft specifications for trial and preliminary data extract	3 days	35
60	Get trial data extract from old billing system	7 days	30
61	Develop trial extract functional specifications	4 days	
62	Develop trial extract per functional specifications	1 day	61
63	Perform trial data extract	2 days	62
64	Assess trial extract	5 days	60
65	Assess data format	2 days	
66	Assess data completeness	2 days	
67	Assess data quality	3 days	
68	Modify draft specifications as needed	2 days	65, 66, 67
69	Develop data quality strategy	7 days	64
70	Develop specifications for quality assessment tools	3 days	
71	Develop quality assessment tools	4 days	70
72	Define QA process and resources	3 days	
73	Get preliminary data extract from old billing system	8 days	68
74	Develop preliminary extract functional specifications	1 day	
75	Develop preliminary extract per functional specifications	5 days	74
76	Perform preliminary data extract	2 days	75
77	Perform quality assessment	12 days	73
78	Run quality assessment tools	1 day	
79	Perform manual QA review per QA process	5 days	
80	Identify and document corrective actions for any quality problems	2 days	79
81	Take corrective actions (recycle with legacy data source or clean up data)	5 days	78, 80
82	Data translations for migration	49 days	
83	Migration analysis/requirements	30 days	
84	Review data structures	1 day	65
85	Define data mappings	10 days	84
86	Review quality assessment results	2 days	77
87	Document data translations requirements	5 days	85, 86
88	Develop translations software	17 days	83

continued on next page

TABLE 40-1 Conversion Project Task List (continued)

#	Task Name	Duration	Pred.
89	Develop functional specifications for data massage software	10 days	
90	Develop data massage software	5 days	89
91	Test data massage software	2 days	90
92	Perform translations on extracted data	2 days	91
93	End-to-end massage and load software test	15 days	92
94	Define scope of end-to-end test	3 days	
95	Develop test scenarios	2 days	94
96	Define test cases	5 days	95
97	Perform test	5 days	96
98	Trial migration	12 days	
99	Perform test loads	5 days	
100	Unit test with migrated data	2 days	
101	Integration test with migrated data	7 days	100
102	Define scope of integration test	2 days	
103	Develop test scenarios	2 days	102
104	Develop test cases	2 days	103
105	Perform test	1 day	104
106	Parallel bill cycle	8 days	100
107	Load usage data for billing cycle	3 days	
108	Run parallel bill cycle for test accounts	5 days	107
109	User acceptance test with migrated data	12 days	
110	Define scope of UA test	2 days	
111	Develop test scenarios	2 days	110
112	Develop test cases	2 days	111
113	Perform UA test	2 days	101, 106, 112
114	Production load	87 days	98
115	Prepare platform for production load	1 day	
116	Load legacy data	15 days	
117	Get final extract from old billing system	1 day	
118	Initiate parallel data feed (usage, transactions)	1 day	
119	Perform QA on final extract	5 days	117
120	Take corrective actions for QA defects	3 days	119
121	Run translations on final extract	1 day	120
122	Resolve translation fall-out	5 days	121

continued on next page

TABLE 40-1 Conversion Project Task List (continued)

#	Task Name	Duration	Pred.
123	Load new system	2 days	121
124	Verify data load	2 days	123
125	Parallel bill cycles	82 days	
126	Define scope of parallel testing	2 days	43
127	Design bill sample	2 days	126
128	Develop review process and materials	5 days	127
129	Run parallel bill production	45 days	127, 116
130	Train reviewers	4 days	
131	Develop training	3 days	128
132	Deliver training	1 day	131
133	Review bills	27 days	
134	Compare bills per review process	15 days	129, 130
135	Identify anomalies	2 days	134
136	Determine causes of anomalies	5 days	135
137	Resolve anomalies	5 days	136
138	User acceptance review	73.5 days	
139	Review migration fallout and resolutions	1 day	122
140	Review results of parallel bill cycles	1 day	133
141	Go/no go decision	0.5 days	139, 140
142	Turn-up for service	5 days	114, 138
143	Update with incremental data if required	4 days	
144	Direct all production feeds to new system	1 day	143
145	Terminate bill distribution from old system	1 day	143

Acquisition and familiarization tasks

Acquisition tasks include the RFI process, the proposal process and negotiating and signing a contract. Familiarization includes all the activities required to organize a team comprised of resources from different organizations within your company and vendor resources. For more detail on acquisition and familiarization tasks, refer to Section III and to Chapter 38 of this book. For the conversion project template, the duration for development of the RFP is shown as 10 days and the duration for analysis of proposals is shown as 15 days, both considerably shorter than for a major platform implementation. However, if the project includes more than one data source, these tasks may require significantly more time.

Platform implementation

In addition to requiring resources on both the old system platform for data extraction and on the new system platform for data loading, conversions typically require significant resources for data reformatting—the functions the project plan calls data "massage." This will involve the capacity for storing and manipulating large quantities of data. Ideally, both "raw" data storage and DBMS capacity will be available for this purpose. If your data center does not have significant spare capacity of this type, you may have to do a separate procurement for temporary or permanent capacity to support this activity. In that case, the Platform Implementation activity may be extended significantly.

Develop migration strategy

A data migration strategy for billing systems should include both business and technical strategy for the data migration. The business strategy should incorporate planning information, including the relationship of conversion activities to other projects and to business activities and constraints. It should also identify standards for data quality, standards for resolution of fallout, standards of success, and identification of the specific work group or special task force that will be responsible for the effort.

Technical strategy should identify architectural premises and tools to be used for the migration, as well as performance requirements for the phases of migration. Technical constraints should be identified.

The data migration strategy should identify the scope of the conversion. For instance, how much, if any, account history will be migrated? Will credit classifications be migrated or recomputed in the new system? Will customer care notations on the account be migrated? Will contract information be migrated?

New system data load software

If your company controls the new billing system platform, work on data load software can begin as soon as the migration strategy is complete. If the data load design identifies expected data formats for use by the translation software, this work is dependent only on the migration strategy and knowledge of the new system.

Many platform vendors have this type of data load software already available for their systems platforms. They may make it available as a standard "bulk-load" capability or only as a service through their consulting organization. If your vendor has tools available and you choose to use them, this entire activity may be reduced to one day for loading the tools.

Develop test plan

Testing activities abound in a data migration. The test plan sets the parameters for those activities—in scope, responsibilities and performance standards. All subsequent testing activities are dependent on this document.

Parallel processing preparation

The ability to run billing cycles in parallel in the old and new systems is a critical feature of billing systems conversions. This allows the comparison of actual bill output for the same customer and bill cycle from one system to the other. Obviously, the bill content and totals should agree! Discrepancies may be as small as a one penny difference due to different rounding algorithms or as significant as applying different tax or rating tables—or applying a discount before or after summarization.

Parallel processing is difficult to set up, a coordination nightmare—and absolutely essential to successful conversions for a host of customer satisfaction, financial, regulatory, and legal reasons.

Implementing parallel processing may require extensive planning and negotiation, so preparation should begin as early as possible in the conversion project.

Develop parallel data feed requirements. What data must be parallel fed, from where and how? If the conversion involves usage-based services, you will need a dual feed of usage files. What about transaction data—service orders and account updates? All required data should be identified and any timing constraints should be specified.

Develop functional specifications for parallel processing support. A parallel feed of usage data may be accomplished by configuring usage data collection or mediation equipment to send the same information to two locations or by having the data forwarded from the legacy system environment. Obviously, this requires significant involvement on the part of the legacy system administrators and network equipment administrators.

Transaction data could be handled many ways. If your new billing system will have an interface from the same sales and/or provisioning systems feeding the legacy system, you may want to make that interface active for the parallel period. If your new billing system will be receiving feeds of this type of information from a new or different front end, hardcopy of those transactions should be supplied so they can be entered—either to the new sales/provisioning environment and fed to the new billing system or to billing personnel for direct entry to the new billing system. Again, this requires significant cooperation, not only from the legacy billing staff, but also from sales and provisioning.

If you have control of all these environments, this can be accomplished with a minimum of overhead. However, if any of these functions are not within your control, add lots of negotiating time to this activity. You may also need an interim work force to accomplish manual inputs, checking of old system versus new system outputs and error resolution.

Implement functional specifications for parallel processing support. "Making it so" often requires lots of coordination. It may require new network connections, administrator involvement and a lot of manual controls. If you can utilize what will become your production billing system data sources, the effort is minimized.

Test parallel processing. If possible, run a test of parallel feeds prior to full parallel processing—a "unit test" for the data interfaces, whether automated or manual.

Data extraction

Extraction of data from a legacy billing system is often literally a process of trial and error, since neither documentation nor data quality can be assumed in a legacy system. Key resources for this activity are the folks who are keeping the old system running.

Draft specifications for trial and preliminary data extract. This activity develops the specifications for both the trial or sample data extract and the preliminary full data extract. Since the technical parameters for both of these extracts are similar, developing the specifications together helps keep them synchronized and also makes efficient use of resources.

Get trial data extract from old billing system. The trial data extract is a sample of data. It provides information about what really is in the legacy system.

Assess trial extract. The trial extract serves to validate extraction logic and it allows an initial assessment of data quality in the legacy system. If the data quality appears to be poor, the time required for resolution of quality assurance defects and translation fallout should be increased in proportion to the quality issues identified. Data format, data completeness and data accuracy within the trial extract should be assessed. If problems are detected with the extract specifications, the draft specifications should be amended as needed.

Develop data quality strategy. Data quality strategy should encompass both automated tools and manual processes. This activity references the trial data extract to develop useful and practical quality assurance methods. It is important to mechanize as much of the quality assurance functionality as possible, since the final data conversion will be handling large quantities of data within time constraints.

Get preliminary data extract from old billing system. The preliminary data extract is essentially a "rehearsal" for the final extract that will load the production environment.

Perform quality assessment. Quality assurance is a critical path activity for conversion activities.

Run quality assessment tools. Quality assurance tools can identify missing data, misformatted data and data that will not parse for conversion.

Perform manual QA review per QA process. Some review tasks require manual procedures to identify issues and root causes.

Identify and document corrective actions for any quality problems. Most resolution work is performed manually. It is critical to capture the corrective actions taken so they can be repeated or automated.

Take corrective actions ("recycle" with legacy data source or clean up data). If the defect is a result of problems with the extract—such as missing data because the extract program did not take into account a reference table—the problem should be corrected in the extract software. If the data are defective at the source, then they should be cleaned up during or as a result of conversion. Some fallout may be documented and clean up delayed until after the bulk of the conversion is completed.

Data translations for migration

Data translation software takes data as they were extracted from the old system (usually in the old system's format) and reformats them for the new system. This may require combining or parsing data fields, using cross-reference tables to map identities from one system to the other, removing special characters or any number of other actions. The end result should be data structured and formatted in the way the data load software expects it to be structured and formatted.

If the conversion team has good information about both the data and the logic of both systems, this is a simple data interface development. If not, this becomes very complex technological detective work. Obviously, that will impact the duration for this activity, for it is usually an iterative process.

Migration analysis/requirements. In order to design data translation software, you will need knowledge of the source data from the source system and from the sample extract. You will also need to understand exactly what the data load software and the receiving system expect in terms of data structure and content.

Develop translation software. The software that "massages" the extracted data into formats expected for data load must perform all the required translations functions and do so within time constraints.

End-to-end "massage" and load software test

This test determines whether the translation software created data that will not only load the "loader" database, but will actually load and act as expected on the new billing system platform.

Trial migration

The trial migration is a rehearsal for the production data load. In addition to verifying that everything works, it provides information on the timing of various activities and allows personnel to gain experience with supporting manual procedures. The trial migration provides the opportunity for testing each module of the conversion, for comprehensive integration tests and for a "practice" parallel bill cycle on specified test accounts. It also provides the opportunity for user testing and user review of the parallel billing test.

Production load

Why do a second load for production? The trial migration allows for optimization of migration processes and software. The production load of data is time constrained. Any activity in the old system after the extract results in some need for reconciliation between the two systems, even if everything is being parallel fed. Eventually, that reconciliation load becomes insurmountable.

Prepare platform for production load. If you have been testing on the same platform that will be used for production, this is the time to clear out all the test data and any non-standard test configurations. If there is any setup required for product and service configuration or account or market configuration, that should be done in this activity. If that setup is part of another project, such as a new billing system implementation, this task should have that project and task as a predecessor.

Load legacy data. Load legacy data include all the activities necessary for converting production billing data from the old platform to the new system.

Get final extract from old billing system. This activity includes the actual extraction and getting it to the translation location.

Initiate parallel data feed (usage, transactions). As soon as the extract is complete, the parallel data feed should be activated. This will minimize reconciliation issues.

Perform QA on final extract. It is critical to document QA results and retain them. This task assumes significant documentation activities.

Take corrective actions for QA defects. Again, it is critical to document QA resolution actions and to retain that documentation. Documentation effort is assumed in this task.

Run translations on final extract. Here you get to run your well tested, highly optimized translation routines against a full production database.

Resolve translation fallout. There will be translation fallout. Your company may decide to defer resolution until the completion of the conversion load. Fallout resolution activities should be tracked as part of the overall project defect tracking process.

Load new system. This task loads the conversion data into the new billing system. Any fallout or other error condition should be fully documented and tracked. The data from the translation fallout may become critical here, as there may be a "snowball" effect that causes you to rethink deferral of resolution of that activity.

Verify data load. This activity inspects the new billing system to ensure that the data loaded and loaded correctly.

Parallel bill cycles. At a minimum, your schedule for parallel billing cycles should encompass the production of two subscriber bills. This allows verification of payment application against a previous month's bill. It also allows the verification of credits and reversal of charges and other activities from one bill cycle to the next.

Define scope of parallel testing. What should your parallel bill review cover? What types of customers, services and transactions should be incorporated in testing? What are the allowable deviations?

Design bill sample. What accounts will be sampled for review? What sampling method will be used? How many accounts will be sampled?

Develop review process and materials. Parallel bill review maybe a manual process or partially automated. It requires good process design and supporting materials.

Run parallel bill production. Bill production should be run on the same dates with the same input in both systems. Then bill copies will be created from the old system for review, since the original bill will be sent to customers in the normal billing process. Bills produced by the new system will not be distributed and can be utilized for the review process.

Train reviewers. Just like any other human function, parallel bill review involves training the people who will review and compare bills. Starting with people who are knowledgeable about the old system is an immense advantage.

Review bills. At a minimum, the bill review activity includes comparing bills from the old and new systems for content and totals, identifying any anomalies, determining the cause of any anomaly and resolving the anomalies. Resolution may consist of fixing a problem or simply accepting small differences in algorithms—with documentation and appropriate accounting accommodation, of course!

User acceptance review

As in any other software contract, you should include a formal user acceptance activity, even though end-users may have been active in various testing and verification activities. This review may include simply reviewing the results of various activities or it may include actual testing tasks.

Review migration fallout and resolutions. Your company's project manager and responsible functional managers should review all process fallout and resolution actions for acceptability to the business.

Review results of parallel bill cycles. Your company's project manager and responsible functional managers should review the results of parallel bill cycles for acceptability to the business.

Go/no go decision. Your company's project manager and responsible functional managers should make a clear Go or No Go decision whether to go live with the new system at this time.

Turn-up for service

Turning up the migrated data for production has several steps. First, any incremental data not incorporated in the new platform should be entered or made available. All production information should now be directed exclusively to the new system and bill distribution for these accounts should be terminated from the old system. Any remaining "test accounts" should be deactivated or removed.

FtA Conversion Project Plan

Table 40-2 shows the higher-level tasks in the FtA conversion plan. The complete task list is available in Table 40-3. Note that the FtA project planner has shown a single activity for developing data load software, but has created separate activities for the extraction and translation development for each of the source systems and that the production loads are scheduled separately. Also note that the BB1 (the smaller broadband company) conversion is allocated minimal time and resources. FtA has decided to manually load customer and service information and only to convert balance and account history information for BB1, which has only 5 customers.

TABLE 40-2 FtA Conversion Project Plan

Task Name	Duration	Pred.
Migrate legacy billing data to new platform	454.5 days	
Implementation phase	234 days	
Platform implementation	27 days	
Platform implementation	4 days	Implementation
Develop migration strategy	16 days	
New system data load software	222 days	11, 12

continued on next page

TABLE 40-2 FtA Conversion Project Plan (continued)

Task Name	Duration	Pred.
Perform BB1 conversion	379.5 days	8
Develop test plan	2 days	8
Parallel processing preparation	6 days	22
Data extraction	19 days	
Data translations for migration	29 days	
End-to-end massage and load software test	6 days	71
Trial migration	7 days	72
Configure for production	10 days	
Production load	59 days	Implementation
User acceptance review	55.5 days	
Turn up for service	2 days	94, 118
Perform BB2 conversion	399.5 days	8
Develop test plan	5 days	
Parallel processing preparation	8 days	129
Data extraction	28 days	
Data translations for migration	40 days	
End-to-end massage and load software test	8 days	178
Trial migration	12 days	179
Production load	70 days	184, 104
User acceptance review	59.5 days	
Turn up for service	5 days	200, 224
Perform ISP conversion	421.5 days	8
Develop test plan	3 days	
Parallel processing preparation	9 days	235
Data extraction	32 days	
Data translations for migration	38 days	
End-to-end massage and load software test	10 days	284
Trial migration	12 days	285
Production load	82 days	290, 209
User acceptance review	68.5 days	
Turn up for service	5 days	306, 330
Perform clec conversion	438.5 days	8
Develop test plan	5 days	
Parallel processing preparation	20 days	341
Data extraction	89 days	

continued on next page

TABLE 40-2 FtA Conversion Project Plan (continued)

Task Name	Duration	Pred.
Data translations for migration	101 days	
End-to-end massage and load software test	15 days	390
Trial migration	14 days	391
Production load	87 days	396, 315
User acceptance review	73.5 days	
Turn up for service	5 days	412, 436

The conversions for BB2 and the ISP are allocated minimum time and resources, since neither company is doing usage-based billing on existing services and their account structure is very simple. The CLEC conversion, on the other hand, involves complex services and accounts and usage-based services. Also, the old billing system is no longer vendor supported and documentation is not available. FtA has decided to defer the CLEC conversion until after the new billing system is in production and stable for other services.

FtA has linked its platform availability to the availability of the new system platform and its production conversion to the "configure for production" activity in the implementation project plan.

TABLE 40-3 FtA Conversion Details

#	Taks Name	Duration	Pred.
1	Complete overall billing project	0 days	2
2	Migrate legacy billing data to new platform	454.5 days	
3	Implementation phase	234 days	
4	Platform implementation	27 days	
5	Platform implementation	4 days	Implementation \42
6	Determine platform requirements	2 days	
7	Allocate platform resources	2 days	6
8	Develop migration strategy	16 days	
9	Review customer and account data	10 days	
10	Review product/service data	10 days	
11	Review customer and account data structures and tools in new system	2 days	9
12	Review product/service data structures and tools in new system	2 days	10
13	Document migration strategy	4 days	9, 10, 11, 12

continued on next page

TABLE 40-3 FtA Conversion Details (continued)

#	Taks Name	Duration	Pred.
14	New system data load software	222 days	11, 12
15	Design data load software	2 days	
16	Develop data load software	1 day	15
17	Synchronize data load software with migration strategy	3 days	16,13
18	Test data load software	2 days	17,5
19	Perform BB1 conversion	379.5 days	8
20	Develop test plan	2 days	8
21	Define scope of data migration software test	2 days	
22	Define scope of parallel processing	1 day	
23	Define testing activities	1 day	
24	Establish testing standards	1 day	
25	Parallel processing preparation	6 days	22
26	Develop parallel data feed requirements	1 day	
27	Develop functional specifications for parallel processing support from old system	1 day	26
28	Implement functional specifications for parallel processing support from old system	1 day	27
29	Develop functional specifications for accepting parallel data in new system	1 day	26
30	Implement functional specifications for accepting parallel data into new system	1 day	29
31	Test parallel processing	5 days	
32	Define test scope	1 day	26
33	Define test scenarios	1 day	32
34	Define test cases	2 days	33
35	Establish test parallel feed	1 day	28, 30
36	Perform test	1 day	34, 35
37	Data extraction	19 days	
38	Draft specifications for trial and preliminary data extract	2 days	13
39	Get trial data extract from old billing system	5 days	8
40	Develop trial extract functional specifications	2 days	
41	Develop trial extract per functional specifications	1 day	40
42	Perform trial data extract	2 days	41
43	Assess trial extract	3 days	39

continued on next page

TABLE 40-3 FtA Conversion Details (continued)

#	Taks Name	Duration	Pred.
44	Assess data format	1 day	
45	Assess data completeness	1 day	
46	Assess data quality	2 days	
47	Modify draft specifications as needed	1 day	44, 45, 46
48	Develop data quality strategy	4 days	43
49	Develop specifications for quality assessment tools	2 days	
50	Develop quality assessment tools	2 days	49
51	Define QA process and resources	2 days	
52	Get preliminary data extract from old billing system	5 days	47
53	Develop preliminary extract functional specifications	1 day	
54	Develop preliminary extract per functional specifications	3 days	53
55	Perform preliminary data extract	1 day	54
56	Perform quality assessment	6 days	52
57	Run quality assessment tools	1 day	
58	Perform manual QA review per QA process	2 days	
59	Identify and document corrective actions for any quality problems	2 days	58
60	Take corrective actions (recycle with legacy data source or clean up data)	2 days	57, 59
61	Data translations for migration	29 days	
62	Migration analysis/requirements	18 days	
63	Review data structures	1 day	44
64	Define data mappings	5 days	63
65	Review quality assessment results	2 days	56
66	Document data translations requirements	3 days	64, 65
67	Develop translations software	10 days	62
68	Develop functional specifications for data massage software	5 days	
69	Develop data massage software	3 days	68
70	Test data massage software	2 days	69
71	Perform translations on extracted data	1 day	70
72	End-to-end massage and load software test	6 days	71
73	Define scope of end-to-end test	2 days	

continued on next page

TABLE 40-3 FtA Conversion Details (continued)

#	Taks Name	Duration	Pred.
74	Develop test scenarios	1 day	73
75	Define test cases	2 days	74
76	Perform test	1 day	75
77	Trial migration	7 days	72
78	Perform test loads	1 day	
79	Unit test with migrated data	1 day	
80	Integration test with migrated data	5 days	79
81	Define scope of integration test	2 days	
82	Develop test scenarios	1 day	81
83	Develop test cases	1 day	82
84	Perform test	1 day	83
85	Parallel bill cycle	2 days	79
86	Load usage data for billing cycle	0 days	
87	Run parallel bill cycle for test accounts	2 days	86
88	User acceptance test with migrated data	7 days	
89	Define scope of UA test	1 day	
90	Develop test scenarios	1 day	89
91	Develop test cases	1 day	90
92	Perform UA test	1 day	80, 85, 91
93	Configure for production	10 days	
94	Production load	59 days	Implementation \118
95	Prepare platform for production load	1 day	
96	Load legacy data	6 days	
97	Get final extract from old billing system	1 day	
98	Initiate parallel data feed (usage, transactions)	1 day	
99	Perform QA on final extract	1 day	97
100	Take corrective actions for QA defects	1 day	99
101	Run translations on final extract	1 day	100
102	Resolve translation fall-out	1 day	101
103	Load new system	1 day	101, 95
104	Verify data load	1 day	103
105	Parallel bill cycles	59 days	
106	Define scope of parallel testing	2 days	22

continued on next page

TABLE 40-3 FtA Conversion Details (continued)

#	Taks Name	Duration	Pred.
107	Design bill sample	2 days	106
108	Develop review process and materials	5 days	107
109	Run parallel bill production	45 days	107, 96
110	Train reviewers	4 days	
111	Develop training	3 days	108
112	Deliver training	1 day	111
113	Review bills	8 days	
114	Compare bills per review process	3 days	109, 110
115	Identify anomalies	1 day	114
116	Determine causes of anomalies	2 days	115
117	Resolve anomalies	2 days	116
118	User acceptance review	55.5 days	
119	Review migration fallout and resolutions	1 day	102
120	Review results of parallel bill cycles	1 day	113
121	Go/no go decision	0.5 days	119, 120
122	Turn up for service	2 days	94, 118
123	Update with incremental data if required	1 day	
124	Direct all production feeds to new system	1 day	123
125	Terminate bill distribution from old system	1 day	123
126	Perform BB2 conversion	399.5 days	8
127	Develop test plan	5 days	
128	Define scope of data migration software test	5 days	
129	Define scope of parallel processing	1 day	
130	Define testing activities	1 day	
131	Establish testing standards	2 days	
132	Parallel processing preparation	8 days	129
133	Develop parallel data feed requirements	2 days	
134	Develop functional specifications for parallel processing support from old system	1 day	133
135	Implement functional specifications for parallel processing support from old system	2 days	134
136	Develop functional specifications for accepting parallel data in new system	1 day	133
137	Implement functional specifications for accepting parallel data into new system	1 day	136
138	Test parallel processing	6 days	

continued on next page

TABLE 40-3 FtA Conversion Details (continued)

#	Taks Name	Duration	Pred.
139	Define test scope	1 day	133
140	Define test scenarios	1 day	139
141	Define test cases	2 days	140
142	Establish test parallel feed	1 day	135, 137
143	Perform test	2 days	141, 142
144	Data extraction	28 days	
145	Draft specifications for trial and preliminary data extract	3 days	
146	Get trial data extract from old billing system	5 days	
147	Develop trial extract functional specifications	2 days	
148	Develop trial extract per functional specifications	1 day	147
149	Perform trial data extract	2 days	148
150	Assess trial extract	5 days	146
151	Assess data format	2 days	
152	Assess data completeness	2 days	
153	Assess data quality	3 days	
154	Modify draft specifications as needed	2 days	151, 152, 153
155	Develop data quality strategy	7 days	150
156	Develop specifications for quality assessment tools	3 days	
157	Develop quality assessment tools	4 days	156
158	Define QA process and resources	3 days	
159	Get preliminary data extract from old billing system	8 days	154
160	Develop preliminary extract functional specifications	1 day	
161	Develop preliminary extract per functional specifications	5 days	160
162	Perform preliminary data extract	2 days	161
163	Perform quality assessment	10 days	159
164	Run quality assessment tools	1 day	
165	Perform manual QA review per QA process	3 days	
166	Identify and document corrective actions for any quality problems	2 days	165
167	Take corrective actions (recycle with legacy data source or clean up data)	5 days	164, 166

continued on next page

TABLE 40-3 FtA Conversion Details (continued)

#	Taks Name	Duration	Pred.
168	Data translations for migration	40 days	
169	Migration analysis/requirements	26 days	
170	Review data structures	1 day	151
171	Define data mappings	6 days	170
172	Review quality assessment results	2 days	163
173	Document data translations requirements	3 days	171, 172
174	Develop translations software	12 days	169
175	Develop functional specifications for data massage software	5 days	
176	Develop data massage software	5 days	175
177	Test data massage software	2 days	176
178	Perform translations on extracted data	2 days	177
179	End-to-end massage and load software test	8 days	178
180	Define scope of end-to-end test	3 days	
181	Develop test scenarios	1 day	180
182	Define test cases	2 days	181
183	Perform test	2 days	182
184	Trial migration	12 days	179
185	Perform test loads	3 days	
186	Unit test with migrated data	2 days	
187	Integration test with migrated data	7 days	186
188	Define scope of integration test	2 days	
189	Develop test scenarios	2 days	188
190	Develop test cases	2 days	189
191	Perform test	1 day	190
192	Parallel bill cycle	8 days	186
193	Load usage data for billing cycle	3 days	
194	Run parallel bill cycle for test accounts	5 days	193
195	User acceptance test with migrated data	12 days	
196	Define scope of UA test	2 days	
197	Develop test scenarios	2 days	196
198	Develop test cases	2 days	197
199	Perform UA test	2 days	187, 192, 198
200	Production load	70 days	184, 104
201	Prepare platform for production load	2 days	

continued on next page

TABLE 40-3 FtA Conversion Details (continued)

#	Taks Name	Duration	Pred.
202	Load legacy data	12 days	
203	Get final extract from old billing system	1 day	
204	Initiate parallel data feed (usage, transactions)	1 day	
205	Perform QA on final extract	4 days	
206	Take corrective actions for QA defects	3 days	205
207	Run translations on final extract	1 day	206
208	Resolve translation fall-out	4 days	207
209	Load new system	2 days	207, 201
210	Verify data load	2 days	209
211	Parallel bill cycles	70 days	
212	Define scope of parallel testing	2 days	129
213	Design bill sample	2 days	212
214	Develop review process and materials	5 days	213
215	Run parallel bill production	45 days	202, 213
216	Train reviewers	4 days	
217	Develop training	3 days	214
218	Deliver training	1 day	217
219	Review bills	13 days	
220	Compare bills per review process	5 days	215, 216
221	Identify anomalies	2 days	220
222	Determine causes of anomalies	3 days	221
223	Resolve anomalies	3 days	222
224	User acceptance review	59.5 days	
225	Review migration fallout and resolutions	1 day	208
226	Review results of parallel bill cycles	1 day	219
227	Go/no go decision	0.5 days	225, 226
228	Turn up for service	5 days	200, 224
229	Update with incremental data if required	4 days	
230	Direct all production feeds to new system	1 day	229
231	Terminate bill distribution from old system	1 day	229
232	Perform ISP conversion	421.5 days	8
233	Develop test plan	3 days	
234	Define scope of data migration software test	3 days	
235	Define scope of parallel processing	1 day	

continued on next page

TABLE 40-3 FtA Conversion Details (continued)

#	Taks Name	Duration	Pred.
236	Define testing activities	1 day	
237	Establish testing standards	2 days	
238	Parallel processing preparation	9 days	235
239	Develop parallel data feed requirements	2 days	
240	Develop functional specifications for parallel processing support from old system	1 day	239
241	Implement functional specifications for parallel processing support from old system	2 days	240
242	Develop functional specifications for accepting parallel data in new system	1 day	239
243	Implement functional specifications for accepting parallel data into new system	1 day	242
244	Test parallel processing	7 days	
245	Define test scope	1 day	239
246	Define test scenarios	1 day	245
247	Define test cases	2 days	246
248	Establish test parallel feed	2 days	241, 243
249	Perform test	2 days	247, 248
250	Data extraction	32 days	
251	Draft specifications for trial and preliminary data extract	3 days	
252	Get trial data extract from old billing system	7 days	
253	Develop trial extract functional specifications	4 days	
254	Develop trial extract per functional specifications	1 day	253
255	Perform trial data extract	2 days	254
256	Assess trial extract	5 days	252
257	Assess data format	2 days	
258	Assess data completeness	2 days	
259	Assess data quality	3 days	
260	Modify draft specifications as needed	2 days	257, 258, 259
261	Develop data quality strategy	5 days	256
262	Develop specifications for QA tools	3 days	
263	Develop QA tools	2 days	262
264	Define QA process and resources	3 days	
265	Get preliminary data extract from old billing system	8 days	260

continued on next page

TABLE 40-3 FtA Conversion Details (continued)

#	Taks Name	Duration	Pred.
266	Develop preliminary extract functional specifications	1 day	
267	Develop preliminary extract per functional specifications	5 days	266
268	Perform preliminary data extract	2 days	267
269	Perform QA	12 days	265
270	Run QA tools	1 day	
271	Perform manual QA review per QA process	5 days	
272	Identify and document corrective actions for any quality problems	2 days	271
273	Take corrective actions (recycle with legacy data source or clean up data)	5 days	270, 272
274	Data translations for migration	38 days	
275	Migration analysis/requirements	27 days	
276	Review data structures	1 day	257
277	Define data mappings	10 days	276
278	Review quality assessment results	2 days	269
279	Document data translations requirements	2 days	277, 278
280	Develop translations software	9 days	275
281	Develop functional specifications for data massage software	5 days	
282	Develop data massage software	2 days	281
283	Test data massage software	2 days	282
284	Perform translations on extracted data	2 days	283
285	End-to-end massage and load software test	10 days	284
286	Define scope of end-to-end test	3 days	
287	Develop test scenarios	1 day	286
288	Define test cases	3 days	287
289	Perform test	3 days	288
290	Trial migration	12 days	285
291	Perform test loads	5 days	
292	Unit test with migrated data	2 days	
293	Integration test with migrated data	7 days	292
294	Define scope of integration test	2 days	
295	Develop test scenarios	2 days	294
296	Develop test cases	2 days	295

continued on next page

TABLE 40-3 FtA Conversion Details (continued)

#	Taks Name	Duration	Pred.
297	Perform test	1 day	296
298	Parallel bill cycle	8 days	292
299	Load usage data for billing cycle	3 days	
300	Run parallel bill cycle for test accounts	5 days	299
301	User acceptance test with migrated data	12 days	
302	Define scope of UA test	2 days	
303	Develop test scenarios	2 days	302
304	Develop test cases	2 days	303
305	Perform UA test	2 days	293, 298, 304
306	Production load	82 days	290, 209
307	Prepare platform for production load	1 day	
308	Load legacy data	15 days	
309	Get final extract from old billing system	1 day	
310	Initiate parallel data feed (usage, transactions)	1 day	
311	Perform QA on final extract	5 days	309
312	Take corrective actions for QA defects	3 days	311
313	Run translations on final extract	1 day	312
314	Resolve translation fall-out	5 days	313
315	Load new system	2 days	313
316	Verify data load	2 days	315
317	Parallel bill cycles	82 days	
318	Define scope of parallel testing	2 days	235
319	Design bill sample	2 days	318
320	Develop review process and materials	5 days	319
321	Run parallel bill production	45 days	308, 319
322	Train reviewers	4 days	
323	Develop training	3 days	320
324	Deliver training	1 day	323
325	Review bills	22 days	
326	Compare bills per review process	10 days	321, 322
327	Identify anomalies	2 days	326
328	Determine causes of anomalies	5 days	327
329	Resolve anomalies	5 days	328
330	User acceptance review	68.5 days	

continued on next page

TABLE 40-3 FtA Conversion Details (continued)

#	Taks Name	Duration	Pred.
331	Review migration fall-out and resolutions	1 day	314
332	Review results of parallel bill cycles	1 day	325
333	Go/no go decision	0.5 days	331, 332
334	Turn up for service	5 days	306, 330
335	Update with incremental data if required	4 days	
336	Direct all production feeds to new system	1 day	335
337	Terminate bill distribution from old system	1 day	335
338	Perform CLEC conversion	438.5 days	8
339	Develop test plan	5 days	
340	Define scope of data migration software test	5 days	
341	Define scope of parallel processing	1 day	
342	Define testing activities	1 day	
343	Establish testing standards	2 days	
344	Parallel processing preparation	20 days	341
345	Develop parallel data feed requirements	5 days	
346	Develop functional specifications for parallel processing support from old system	2 days	345
347	Implement functional specifications for parallel processing support from old system	5 days	346
348	Develop functional specifications for accepting parallel data in new system	2 days	345
349	Implement functional specifications for accepting parallel data into new system	2 days	348
350	Test parallel processing	15 days	
351	Define test scope	1 day	345
352	Define test scenarios	4 days	351
353	Define test cases	6 days	352
354	Establish test parallel feed	4 days	347, 349
355	Perform test	4 days	353, 354
356	Data extraction	89 days	
357	Draft specifications for trial and preliminary data extract	3 days	
358	Get trial data extract from old billing system	22 days	
359	Develop trial extract functional specifications	8 days	
360	Develop trial extract per functional specifications	10 days	359
361	Perform trial data extract	4 days	360

continued on next page

TABLE 40-3 FtA Conversion Details (continued)

#	Taks Name	Duration	Pred.
362	Assess trial extract	8 days	358
363	Assess data format	4 days	
364	Assess data completeness	4 days	
365	Assess data quality	5 days	
366	Modify draft specifications as needed	3 days	363, 364, 365
367	Develop data quality strategy	11 days	362
368	Develop specifications for QA tools	6 days	
369	Develop QA tools	5 days	368
370	Define QA process and resources	5 days	
371	Get preliminary data extract from old billing system	24 days	366
372	Develop preliminary extract functional specifications	10 days	
373	Develop preliminary extract per functional specifications	10 days	372
374	Perform preliminary data extract	4 days	373
375	Perform quality assessment	35 days	371
376	Run quality assessment tools	2 days	
377	Perform manual QA review per QA process	15 days	
378	Identify and document corrective actions for any quality problems	10 days	377
379	Take corrective actions (recycle with legacy data source or clean up data)	10 days	376, 378
380	Data translations for migration	101 days	
381	Migration analysis/requirements	73 days	
382	Review data structures	1 day	363
383	Define data mappings	15 days	382
384	Review quality assessment results	5 days	375
385	Document data translations requirements	5 days	383, 384
386	Develop translations software	25 days	381
387	Develop functional specifications for data massage software	12 days	
388	Develop data massage software	10 days	387
389	Test data massage software	3 days	388
390	Perform translations on extracted data	3 days	389
391	End-to-end massage and load software test	15 days	390
392	Define scope of end-to-end test	3 days	

continued on next page

TABLE 40-3 FtA Conversion Details (continued)

#	Taks Name	Duration	Pred.
393	Develop test scenarios	2 days	392
394	Define test cases	5 days	393
395	Perform test	5 days	394
396	Trial migration	14 days	391
397	Perform test loads	5 days	
398	Unit test with migrated data	2 days	
399	Integration test with migrated data	7 days	398
400	Define scope of integration test	2 days	
401	Develop test scenarios	2 days	400
402	Develop test cases	2 days	401
403	Perform test	1 day	402
404	Parallel bill cycle	8 days	398
405	Load usage data for billing cycle	3 days	
406	Run parallel bill cycle for test accounts	5 days	405
407	User acceptance test with migrated data	14 days	
408	Define scope of UA test	2 days	
409	Develop test scenarios	2 days	408
410	Develop test cases	4 days	409
411	Perform UA test	4 days	399, 404, 410
412	Production load	87 days	396, 315
413	Prepare platform for production load	1 day	
414	Load legacy data	15 days	
415	Get final extract from old billing system	1 day	
416	Initiate parallel data feed (usage, transactions)	1 day	
417	Perform QA on final extract	5 days	415
418	Take corrective actions for QA defects	3 days	417
419	Run translations on final extract	1 day	418
420	Resolve translation fall-out	5 days	419
421	Load new system	2 days	419
422	Verify data load	2 days	421
423	Parallel bill cycles	87 days	
424	Define scope of parallel testing	2 days	341
425	Design bill sample	2 days	424
426	Develop review process and materials	5 days	425

continued on next page

TABLE 40-3 FtA Conversion Details (continued)

#	Taks Name	Duration	Pred.
427	Run parallel bill production	45 days	414, 425
428	Train reviewers	4 days	
429	Develop training	3 days	426
430	Deliver training	1 day	429
431	Review bills	27 days	
432	Compare bills per review process	15 days	427, 428
433	Identify anomalies	2 days	432
434	Determine causes of anomalies	5 days	433
435	Resolve anomalies	5 days	434
436	User acceptance review	73.5 days	
437	Review migration fallout and resolutions	1 day	420
438	Review results of parallel bill cycles	1 day	431
439	Go/no go decision	0.5 days	437, 438
440	Turn-up for service	5 days	412, 436
441	Update with incremental data if required	4 days	
442	Direct all production feeds to new system	1 day	441
443	Terminate bill distribution from old system	1 day	441

Summary

In this chapter, we looked at the activities involved in a billing system data conversion. Data conversion is a time- and resource-consuming activity that should be a significant part of your planning if your project includes conversion activities. Throughout the course of this chapter and all of Section IV, we have identified significant testing activities and reviewed the importance of allocating adequate time and resources to testing. In Chapter 41, we pull together much of this information and other testing topics related to billing systems.

Chapter

41

Testing

Extensive test activities are specified in each of the subproject plans presented in the preceding chapters. Why then have we devoted this final chapter to the topic of testing?

Billing systems create some unique challenges for testing. They also require well planned and comprehensive testing, because of the complexity of the environment and because of the criticality of the function to your business. This is especially true when you are moving from one system to another; in this case, your customers may be your most critical testers!

Testing Expertise

If your company does not have resources with significant experience in testing billing systems, you should seriously consider buying that expertise—either by hiring folks with that background or by contracting with a consulting or services company that has that expertise. If you want to develop that capability, include a knowledge transfer activity focusing on testing in the contract.

Testing includes test planning, system test, and manual testing. Here is a quick look at all the activities requiring testing expertise in the linked project plans.

Test planning

The following testing activities require test planning expertise:

- Definition of the scope for each variety of testing—acceptance, operational, interface integration, conversion, and user acceptance
- Schedule development for the test plans, including which activities can be accomplished concurrently and which must be done sequentially
- Creating and managing parallel tests for conversions that are efficient for the company and transparent to customers.

System test

The following testing activities require system test expertise:

- Specification and creation of the test environment(s), providing the ability to both test and train without negative impact on production
- Management of the test environment(s), including such items as resetting dates to enable accelerated testing cycles and running "stress" or volume tests
- Design and use of control reports to ensure synchronicity of parallel testing inputs
- Troubleshooting error conditions, particularly during integration and conversion testing.

Manual testing

The following testing activities require manual testing expertise:

- Definition of test scenarios
- Creation of test cases
- Troubleshooting error conditions, particularly when manual processes or input conditions need to be identified and analyzed

Overall

Test management will provide the balanced and—it is to be hoped—unbiased view of the accuracy of the tested systems and processes as input to "go-no go" decisions and end-of-project analysis documents.

Testing Constraints

The most frequently seen testing constraint is time for testing. Because the testing step is at the end of the process and earlier steps may exceed scheduled time frames, project managers often slash the time allotted for testing activities. It is not unheard of for companies to eliminate some test steps altogether—and then find that there are significant problems that require far more time to fix after the fact than would have been needed before final acceptance.

Another significant constraint is the shortage of knowledgeable personnel. Especially in a conversion situation, people who are conversant with both old and new systems are highly prized. Luring this type of person from his or her home department or discipline to spend a fair amount of time doing testing is difficult.

Expense is another factor and is regularly cited in discussions of testing time and provision of knowledgeable testers. The cost/benefit argument is useful here: it is much more costly to have to rerun a conversion or implementation than to do it correctly the first time.

Occasionally, an objection raised to testing is that production files, either billing or financial, will be negatively impacted if test accounts or media are entered into the production environments. Test media can be prepared and managed in a number of different ways to avoid such negative impacts. A method used successfully by many companies is the establishment of unique elements (e.g., segments, accounts, telephone numbers, accounting codes) identified as test or auditor elements. These transactions and accounts can then be tracked on a regular basis from end to end, thereby exercising all aspects of the processes and systems.

Permanent Test Partitions

Testing should not be viewed as a one-time step. Most modern billing systems use a multitude of tables and configurations that are controlled by users within the company. Changes to this type of system tool should be carefully monitored and controlled. Establishing a permanent test partition is something that the authors strongly recommend. Such a separate environment allows for both testing and training, giving you the assurance that the production system is protected from trial and error changes.

Summary

In this chapter, we have emphasized the value of including and maintaining testing and testing expertise as part of any billing system conversion or implementation. Time spent in good test planning shortens the overall implementation of the project. By the same token, testing fully and intelligently enhances the end product with a minimum of rework—and customer dissatisfaction. This is true for both major changes, such as the installation of a complete billing system, as well as the small changes every month. Testing can help us realize that all the efforts we discuss in this book produce the exact billing system we contracted to get. What value!

Appendix A

Data Gathering Forms

In Section II of this book, we looked at data gathering for a billing systems project. This appendix contains sample data gathering forms the authors find useful. The accompanying CD contains MS Word® versions of these forms that you can copy and use or modify as needed. If your environment is large and complex, you may want to convert the forms to be a front end for databases of collected information.

Data entry fields are shown as shaded boxes. Yes/no conditions are shown as a checkbox. Where a data entry field displays text, the box is a drop-down field with multiple fixed choices.

Figure A-1 corresponds to the product and service data gathering in Chapter 15.

Figure A-2 is designed for use in consolidating information across products and services, as discussed in Chapter 16.

Figure A-3 corresponds to the customer and account data gathering in Chapter 17.

Product/Service Worksheet

Analyst

Date

Product/Service Code

What is the product/service code?

Service status

Embedded intelligence? ☐ Describe

Subfields—describe

Field length and type

Field 1

Field 2

Field 3

Character set

Alpha

Numeric

Allowed special characters

Generated within the company? ☐

Product/Service Name/Description

What is the product/service name or description to be used on the bill?
Will this be different in the new system from the old system? ☐
If so, explain.

Embedded intelligence? ☐ Describe

Subfields—describe

Field length and type

Field 1

Field 2

Field 3

Character set

Alpha

Numeric

Allowed special characters

Generated within the company? ☐

Product/Service Detail

Type of service (switched, dedicated, Internet access, etc.)

Describe the product/service in plain language.

How is this different from other products or services?

Figure A-1 Product and service data gathering.

Product/Service Worksheet

Product/Service Identifier Format

Format

Embedded intelligence? ☐ Describe

Subfields—describe

Field length and type

Field 1

Field 2

Field 3

Character set

Alpha

Numeric

Allowed special characters

Generated within the company? ☐

If not internally generated, identify source or control

Billing Basis

Recurring, one-time or usage-based?

Units Format for units

Advance/Arrears

Is the amount of the charge calculated in the billing system or provided by another system?

Geographic Market

In the United States, which states?

What countries?

Customer Market Segment

Market segment

Price Structure

Describe the price structure

Provide the formula for calculating price

Currency

Discount structure

Discounts

Eligible for what discounts?

Relationship of discounts?

Formulas?

Precedence?

Tax impact?

Figure A-1 Product and service data gathering (continued).

Product/Service Worksheet

Promotions

Describe any promotions

Relationship to other services and account

Special Contracts

How many contracts for this service?

What percentage of this service base has been sold under contract?

Describe customers under contract

How is contract information captured and retained?

How are expirations and renewals managed?

How are contracts and contract compliance audited?

Related Products, Services, and Features

Related Product	Relationship	Markets/Jurisdictions

Warranty and Support Information

Does the product include warranty or support? ☐ Describe.

Identify options and additional billing elements for warranty and support.

Terms and conditions

Accounting Classification

Identify accounting classification code

Embedded intelligence? ☐ Describe

Subfields—describe

Field length and type

Field 1

Field 2

Field 3

Character set

Alpha

Numeric

Allowed special characters

Generated within the company? ☐

Quantity

How many of this product or service are in service?

How many are anticipated over the life of the new billing system?

Order Volume

How many orders per month (history)?

How many orders per month (projected)?

Order Source Quantity

Figure A-1 Product and service data gathering (continued).

Product/Service Fields and Formats Consolidated Data

Analyst

Date

Product Line

Market Segment

1. Product/Service Code

Description

Source

Use

Length

Format

Character Set

Volatility

Volume

Unique?

2. Product/Service Name/Description

Description

Source

Use

Length

Format

Character Set

Volatility

Volume

Unique?

3. Product/Service Detail

Description

Source

Use

Length

Format

Character Set

Volatility

Volume

Unique?

Figure A-2 Product and service data consolidation.

Product/Service Fields and Formats Consolidated Data

4. Product/Service Identifier Format

Description
Source
Use
Length
Format
Character Set
Volatility
Volume
Unique?

5. Billing Basis

Description
Source
Use
Length
Format
Character Set
Volatility
Volume
Unique?

6. Billing Information Sources

Description
Source
Use
Length
Format
Character Set
Volatility
Volume
Unique?

7. Geographic Market

Description
Source
Use
Length

Figure A-2 Product and service data consolidation (continued).

Product/Service Fields and Formats Consolidated Data

- Format
- Character Set
- Volatility
- Volume
- Unique?

8. **Customer Market Segment**
   - Description
   - Source
   - Use
   - Length
   - Format
   - Character Set
   - Volatility
   - Volume
   - Unique?
9. **Price Structure**
   - Description
   - Source
   - Use
   - Length
   - Format
   - Character Set
   - Volatility
   - Volume
   - Unique?
10. **Discounts**
    - Description
    - Source
    - Use
    - Length
    - Format
    - Character Set
    - Volatility
    - Volume
    - Unique?

Figure A-2 Product and service data consolidation (continued).

Product/Service Fields and Formats Consolidated Data

11. Promotions

- Description
- Source
- Use
- Length
- Format
- Character Set
- Volatility
- Volume
- Unique?

12. Special Contracts

- Description
- Source
- Use
- Length
- Format
- Character Set
- Volatility
- Volume
- Unique?

13. Related Products, Services and Features Indicators

- Description
- Source
- Use
- Length
- Format
- Character Set
- Volatility
- Volume
- Unique?

14. Warranty and Support Indicator

- Description
- Source
- Use
- Length

Figure A-2 Product and service data consolidation (continued).

Product/Service Fields and Formats Consolidated Data

Format

Character Set

Volatility

Volume

Unique?

15. Accounting Classification

Description

Source

Use

Length

Format

Character Set

Volatility

Volume

Unique?

Figure A-2 Product and service data consolidation (continued).

Customer Information Worksheet

Documented by

Date

1. **Account Identifier/Account Number**

   Format

   Character Set Alpha

   Numeric

   Allowed special characters

   Length

   Subfields

   Intelligence

   System Assigned? ☐ If so, in which system?

   Comments:

2. **Account Name**

   Parsed? ☐

   Rules

Figure A-3 Customer and account data gathering.

Customer Information Worksheet

Fields	Length	Char. Set	Edit Rules	Source

Comments

3. Billing Contact

Parsed? ☐

Rules

Fields	Length	Char. Set	Edit Rules	Source

Comments

4. Billing Address

Parsed? ☐ If so, in which system(s)

Rules

Fields	Length	Char. Set	Edit Rules	Source
Street Number				
Street Directional				
Street Name				
Apt/Office No.				
PO Box				
Community				
State				
Zip				

Comments

5. Directory Listing Appearance

Parsed? ☐

Rules

Fields	Length	Char. Set	Edit Rules	Source
Last Name or Company Name				
First Name				
Middle Name or Initial				
Title				
Listing Address				
Listing City				
Additional lines				
Additional lines				

Figure A-3 Customer and account data gathering (continued).

Customer Information Worksheet

Additional lines

Additional lines

Comments

6. Final Billing Address

Parsed? ☐ If so, in which system(s)

Rules

Fields	Length	Char. Set	Edit Rules	Source
Street Number				
Street Directional				
Street Name				
Apt/Office No.				
PO Box				
Community				
State				
Zip				

Comments

7. Credit Information

Rules

Fields	Length	Char. Set	Edit Rules	Source

8. Credit Classification

Rules

Fields	Length	Char. Set	Edit Rules	Source

Credit Class

Comments:

9. Credit History

Rules

Fields	Length	Char. Set	Edit Rules	Source
Current Month				

Comments:

10. Deposit Information

Deposit indicator ☐

Deposit taken date Field Length: Subfields:

Figure A-3 Customer and account data gathering (continued).

Customer Information Worksheet

	Character Set	
Deposit Amount	Field Length:	Subfields:
	Character Set	Allowed special characters
Deposit return date	Field Length:	Subfields:
	Character Set	
Interest	To be paid? ☐	Interest rate:
Date interest to be paid		
	Field Length:	Subfields:
	Character Set	
Deposit return medium	Applied to bill? ☐	Held for life of account?
	Check/draft to customer? ☐	

Comments:

11. Bill Cycle

Rules

Fields	Length	Char. Set	Edit Rules	Source
Current Month				
Bill Cycle Ind.				

Comments:

12. Balance and Payment History

Rules

Fields	Length	Char. Set	Edit Rules	Source
Total of Last Bill:				

For Current Period

Bal. from Previous Period

Payments Received Since Last Bill:

Date:

Amount:

Balance Forward:

Date:

Amount:

Balance Forward:

Date:

Amount:

Balance Forward:

Figure A-3 Customer and account data gathering (continued).

Customer Information Worksheet

Date:

Amount:

Balance Forward:

Previous Months Payment History

Comments:

13. Primary Bill Medium

Rules

Fields	Length	Char. Set	Edit Rules	Source
Primary Medium				
Additional Bill				
Allowed?				
Additional Bill				
Allowed?				

Comments:

14. Preferred Payment Method

Rules

Fields	Length	Char. Set	Edit Rules	Source

15. Market Segment

Rules

Segment Ind.

16. Tax Jurisdiction

Rules

Tax Jurisdiction

17. Tax Exemption Information

Tax Exempt? ☐

Certification Received? ☐

Rules

Fields	Length	Char. Set	Edit Rules	Source
Reason for Tax Exemption				
Date Certification Received				
Location Certificate Filed				

Comments:

18. Third Party Information

Third Party Notification Indicator? ☐

Figure A-3 Customer and account data gathering (continued).

Customer Information Worksheet

Rules

Fields	Length	Char. Set	Edit Rules	Source
Name				
Address				
Telephone No.				
Relationship				

Comments:

19. Account Structure

Rules

Fields	Length	Char. Set	Edit Rules	Source
Master Account Indicator				
Slave Account Indicator				

Comments:

Figure A-3 Customer and account data gathering (continued).

Appendix

B

Requirements Document Models

The enclosed CD contains "fill-in-the-blanks" tracking forms for billing systems business requirements documents. The documents are in MS Word 2000® format with standard document formatting and fonts. These documents provide an outline for each type of requirements document, along with prompts for typically important features, functions and controls.

Not surprisingly, the requirements forms include the requirements checklist illustrated in Section III and follow the structure of the requirements checklist and evaluation. The requirements included are as follows:

- Customer level
  - Fields/formats
  - Market segments
  - Media options
  - Billing options
  - Geographic markets
  - Account structures
- Product/service level
  - Service categories
  - Fields/formats
  - Billing basis
  - Volatility
- Operational and systems level
  - Functionality

  - Ease of use
  - Ease of management
  - Exception handling
  - Auditability
  - Sizing/volumes
  - Billing cycles
  - Platform
- Interface level
  - Types of interfaces
  - Specifications/tools
  - Quality tools
  - Capacity
- Standards level
  - Call detail standards
  - Regulatory requirements
  - Billing output standards
  - Payment processing standards

Appendix C

Document Models

The enclosed CD contains "fill-in-the-blanks" models for billing systems procurements documents. The documents are in MS Word 2000® format with standard document formatting and fonts. These documents provide an outline for each type of procurement document, along with prompts for important decision points.

The procurement document models coordinate with the discussions in Section III. Included are the following document models:

- Request for information (RFI)
- Request for proposal (RFP)
- Contract

The contract model includes business, financial, technical and risk management material. It explicitly does not include any guidance on legal terminology, contract law or relevant boilerplate material. Your corporate lawyers should provide guidance appropriate to your business and jurisdiction.

Appendix D

Resources

The enclosed CD contains information about resources that may be of help in your billing process. This information is contained in MS Access® databases. One database includes information on a broad spectrum of vendors of billing products and/or services. A second contains information of billing-related publications, conferences and training. The authors do not endorse any of the products or services referenced in these databases. Indeed, in many cases, all we know about the vendor is what their Web site proclaims about them. However, these databases provide both some starting information and ways of organizing, querying and analyzing high-level vendor information.

The billing vendor database includes two tables—one for vendor information and one for product information. The vendor information provides all the contact information we could identify from public sources for the vendor. The tables are linked and sample queries and forms are provided.

The product information, based on vendor sales literature, includes the name of the product or product line, check off fields for basic billing functionality—rating, rendering, support for electronic presentment, support for electronic payment, tax processing and databases, payment processing, usage collection and mediation, integration middleware and other unspecified. There are also indicators for service bureau functions (e.g., bill printing contractors, tax processing services, full service billing outsourcers) and for purchased products. The database indicates if the product targets small, medium or large applications. The database includes indicators for types of services supported in the product.

The resources database simply lists the type of resource and contact information.

Glossary

Items have been included in this glossary when they appear in the text as acronyms or when a term is used that has a unique meaning in the context of billing or telecommunications and is not explicitly defined in the text.

Adjustment
A change made to charges, billed or unbilled, after they have been assigned to an account

AMA
Automated Message Accounting

ANSI
American National Standards Institute: a private, non-profit organization that administers and coordinates the United States voluntary standardization and conformity assessment system

API
Application programming interface: specific method prescribed by an application program by which a programmer writing an application program or interface routine can make requests of another application

Application services
Value-added content and processing services usually in the context of the World Wide Web and convergent technology

ASP
Application Service Provider

ATIS
Alliance for Telecommunications Industry Solutions: standards body that is leading the development of telecommunications standards, operating procedures and guidelines in North America

BAF
Billing AMA Format: message accounting exchange format by many telecommunications companies in North America

Billing element
Charge that is shown on a bill for a service or product

Billing medium
One of the vehicles by which a customer is sent a bill or invoice statement paper (e.g., printed on paper, email, Web, CD)

Book of business
A selected group of customers and associated services

BTN
Billing Telephone Number: the number to which a call or product/service will be billed

CAMA
Computerized Automated Message Accounting

CARE
Customer Account Record Exchange: a forum sponsored by the Subscription Committee of the OBF to provide a forum for customers and providers to develop common definitions and recommendations for resolution of national subscription issues

CARS
CABS (Carrier Access Billing) Auxiliary Report Specifications: OBF committee

CCB
Common Carrier Bureau

CD
Compact Disc

CDR
Call Detail Record

CD-ROM
Compact Disc-Read Only Memory

Centrex
Central Office Exchange System—Provides sophisticated multiline call management features (transfer, conferencing, hold, ACD, etc.) on a telephone company switch. Considered the alternative to PBX systems.

CFO
Chief Financial Officer

CIO
Chief Information Officer

CLCI
Common Language Circuit Identifier®: a product of Telcordia

CLEC
Competitive Local Exchange Carrier

Convergence
Intersection of wireless services, content-delivery services, IP-based services, Internet-specific services, cable services

CORBA
Common Object Request Broker Architecture: an OMG middleware platform, which includes the Interface Definition Language OMG IDL, and protocol IIOP

CPE
Customer Premises Equipment: pieces of equipment on the customer's premises that are connected to telecommunications service

Cron files
Executables run at a specified time by the UNIX scheduler program "cron"

CTO
Chief Technology Officer

CWM
Common Warehouse Metamodel: Standard interfaces that can be used to enable easy interchange of warehouse and business intelligence metadata between warehouse tools, warehouse platforms and warehouse metadata repositories in distributed heterogeneous environments

DBMS
Data Base Management System

Dedicated services
Telecommunications services (connections, bandwidth) leased to a single subscriber end-to-end

Dispatch systems
Systems that support the dispatch of technicians to a customer premises or other remote location

DSL
Digital Subscriber Line

Dun & Bradstreet
A U.S. firm that provides assessments of companies including financial information, size (employees and annual sales), type of ownership, principal executives, and biographies)

EBPP
Electronic Bill Presentment and Payment

EDGE
Enhanced Data rates for GSM Evolution: a high-bandwidth dial service and one of the GSM family of technologies

EDI
Electronic Data Interchange

EFT
Electronic Funds Transfer

EFTA
European Fair Trade Association

EMI
Exchange Message Interface: maintained by the Message Processing Committee of the OBF

ETSI
European Telecommunications Standards Institute

EU
European Union

Facilities-based provider
Telecommunications company that operates its own switches

FCC
Federal Communications Commission

General ledger
Record of the financial transactions of a company

GPRS
General Packet Radio Service: one of the GSM family of technologies

GSM
Global System for Mobile Communications: set of wireless telecommunications standards

GSMA
Global System for Mobile Communications Association: responsible for the deployment and evolution of standards for the GSM family of technologies

GUI
Graphic User Interface

HTML
HyperText Markup Language: language used for creation of Web sites

IANA
Internet Assigned Numbers Authority

IETF
Internet Engineering Task Force

IFAST
International Forum for ANSI-41 Standards Technology: ATIS-sponsored subgroup concerned with family of standards for international wireless roaming

IFX
Interactive Financial Exchange: Extensible Markup Language-based (XML) incarnation for EBPP

IIOP
Internet Inter-ORB Protocol: a protocol included in the CORBA middleware platform

ILEC
Incumbent Local Exchange Carrier

IP
Internet Provider

IPDR.org
Internet Protocol Detail Record Organization: an organization to define the essential elements of data exchange between network elements, operation support systems and business support systems

IPP
Internet Presence Provider

ISO
International Organization for Standardization: a worldwide federation of national standards bodies

ISO 9000
Series of standards provide a framework for quality management promulgated by the ISO

ISP
Internet Service Provider

IT
Information Technology

ITU
International Telecommunications Union: United Nations-sponsored organization with membership of telecommunications policy makers and regulators, network operators, equipment manufacturers, hardware and software developers, regional standards-making organizations, and financing institutions

ITU-T
Telecommunication Standardization division of ITU: experts prepare the technical specifications for telecommunications systems, networks and services

Jurisdiction
The geographical area regulated by a governmental body

LAN/WAN
Local Area Network/Wide Area Network

LATA
Local Access Transport Area

LEC
Local Exchange Carrier

LINUX
An open-source operating system for multiple platforms; marketed by Red Hat

LNP
Local Number Portability

Lockbox bank
Bank to whose address customers direct payments, which the bank processes for the company under contract

Marketing Service Description
Information on a service offering from the market perspective, including: how the customer will see the service, billing basis, price and discount structure, market structure and market sensitivities, as well as projected revenues, sales compensation structure and marketing and advertising costs

Marketing Service Plan
Marketing service description information for planned services

MDA
Model Driven Architecture: multiplatform specification of the OMG

MECAB
Multiple Exchange Carrier Access Billing: OBF committee

Mediation
Call Detail Record (CDR) collection, formatting and translation

Middleware
Transaction management, timing and regeneration support for an interface

MOF
Meta-Object Facility: standard interfaces that can be used to define and manipulate a set of interoperable metamodels and their corresponding models

MSP
Management Service Providers

NDM-U
Network Data Management-Usage: a specification developed by the IPDR

NPA
Number Plan Area

NPA/NXX
Number Plan Area and the local exchange within the NPA

OBF
Operations and Billing Forum of the Alliance for Telecommunications Industry Solutions: consists of six subcommittees that affect ordering, billing, provisioning and exchange of information about access services, other connectivity and related matters

OFX
Open Financial Exchange: the financial transaction standard originated by CheckFree Corp., Microsoft Corp. and Intuit Inc.

OMA
Object Management Architecture: defines standard services that will carry over into MDA work

OMG

Object Management Group: an open membership, not-for-profit consortium that produces and maintains computer industry specifications for interoperable enterprise applications

OMG IDL

Object Management Group Interface Definition Language: included in the CORBA middleware platform

ORB

Object Request Broker

OS

Operating System: may be in place for any of the major systems, such as network or billing

Override pricing

A method to adjust the regular price of an item (product or service) to provide a discount or special pricing arrangement

PBX

Private Branch Exchange: a piece of equipment for switching calls on a customer premise

PC

Personal Computer

PDA

Personal Digital Assistant

PIN

Personal Identification Number

POTS

Plain Old Telephone Service: long distance voice services plain old telephone service

Provisioning systems

Systems that support the physical implementation of a service, including engineering, installation, testing and management

Q interface

Q interface exists between two TMN-conformant functional blocks that are within the same TMN domain. Any functional component that interfaces directly to the OS uses the Q3 interface

Q.825 formats

An ITU Specification of TMN applications at the Q3 interface: Call detail recording

QoS

Quality of Service: variable parameter of service as a result of dynamic bandwidth assignment as a network capability

Qunix
A fictitious operating system, used to illustrate development of a statement in an RFP

R&D
Research and Development

RAO
Revenue Accounting Office

Rate element
Charge that is shown on a bill for a service or product

RBOC
Regional Bell Operating Company: a company providing local telecommunications services that was formed when AT&T was broken up in 1984

RFI
Request for Information

RFP
Request for a Proposal

SDH
Synchronous Digital Hierarchy

SECAB
Small Exchange Carrier Access Billing: OBF committee

Settlements
An arrangement between companies and/or governments to compensate the non-billing companies for terminating, originating or transporting communications traffic (calls)

SLC
Subscriber Line Charge

SME
Subject matter expert

Solaris
Sun Microsystems' implementation of the UNIX operating system

SONET
Synchronous Optical Network

Spooling
The activity of placing information into a special area (e.g., a buffer, disk, or in memory) where it can be accessed at a later time; generally applied to storage of formatted data awaiting printer or other output resource

SQL
Standard Query Language

SSA
Standard Metropolitan Statistical Areas: areas identified by the United States government for data gathering and statistical reporting

Switched services

Telecommunications services that rely on switching systems, such as local calls from one telephone to another or long distance calls

TAP

Transferred Account Procedure: standards for roaming billing information set by the GSMA

Technical Service Description

Information on the technical structure of a service, as well as information on the cost of delivering the service. This may also contain information on the formatting of information in and out of the provisioning and sales systems

Technical Service Plan

Technical service description information for planned services

Telco

Telecommunications company that provides local and/or long distance services

TMN

Telecommunications Management Network

10-K

A report form required annually by the U.S. Securities and Exchange Commission for publicly held corporations

3G wireless

Third Generation or "always-on" data transaction services for digital wireless communications. One of the GSM family of technologies

TIPHON

Telecommunications and Internet Protocol Harmonization Over Networks: sponsored by ETSI

TN

Telephone number

Toll switch

Equipment that receives a call from a local switch (or directly from a dedicated line) and routes it to another toll switch

Treatment

Activities to affect collection of outstanding billing balances, including service suspension

TSACC

Telecommunications Standards Advisory Council of Canada

2.5G

Another term used for EDGE, a high-bandwidth dial service and one of the GSM family of technologies

UDT
Universal Date Time: structured date/hour/minutes/second format used in CDRs

UML
Unified Modeling Language: A specification defining a graphical language for visualizing, specifying, constructing and documenting the artifacts of distributed object systems

UNIX
A widely-used computer operating system

URL
Universal Resource Locator (Internet address)

USF
Universal Service Fund Surcharge

V&H coordinates
Vertical and Horizontal: telecommunications-unique parameters to identify service endpoint locations, used in many older tariffs and legacy services, for private line services

VoIP
Voice over Internet Provider

VP
Vice President

W3C
World Wide Web Consortium: develops interoperable technologies for the Web

WAN
Wide Area Network

WLAN
Commercial wireless LAN

XMI
XML Metadata Interchange: A specification enabling easy interchange of metadata between modeling tools (based on the OMG-UML) and metadata repositories (OMG-MOF-based) in distributed heterogeneous environments.

XML
Extensible Markup Language: language that is more powerful than the older HTML, used for newer Web applications such as EBPP. XML is a project of W3C and is a public format

Y2K
Year 2000: a term used extensively as preparations were made to prepare software for transition from 1999 to 2000

Index

NOTE: Boldface numbers indicate illustrations.

ABOUT THE AUTHORS

JANE HUNTER has over 25 years' experience in telecommunications and systems management. For the past 9 years, she has been an independent telecommunications and information services consultant. She has in-depth experience with telecommunications billing and financial systems, Operations Systems (OSs), customer care environments, operations processes and procedures, and field operations in a variety of environments. Ms. Hunter's experience includes implementation and conversion of billing and provisioning systems for interexchange and local phone companies.

MAUD THIEBAUD is a consultant to local communications carriers and to suppliers of local telecommunications provisioning and management systems. She has extensive experience in the development and implementation of quality processes and metrics for billing services. Her experience includes management of the development of acceptance criteria and acceptance testing for new telecommunications billing systems. Prior to becoming a consultant, Ms. Thiebaud held a wide range of corporate and line and staff assignments for Pacific Bell and AT&T in billing methods, billing operations, finance, customer care, and regulatory affairs in both the local exchange and long distance segments of the industry.

SOFTWARE AND INFORMATION LICENSE

The software and information on this diskette (collectively referred to as the "Product") are the property of The McGraw-Hill Companies, Inc. ("McGraw-Hill") and are protected by both United States copyright law and international copyright treaty provision. You must treat this Product just like a book, except that you may copy it into a computer to be used and you may make archival copies of the Products for the sole purpose of backing up our software and protecting your investment from loss.

By saying "just like a book," McGraw-Hill means, for example, that the Product may be used by any number of people and may be freely moved from one computer location to another, so long as there is no possibility of the Product (or any part of the Product) being used at one location or on one computer while it is being used at another. Just as a book cannot be read by two different people in two different places at the same time, neither can the Product be used by two different people in two different places at the same time (unless, of course, McGraw-Hill's rights are being violated).

McGraw-Hill reserves the right to alter or modify the contents of the Product at any time.

This agreement is effective until terminated. The Agreement will terminate automatically without notice if you fail to comply with any provisions of this Agreement. In the event of termination by reason of your breach, you will destroy or erase all copies of the Product installed on any computer system or made for backup purposes and shall expunge the Product from your data storage facilities.

LIMITED WARRANTY

McGraw-Hill warrants the physical diskette(s) enclosed herein to be free of defects in materials and workmanship for a period of sixty days from the purchase date. If McGraw-Hill receives written notification within the warranty period of defects in material or workmanship, and such notification is determined by McGraw-Hill to be correct, McGraw-Hill will replace the defective diskette(s). Send request to:

Customer Service
McGraw-Hill
Gahanna Industrial Park
860 Taylor Station Road
Blacklick, OH 43004-9615

The entire and exclusive liability and remedy for breach of this Limited Warranty shall be limited to replacement of defective diskette(s) and shall not include or extend to any claim for or right to cover any other damages, including but not limited to, loss of profit, data, or use of the software, or special, incidental, or consequential damages or other similar claims, even if McGraw-Hill has been specifically advised as to the possibility of such damages. In no event will McGraw-Hill's liability for any damages to you or any other person ever exceed the lower of suggested list price or actual price paid for the license to use the Product, regardless of any form of the claim.

THE McGRAW-HILL COMPANIES, INC. SPECIFICALLY DISCLAIMS ALL OTHER WARRANTIES, EXPRESS OR IMPLIED, INCLUDING BUT NOT LIMITED TO, ANY IMPLIED WARRANT OF MERCHANTABILITY OR FITNESS FOR A PARTICULAR PURPOSE. Specifically, McGraw-Hill makes no representation or warranty that the Product is fit for any particular purpose and any implied warranty of merchantability is limited to the sixty day duration of the Limited Warranty covering the physical diskette(s) only (and not the software or information) and is otherwise expressly and specifically disclaimed.

This Limited Warranty gives you specific legal rights, you may have others which may vary from state to state. Some states do not allow the exclusion of incidental or consequential damages, or the limitation on how long an implied warranty lasts, so some of the above may not apply to you.

This Agreement constitutes the entire agreement between the parties relating to use of the Product. The terms of any purchase order shall have no effect on the terms of this Agreement. Failure of McGraw-Hill to insist at any time on strict compliance with this Agreement shall not constitute a waiver of any rights under this Agreement. This Agreement shall be construed and governed in accordance with the laws of New York. If any provision of this Agreement is held to be contrary to law, that provision will be enforced to the maximum extent permissible and the remaining provisions will remain in force and effect.